'Wonderfully researched, vividly written ... an example of medical history at its absolute best.' – *Michael Neve, medical historian and Wellcome Trust Book Prize panellist, 2010*

'Williams recounts the history of smallpox in a breezy, accessible style. And what a history it is.' – *Clive Anderson, journalist and broadcaster, Big Read 2010, New Scientist*

'Williams's account of our battle with the disease revisits historical accounts of its horrendous impact and the fascinating story of medical progress – including the pioneering use of vaccination by a country doctor in 1796 – and its relevance in the fight against modern epidemics.' – *The Times*

'Williams writes with a great command of the English language ... an excellent holiday read.' – *Royal College of Physicians, 2010*

'A meticulously researched story with pace and flair ... Both the history and the science are terrific.' – *Medical Journalists Association Open Book Award, 2011*

'In lively prose with unpatronising insight into past medical dilemmas, he dramatises the scourge and its treatment first by variolation (immunisation with live smallpox virus) then vaccination, but also shows how controversial smallpox vaccination was during the 19th century.' – *The Lancet*

'... an engaging narrative, in which medical history is interweaved with social history and reflections on contemporary issues.' – *BBC History Magazine*

'... the author explores one of the most exciting success stories in the history of medicine. His book also gives original and engaging insights into the anti-vaccination campaigns which remain active today.' – *The Guild of Health Writers*

'Williams has managed to bring to life one of the most enthralling, life changing success stories in the history of medicine' – *Laboratory News*

'A breezy, accessible account.' – *The Week*

'This extraordinary book brings alive the sheer horrors of smallpox and how mankind has managed to wipe it out using vaccination, pioneered by a Gloucestershire country doctor in 1796. This history has a very modern message, and this book needs to be read by everyone interested in public health today.' – *Mark Horton, presenter of BBC TV's Coast and Professor of Archaeology, University of Bristol*

'*The Angel of Death* is a fascinating account of the most terrible disease to afflict mankind. Smallpox showed no mercy: the young, old, poor and royalty all equally at risk; whole societies almost wiped out in its inexorable wake. 2010 marks the 30th anniversary of its final eradication; Gareth Williams charts this compelling story with a plot that weaves seamlessly between cultures and centuries. Written in a wonderfully flowing and engaging style, this is a must read for all lovers of history. Highly relevant for today as the fight lives on to banish other deadly diseases from the world.'
– *Sarah Parker, Director, Edward Jenner Museum, Gloucestershire, UK*

'Filled with fascinating historical detail, this story resonates with contemporary concerns about epidemics and the fight against them. Gareth Williams effortlessly weaves together medical science writing and social history to tell the compelling tale of a battle against a deadly disease.' – *Alice Roberts, writer and broadcaster*

Angel of Death

The Story of Smallpox

Gareth Williams
Professor of Medicine
University of Bristol

Illustrations by Ray Loadman

palgrave
macmillan

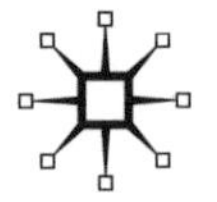

First published in hardback 2010 and in paperback 2011 by
PALGRAVE MACMILLAN

Palgrave Macmillan in the UK is an imprint of Macmillan Publishers Limited, registered in England, company number 785998, of Houndmills, Basingstoke, Hampshire RG21 6XS.

Palgrave Macmillan in the US is a division of St Martin's Press LLC, 175 Fifth Avenue, New York, NY 10010.

Palgrave Macmillan is the global academic imprint of the above companies and has companies and representatives throughout the world.

ISBN 978–0–230–27471–6 hardback
ISBN 978–0–230–30231–0 paperback

This book is printed on paper suitable for recycling and made from fully managed and sustained forest sources. Logging, pulping and manufacturing processes are expected to conform to the environmental regulations of the country of origin.

A catalogue record for this book is available from the British Library.

A catalog record for this book is available from the Library of Congress.

10 9 8 7 6 5 4 3 2 1
20 19 18 17 16 15 14 13 12 11

Printed and bound in Great Britain by
CPI Antony Rowe, Chippenham and Eastbourne

To Caroline, Tim, Jo, Sally and Pippa

Joan and June

And in memory of Alwyn and Geoff

'Now within a few months ... the smallpox was brought into Boston, and within a few months more the besom of destruction swept away near a thousand people. And how strangely was way made for the Destroying Angel to do his execution!'

Reverend Increase Mather (1639–1723)
Sermon preached in Boston, September 1720

'It was difficult for the Angel of Death to kill everybody in the whole world, so he appointed doctors to assist him.'

Rabbi Nachman of Breslov (1772–1810)
Sichot Ha-Ran, no. 50

Contents

List of Illustrations

Plates are shown between pages 204 and 205.

Plates

Illustrations in text pages

Preface to the Paperback Edition

Most books are ephemeral things, as was made clear by the statement at the front of the hardback edition of *Angel of Death* that it was 'printed on paper suitable for recycling'. Fortunately, the book has made it into paperback, and this has also provided me with the opportunity to correct a number of mistakes in the original, and to thank again all those who have helped this particular Angel to take to the skies.

Dr. Derrick Baxby, Dr. Ruth Richardson, Pauline Hemingway, Dr. David Powell, Susan Hood and Dr. Michael Bamber have all kindly pointed out errors, and I'm very grateful to them for their wisdom and knowledge in this regard. After their interventions, there really shouldn't be any mistakes left – but if there are, then they are entirely my responsibility.

In the last year, *Angel of Death* has taken me on an exciting journey that has also been great fun. Much of this has been thanks to the excellent team at Palgrave, who have been wonderful fellow travellers. I remain indebted in particular to Michael Strang, far-sighted publisher and wise counsel, and Abby Coften, the one-woman publicity machine whose tentacles know no bounds.

Gareth Williams
Rockhampton, Gloucestershire

Acknowledgements

Writing a book is a journey best made in good company. This one has been huge fun, largely because I've been lucky enough to acquire such excellent fellow-travellers along the way. I've greatly appreciated the quality of their advice, the strength of their support and, in several cases, their superhuman patience and inexplicably good humour.

Top of the list is my long-suffering wife Caroline, who still manages a tolerant smile whenever I tell her that a book is "nearly finished" (more appropriate responses would include grabbing me by the throat or running screaming from the house). Thanks also to Tim and Jo, and to Sally and Pippa, for putting up with a distracted Dad and a dysfunctional pack-leader, respectively.

This particular Angel has led me deep into unfamiliar territory. Luckily, Dr. Robert Spencer, Dr. Derrick Baxby, Professor Alasdair Geddes and Dr. Chris Burns-Cox were on hand to guide me safely through its numerous minefields. Their collective expertise covers an astonishingly wide area – everything from detailed critiques of Jenner's *Inquiry* and poxvirus biology, to germ warfare and personal insights into the smallpox eradication campaign. I'm indebted to them for being so generous with their time and knowledge. The staff of the Jenner Museum in Berkeley have also been fantastic, notably Sarah Parker, Steven Deproost, Fiona Kam Meadley and (tucked away in the archives behind the DANGER SMALLPOX sign) Jan Leach and Glenys Hannam. I'm grateful to them and to Rod Sayers and Jan Bell for their consistently warm welcome – including the special treat of grapes from Dr. Jenner's own vines. With such expert backup, it goes without saying that any errors in the book are entirely of my own making.

Thanks to the internet and search engines, chasing up sources is now easier than ever before and Google has rescued me countless times by spiriting up half-remembered references from the ether. However, there is nothing like the real thing and I would have struggled without the expert assistance of Patsy Williams and Nial Busby (St. Deiniol's Library), Pamela Forde and Peter Basham (Heritage Centre, Royal College of Physicians of London), Stella Calvert-Smith (Wellcome Library), Sarah Pearson and Louise King (Hunterian Museum, Royal College of Surgeons of London), Stéphanie Bolliger (WHO, Geneva), Keith Moore (Royal Society), Ruth Thomas (Middlesbrough Reference Library), Betty Smallwood (Medical Society of London), Katya Gracheva (RIA Novosti, Moscow) and Meg Wise (Thornbury Museum). My grateful thanks to all of them.

With this book sailing perilously close to its word limit, it's fortunate that a picture is worth a thousand words. Ray Loadman deserves particular praise for the elegance of his artwork (especially given the disgraceful roughness of the original sketches) and for maintaining his customary forgiving nature throughout. My gratitude also goes to Bhavna Singham for hunting down elusive illustrations; to D.A. Henderson for kindly providing a wonderful photo of man, dog and Temple; and to the world's most advanced, user-friendly voice recognition system, namely Christine Greenwood, for once more translating my incoherent ramblings into clean text.

Readers may be reassured to know that some brave souls have already read chunks of the manuscript, apparently without lasting harm. The roll of honour is headed by Joan Williams and June Evans, my mother and mother-in-law (both friendly critics and tough old birds), the long-suffering Caroline (see above), Bob (alias Dr. Robert) and Tracy Spencer, Jenifer Roberts and Paul

Beck, and Charlotte Thomas. Close behind are Dr. Alice Roberts, Katharine Whitehorn, Professor David Punter, Professor Robert and Dr. Paula Sells, Cherry Lewis, Chris Wakling, Professor Parveen Kumar and Dr. Trevor Thompson. I'm deeply grateful to all of them for their considerable investments of time and energy as well as their invaluable comments and suggestions.

This brings me to the team at Palgrave Macmillan, who have done a brilliant job of transforming an electronic manuscript into such a solid and attractive book. Above all, I'm indebted to Michael Strang for having fought the Angel's corner so effectively in tough times and for having pushed the book through into production in record time. It's been a great pleasure working with him and his colleagues, especially Abby Coften, the high-energy head of publicity and marketing, and Ray Addicott and Tracey Day, who succeeded in making that last phase of copy-editing and proofs an entirely pain-free experience.

Finally, I owe a great deal to all those who kept me buoyant throughout the lifetime of this book, by providing encouragement, good cheer, stiff drinks and the occasional kick in the pants. Under this heading come several people already mentioned – Caroline, Joan, June, Bob and Tracy, Jenifer and Paul – and I'm also grateful to Colin Gardner, Jean Crispin and Sean Cusack in this regard.

Setting the Scene

On 9 December 1979, a group of 20 international experts assembled in Geneva to sign a ceremonial parchment that celebrated the simultaneous end of two eras: a huge public health campaign that had run for just over a decade and a reign of terror that had gripped mankind for several millennia. The 20-word declaration on the parchment, in six languages that span the planet, certified that smallpox had been eradicated from the world. Five months later, on 8 May 1980, the World Health Organisation ratified that declaration and, apart from some samples locked away in half a dozen research laboratories, the scourge of smallpox was officially extinct.

This should have been momentous news. At the peak of its powers – from the Middle Ages to the end of the nineteenth century – smallpox had been a global scourge of biblical proportions. Some of the larger outbreaks turned the concept of 'decimation' on its head by wiping out 90 per cent of a population, not just one in ten. Even in the quiet years between epidemics, it continued to kill people in their millions. The extermination of this monster was therefore a huge triumph of preventative medicine and arguably one of the greatest medical coups of all time.

Strangely, however, there was little reaction from the peoples of the newly cleansed planet. By then, smallpox was not just on its way out: it had already joined moon landings and the Vietnam War on the list of things that had been and gone and no longer needed to be thought or worried about. Medicine too had moved on. Dead diseases had to give way to new concerns – such as the improbable plague that seemed to target homosexual men in America, robbing them of their power to fight off infections, and which was soon to be given the clumsy title of 'acquired immunodeficiency syndrome'.

Like many doctors, I barely noticed these ground-breaking announcements. In my defence, it was early in my career and I was busy trying to persuade my first research experiments to work. I never saw a case of smallpox and my knowledge of the disease was survival level at best – but then the only realistic risk of meeting it was a rather predictable multiple-choice question about how to distinguish its rash from that of chickenpox.

I doubt that smallpox ever made me stop and think until 2003 when my career brought me to Bristol and the family to a house in the village of Rockhampton, set in flat farmland on the edge of the Severn basin. This is rural Gloucestershire, one of England's traditional dairy counties – although cows are much less obvious now thanks to the European Union, foot-and-mouth disease and supermarket pricing policies.

This is also Jenner country. The list of the Rectors of Rockhampton displayed in the village church goes back to 1270 and includes three Jenners in the mid-eighteenth century. They were the father and two elder brothers of Dr. Edward Jenner of Berkeley, acclaimed across the world as the father of vaccination, which eventually – as certified in December 1979 – fulfilled his expectation that it would rid the world of smallpox. Rockhampton fell within Jenner's practice and the road between here and Berkeley, looping alongside a wooded ridgeback hill that was the deer-park of the Berkeley Estate, was one of his favourite outings.

In 2008, I was given a year of sabbatical leave – arguably overdue (in the golden age of universities, every seventh year was ring-fenced for personal study) but also a rare privilege nowadays in the cash-strapped Groves of Academe. Looking from my study window along the road where Jenner used to ride on his rounds, I found myself thinking about the history of vaccination – which was rather odd for someone whose career is firmly planted in diabetes and obesity. Shortly after that, I was in Jenner's own house in Berkeley – now the Edward Jenner Museum – and behind a door carrying a metal sign (rescued from a derelict isolation hospital) with the fearsome block capitals: DANGER. SMALLPOX. KEEP OUT. The sign guards the Museum's archives where I rapidly discovered that there was much more to the story of smallpox than I had ever imagined.

It was created in prehistory by a freak mutation, straight from fantasy fiction. This transformed a dull virus that infected gerbils into one of the most savage killers in the history of the human race. From the birth of civilisation, smallpox was a destructive force so powerful that it was given the status of a god. Until the nineteenth century, it was the Angel of Death that stood over the lottery of life, sweeping down on one person in three and killing up to half of its victims. During the twentieth century, it remained untreatable and so dangerous that the corpses of its victims were wrapped in heavy polythene and cremated in a sealed coffin filled to the brim with sawdust and gallons of industrial disinfectant. Throughout its lifetime, smallpox was loathed and feared for the terrible legacies of disfigurement and blindness that it inflicted on those who escaped death. Some survivors killed themselves rather than live with their mutilated faces; even in the 1960s, mirrors were routinely removed from the rooms of female patients suffering a severe attack. Finally, smallpox has left behind a memory so terrifying that the world's last experimental samples of the virus are locked away behind the same level of security precautions that protect against the unleashing of a nuclear strike.

Smallpox was a curse that also brought blessings. As the only disease that mankind has so far succeeded in wiping off the face of the planet, it gives us hope that other infections can also be eradicated. It was smallpox that triggered the discovery of vaccination, one of the most valuable transferable technologies in the history of medicine and one that has saved hundreds of millions of lives. The struggle to understand how people could be protected against smallpox spun off the science of immunology, one of the broadest disciplines in medicine which has laid bare the secrets of a vast range of diseases.

Above all, the story of smallpox was about people: how they lived under the shadow of the Angel of Death, how they coped with the disease and its aftermath, how some rose above it while others were crushed, and how – for good or bad – some managed to exploit it. Smallpox drew out the extremes of human behaviour, from courage, compassion and inventiveness at their most noble to the worst excesses of evil, malice and dishonesty.

The people swept up by smallpox include some of the most colourful and engaging characters in the history of medicine. Near centre-stage is Edward Jenner, rightly hailed for having brought vaccination into practice – even though he was not the first to try it out and took over 25 years to finish his experiments. Another conspicuous member of the cast was a pushy English Lady who went native in Constantinople and used her children to test a bizarre Turkish folk-custom supposed to protect against smallpox. Working in parallel with her was a witch-hanging Puritan preacher in Boston, on a crusade to save the town from smallpox and New England from Satan. Later came an American mother, devastated by the death of her only child after vaccination and hell-bent on ridding the world of Jenner's despicable invention. Finally, during the dying throes of the Angel of Death, the front line of the global eradication campaign included charismatic and persuasive evangelists who had to fight bureaucracy, superstition and politics as well as smallpox itself.

Smallpox was a great leveller and the people who encountered it came from all sections of society and from across the planet. All the players in this history, whether major or minor, have one characteristic in common. In every case, and in one way or another, smallpox changed their lives forever. Their stories come together to make an narrative of huge emotional power as well as one of the most gripping epics in the history of medicine. They also show us how mankind's battle against smallpox only succeeded by overcoming problems caused by human nature and people as well as by the virus itself.

This was how I came to write this book. Quite simply, the story of smallpox hooked me and I hope that it will do the same for you.

1 Know the Enemy

During its lifetime, smallpox was many things. The 'Angel of Death' that wiped out millions was also a godsend to Conquistadores, a useful tool for making people obey the Scriptures and the lesson of nature that brought us vaccination. To make sense of these paradoxes, we first need to meet the variola virus that causes smallpox and understand how it maimed and killed. This chapter lays out a brief *curriculum vitae* of smallpox, beginning with the puzzle of how the 'greatest killer' could possibly be described as 'small'.

Also known as

"What's in a name? that which we call a rose
By any other name would smell as sweet."
William Shakespeare *Romeo and Juliet*, Act 2 scene 2, 1597

'Pocca' is an Anglo-Saxon word meaning a pouch or blister, while 'pocks' or 'pox' were terms long used to describe any unpleasant skin eruption. The seemingly inappropriate 'small pocks' (later small-pox and then smallpox) was coined in the late fifteenth century to distinguish the disease from the 'great pox' of syphilis, one of the bounties brought home from South America by Columbus' sailors. Terrible though it was, smallpox was an established fact of life in Europe so the new curse of syphilis – mutilation, paralysis and a lingering death – may well have seemed a bigger threat.

German used the same root as English to name the disease: *Kinderpocken*, which also reminds us that smallpox targeted children. The French *variole* and the Spanish *viruela* come from the Latin *varius*, meaning 'speckled'. This also gave rise to *variola*, the Latinised name of smallpox and now the scientific name of the virus that causes it. The original Italian *vajuolo umano* (nowadays *vaiolo*) comes from the Latin *vacuola* ('cavity'), while one of the Mandarin Chinese words for smallpox, *dou chouang*, describes a bean-like eruption with ulcers. These descriptive names do no justice to the horror of the disease. North American Indians, who met smallpox late in their history, simply called it "rotting face".[1]

Smallpox also had graphic figurative names which reflected dread and awe: the 'Angel of Death', the 'Destroying Angel' and Macaulay's "most terrible of the Ministers of Death".[2] The supernatural theme was picked up in the common Mandarin name *tian hua* ('heavenly flowers', possibly a euphemism for the rash and an allusion to Chinese smallpox gods) and *boginje* ('goddesses') in Serbo-Croat. More enigmatic are *eulogiá* in modern Greek, subtly different from *eulogía* (blessing), and *koze* (also meaning 'goats') in Slovenian.

Smallpox gave its name to the family of viruses to which it belongs – the poxviruses. Close family members include cowpox (*Kuhpocken* in German and *variole bovine* in French) and the mysterious vaccinia virus, the active ingredient of the smallpox vaccine (the mystery being that nobody knows where it came from). Chickenpox produces a rash that can be difficult to distinguish from that of smallpox but is caused by the varicella virus which has nothing to do with poxviruses (or chickens).

Fellow travellers

"A vast amount of erudition has been expended to little purpose in debating whether or not small-pox was known to the Greeks and Romans."
British Medical Journal, 1896

Smallpox and mankind go back a long way together. There are many gaps in its early history and attempts have been made to fill these in with, for example, pockmarks on the face of a mummified Pharaoh and reading between the lines in the Bible.

The Pharaoh in question was Ramses V, whose death from an acute illness in his early thirties was recorded in 1157 BC.[3] The ravages of smallpox are more durable than those of mummification and the passage of over 3,000 years, as typical pitted scars still adorn his face. Two other mummies, one a couple of hundred years older, show similar pockmarks. Surviving Ancient Egyptian medical papyri contain recognisable descriptions of diseases such as diabetes, but strangely make only passing mention of a skin eruption that may or may not have been smallpox.

The Bible might also be expected to refer to such an obvious and lethal condition, especially as other diseases such as leprosy and plague figure prominently. The Ancient Hebrew terms *yallepheth* (an eruption or scab) and *gārābh* (an itch, scab or festering sore) would both have fitted the bill for smallpox but nasty skin eruptions merited only a couple of brief mentions.[4] In Job 2:7–8, the Devil smote the unfortunate Job with a curse of boils, an act later emulated by Moses when he punished the Egyptians (Exodus 9:8–11). Illustrators of early Bibles often portrayed the Egyptians' affliction as something like smallpox – a horror familiar to all that would encourage compliance with scriptural directives. However, Job's misery might have been some other disease. The condition of recurrent abscesses known today as Job's syndrome has nothing to do with smallpox; its cause is a genetic defect which disables the white blood cells that normally kill off the bacteria responsible for these infections.

Smallpox figured more prominently in other religions, especially centuries later when the disease was more firmly established. Smallpox gods sprang up in parts of Asia and Africa, commanding fear and respect and being available for appeasement to spare the faithful from the disease.[5] They included Shapona, the god of smallpox revered by the Yoruba in West Africa, and the goddesses Shitala Mata (Indian Hindu)

Figure 1.1 Shitala Mata, goddess of smallpox. Drawing by Ray Loadman.

and Chu'an Hsing Hua Chieh (Chinese). Statuettes of Shapona, who doubled as the god controlling the earth, show a sinister, squat figure with grim features. The two goddesses are usually depicted sitting on a donkey (Shitala) or a horse (Chu'an Hsing). Shitala, often sporting an appropriately spotted robe, epitomises the evil-twin split personality of these deities. Her name means 'cool' and she is the cure, sprinkling cold water to relieve the unbearable heat and pain of the eruption – but it was she who inflicted the disease in the first place. In all these cultures, smallpox was the most terrible of punishments, reserved for the worst transgressions.

Shapona and Shitala are more recent creations than they might appear, with robust references only since the seventeenth and sixteenth centuries. Chu'an Hsing is older, probably dating from the eleventh century. The smallpox gods were a remarkably durable bunch. In the sixteenth century, African expatriates took Shapona across the Atlantic and continued to worship him (sometimes integrated with Christianity) in their new homes. This ritual probably provided much-needed comfort, as smallpox was a frequent stowaway on the packed ships that carried them into slavery in South America. Four centuries later, two-faced Shitala, alternately tossing the pox and its relief from her donkey, rode out in the 1970s to undermine the World Health Organisation's campaign to eradicate her disease from various Hindu regions of India. For fear of offending Shitala, her devotees refused vaccination for themselves and their children and threw the corpses of smallpox victims into rivers, rather than defile her with the traditional cremation.

Returning to Europe, the classical Greek and Roman medical traditions followed the biblical example and gave surprisingly little attention to smallpox. There is no specific mention in the writings of Hippocrates (c. 440–c. 340 BC), the father of the Oath which still embodies the aspirations and ideally the actions of the doctor. Five hundred years later, his compatriot and conceptual successor Galen (AD 129–216) referred only briefly to a lethal sickness that made the skin bubble up, turn black and ulcerate. Galen was physician to the Roman Emperor Marcus Aurelius Antonius, and the brevity may be explained by his hasty departure from Rome where this 'Antonine Plague' was beginning its 15-year reign of terror. Scholars have agonised over whether smallpox caused this and other catastrophic diseases that struck the Greek and Roman Empires, notably the horrific 'Plague of Athens' reported in 430 BC by Thucydides.[6] On balance, they were probably not and perhaps significantly, there was no specific word for smallpox in either classical Greek or Latin. When Bishop Marius in the Swiss city of Avenches coined the term 'variola' in AD 570, this was 100 years after the fall of the Western Roman Empire. He presumably applied this to the disease that was unmistakably smallpox, described a decade later by another Bishop, Gregory of Tours.

Over 300 years elapsed before the next significant milestone in the history of smallpox was planted, in tenth-century Baghdad by the Persian physician Rhazes (865–925).[7] Rhazes wrote several comprehensive medical texts that provided continuity with Hippocrates and Galen yet also broke new ground – a winning combination which ensured that his influence outlived him by several hundred years. In 910, he wrote *A Treatise on Smallpox and Measles*. The title could not be bettered: he described both diseases clearly, as well as the diagnostic features that distinguished them. Intriguingly, he reported measles to be worse than smallpox.

During the first millennium AD, smallpox tightened its grip across Europe, the Middle East and Northern and Western Africa, carried along by those most human of behaviours: invasion, trade and religious crusades of all persuasions.[8] China was invaded several times by smallpox, initially imported across the unfinished Great Wall by the Huns in about 250 BC (hence the Chinese nickname of 'Hun pox'). Over several hundred years it became self-sustaining and then spread via Korea to Japan towards the end of the sixth century. Smallpox was also firmly established in India, a putative origin of the disease. Ancient Sanskrit medical texts written before AD 400 (and possibly centuries before that) clearly described an unpleasant and often fatal skin eruption that went through stages like those of smallpox.

Across the territory over which smallpox held dominion, outbreaks continued to flare up and die down like a firework display in extreme slow motion. By AD 1000, however, large swathes of the planet were still virgin territory for the disease: Russia, Scandinavia, Iceland, the whole of the Americas and the Antipodes. Over the next few hundred years, this happy state of innocence ended for each of these regions, often catastrophically. When the virus hit previously untouched societies which had no immunity against it, the effects were devastating. These experiments of nature – not all accidental – show smallpox at its most efficient and brutal. The Danish ship which brought smallpox to Iceland in 1241 triggered an outbreak that eventually killed 20,000 people, about 40 per cent of its inhabitants.[9] And in 1541, one infected African slave on the Spanish Conquistadores' expedition to Mexico unleashed an epidemic that ultimately killed over 20 million people, some 90 per cent of the initial population, and toppled the Aztec civilisation (see Chapter 2).

The story so far: smallpox has made its debut and is marching out across the planet. But what was the cause of the disease?

Pinning the blame

"It is generally (but not universally) accepted that smallpox is due to contagion rather than to some cataclysmic changes in the atmosphere or bowels of the earth."

Royal Commission on Vaccination, *Final Report*, 1896

Every child in the developed world knows that germs cause disease, having been told from infancy that both are spread by coughs, sneezes and unwashed hands. We also know that diseases such as measles, chickenpox and smallpox are infections, even though we may not be sure whether they are caused by bacteria or viruses.

This understanding has only crystallised during the last hundred years or so. At the turn of the twentieth century, the notion that smallpox was an infection (contagion) was still actively resisted. The main opponents were believers in 'miasma theory', namely that sufferers from diseases such as cholera and smallpox were caught in the crossfire between clashing elements of nature – the "catalcysmic changes" in air and earth mentioned in the quotation. This sounds ridiculous now but miasmatists were powerful in medicine and society and their stand-off against the 'germ theorists' led by Louis Pasteur and the German Robert Koch was bitter and lasted for decades. Lined up alongside the miasmatists were various anti-Pasteurian scientists and natural healers, who maintained that smallpox was simply the body eliminating wastes through the skin and therefore a benign process that must be encouraged, not 'treated' (see Chapter 12). Given that smallpox was both hideous and murderous, this was a huge triumph of gullibility over common sense. Nevertheless, this view had widespread support up to the first quarter of the twentieth century and its adherents can still be found lurking on the internet today.

The early 'medical' concepts of infection were abstract. Hippocrates saw diseases as imbalances between the four 'humours' – blood, yellow bile, black bile and phlegm – that he suggested regulated all bodily functions.[10] His theory obviously cuts across every law of physiology but at the time its neatness caught the imagination and filled a conceptual vacuum. It also paralysed medical thinking for centuries. Its last vestiges still cling to the fringes of medicine, in archaic terms such as 'plethoric' (a morbidly ruddy complexion, attributed by Hippocrates to a plethora, or excess of blood) and 'melancholy', the depression caused by too much black bile (*melas kholi* in Greek). Hippocrates and his humours had unfortunate consequences for smallpox sufferers that lasted until well into the nineteenth century. According to him, fevers were due to a harmful plethora and must therefore be treated by bleeding. The imaginative and unpleasant ways in which this dogma was applied in smallpox are described later.

Rhazes from Baghdad developed a different theory about the cause of diseases such as smallpox.[11] He believed that everyone was born with latent diseases in the blood and when a particular 'contagion' built up to a critical level it was expelled by the body through the appropriate route – the gut (diarrhoea), lungs (coughing up phlegm) or skin (an eruption). Smallpox was therefore a natural process; if the victim died, it was because the process of elimination had not been successful.

Rhazes' teaching was hugely influential and the embers of Persian medicine continued to warm European practice for centuries – ironically, up to the age when English medicine held itself to be the best in the world and sneered at ideas coming from the East. A great seventeenth-century English physician, Thomas Sydenham, echoed Rhazes' view that the eruption of smallpox was evidence of the body's commendable efforts to rid itself of the disease: "nothing but Nature's Endeavour to thrust forth with all her Might the Morbific Matter".[12] Sydenham was a powerful opinion and his views were parroted up to 200 years later by natural healers and anti-vaccinationists. He also suggested that smallpox epidemics were due to "some peculiar constitution of the atmosphere" and so helped to lay the ground for the miasma theory of disease.

By contrast, the Dutch physician Hermann Boerhaave stated in the early 1700s that "smallpox only arises from contagion" (person-to-person spread). Another Dutchman, Antony van Leeuwenhoek, had already reported seeing tiny moving 'animalcules' when he looked at scrapings from his teeth under his newly-invented microscope. However, the concept of contagion remained nebulous and did not stretch to the notion that particular life-forms invisible to the doctor's eye could produce diseases – even though the Italian Girolamo Fracastoro had suggested exactly that in the mid-sixteenth century with his theory that 'seminaria' (tiny seeds) caused smallpox and measles.[13]

The dichotomy between Sydenham and Boerhaave set the scene for the messy confrontation between germs and miasma, which reached a peak around the turn of the twentieth century. This was a titanic clash of cultures and personalities. In one corner were traditionalists who dared not challenge the wisdom that had held sway for 1,800 years. In the other was the new breed of experimentalists, adept in the 'seeing-is-believing' wizardry of the microscope and unafraid to tear up their own results and accepted wisdom if they did not make sense.

The impasse is nicely illustrated by the battle over the nature of cholera – a common condition causing torrential and often fatal diarrhoea – between Max Josef von Pettenkofer, head of the world-famous Institute of Hygiene in Munich, and the younger Robert Koch, a microbe-hunter to rival Pasteur himself.[14] Von Pettenkofer ridiculed Koch's claims that the comma-shaped 'microbes' seen down his microscope caused cholera, and that sheep given a culture of the microbes promptly developed severe diarrhoea and died. By way of evidence, Koch sent von Pettenkofer a flask of cholera culture broth for him to experiment with as he saw fit. Von Pettenkofer certainly rose to the challenge: he drank the lot in front of his students and then delighted in reporting back to Koch that his microbial theory of cholera was complete rubbish. However, von Pettenkofer omitted to mention that he had already had cholera, which evidently provided enough immunity to prevent Koch's brew from killing him. His victory was short-lived. More bacteria fell victim to Pasteur, Koch and their allies and the realisation that miasma theory was doomed finally caught up with von Pettenkofer in early February 1901. He locked himself away, surrounded by all the works that had made him famous, and shot himself in the head.

It was assumed that bacteria also caused smallpox and many potential culprits were paraded in triumph towards the end of the nineteenth century.[15] They all turned out to be only opportunistic fellow-travellers that had taken up the invitation to grow in ravaged skin. In 1892, the Italian microbe expert Giuseppe Guarnieri stirred up great excitement with his discovery in thin tissue sections of tiny particles clustered inside damaged cells.[16] Confident that these were the causative bacteria, Guarnieri named them *Cytorrhyctes variolae*, meaning 'the cell destroyer of smallpox'. He was both right and wrong. These 'Guarnieri bodies', later used as a diagnostic feature of smallpox, were indeed the cause but were not bacteria. Instead, they were the 'virus factories' that churn out thousands of replicated viruses, which are individually too small to see clearly with the light microscope. The 1898 edition of William Osler's monumental

Principles and Practice of Medicine gives detailed descriptions of the bacteria responsible for anthrax, diphtheria, cholera and tuberculosis, but admits that "the nature of the contagion of small-pox is unknown".[17] It was another 50 years before the variola virus was isolated and finally visualised.

At this point, a brief sortie into the peculiar world of viruses will help to explain what smallpox did. Like a good dental procedure, this account aims to be quick and painless. Unlike a dental procedure, it may leave you wanting more – in which case, further information can be found in the references cited for this chapter.

Nature's triumph of nanotechnology

"So, naturalists observe, a flea
Hath smaller fleas that on him prey;
And these have smaller fleas to bite 'em,
And so proceed *ad infinitum*."

Jonathan Swift 'On Poetry', 1733

If the discovery of bacteria broke the mould of how people thought about disease, viruses were to stretch imagination and credulity even further. Their existence had been deduced for decades before they were eventually seen.

Viruses were a concept that filled a diagnostic vacuum: something that caused disease in animals or plants, when no bacteria could be found to incriminate. The first extensive studies in plants revealed some remarkable properties. The infection could pass through a filter made of unglazed porcelain, fine enough to trap bacteria. Bizarrely, some extracts could be crystallised like a salt, initially suggesting that toxic chemicals might be responsible – but unlike a chemical poison, the filterable agent somehow managed to produce more of itself when given access to the plants on which it preyed. 'Virus' ('poison' in Latin) was adopted as a helpfully non-specific term. The word had been in use for centuries – including by Edward Jenner – to denote the active principle, whatever it was, that caused various diseases.

Viruses remained invisible and therefore mysterious until the invention in 1931 of the electron microscope, which expanded the power of seeing by several hundred-fold beyond the limits of the light microscope. Viruses turned out to be hugely diverse in shape and size. The first to be visualised by electron microscopy was the tobacco mosaic virus (TMV) which causes unsightly yellow speckling of tobacco leaves and economic ruin for those who grow the weed for a living. TMV was found to be rod-shaped, with a surface covered in tiny, tightly-packed beads, rather like a stylised cob of corn. The electron microscopists soon built up a rogues' gallery of human viruses. The adenoviruses that often produce cold-like respiratory symptoms turned out to be perfect 20-sided crystal-like particles with spikes sticking out of the corners, while various 'small round viruses' were isolated from cases of severe diarrhoea and rabies was found to be due to the sinister bullet-shaped rhabdovirus. And in 1948, the electron microscope finally showed us the cause of smallpox.[18]

Big and ugly

"Forget the outside world. Life has different laws in here."

Alexander Solzhenitsyn *The Love-Girl and the Innocent*, Act 1 scene 3, 1969

Variola, the virus that caused smallpox, turned out to have an undistinguished rounded brick shape, about 270 nanometres (nm) across and 400 nm long and with a granular surface (1 nm is 10^{-9} of a metre, or one-millionth of a millimetre).[19] Other members of the main poxvirus family (orthopoxviruses), of which variola is the archetype, include cowpox, vaccinia, monkeypox and taterapox which infects gerbils. These are all similar in appearance, while the related parapoxviruses show structural differences that allow them to be told apart by electron microscopy. Variola is one of the biggest viruses; the wire-like filoviruses that cause the even deadlier Marburg and Ebola fevers can be 40 times longer, but are so thin as to be invisible on light microscopy.

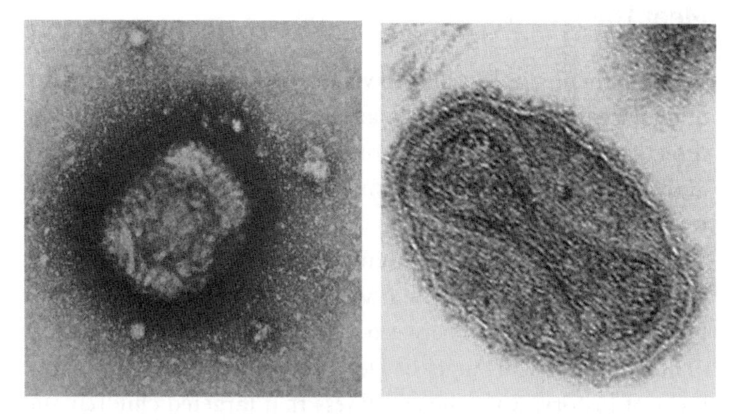

Figure 1.2 The variola (smallpox) virus. Electron micrographs showing the surface (left) and internal structure (right). Images reproduced by kind permission of the Edward Jenner Museum, Berkeley, Gloucestershire.

Smallpox was terrifyingly successful but, like all viruses, it clung to its version of existence somewhere between organic chemistry and the simplest living organisms. Viruses are not alive: they lack all the essential machinery of life and are totally dependent on their host for survival. They are the matter of reproduction stripped to its barest essentials, with a dab of genetic material wrapped up in a protective coat of various proteins. The genetic material is either DNA, which carries the genes in living organisms including man, or the related nucleic acid, RNA. DNA viruses include the poxviruses, varicella (chickenpox) and adenoviruses, while HIV (AIDS), Ebola, rabies and influenza are all examples of RNA viruses.

There is nothing evil about viruses, even ones like smallpox that inflict horrific damage. A virus does only what it is programmed to do: to plug into the machinery of the host's cells and churn out countless copies of itself. It is unfortunate but irrelevant if that process happens to tear the guts out of the cells, wreck tissues or kill people.

Meet the family

Between them, the 20-plus poxviruses can infect a wide range of animals, from insects to goats and from canaries to man. Those directly involved in the story of smallpox are variola, cowpox and vaccinia, while the taterapox that infects gerbils plays a brief but crucial historical role.[20]

Smallpox came in two main varieties, caused by distinct strains of variola virus which were identical in size and shape but showed clear water between their biological effects.[21] *Variola major*, the classical Angel of Death that held sway throughout most of history, was a severe illness that targeted children and killed up to 50 per cent of its victims. Much more benign was *Variola minor* which usually produced a milder rash and carried a mortality of only 0.2 per cent. *Variola minor* was seen initially as a clinical curiosity but eventually spread widely throughout America, displacing *major* in some regions. Catching *minor* was a double stroke of good fortune: the victim usually escaped with a slight illness and trivial scarring and was then left with lifelong immunity against *major* as well.

Variola, whether *major* or *minor*, affected only humans. Other poxviruses are more promiscuous. Despite its name, cowpox is equally at home in certain rodents such as field mice and voles (which were probably its orginal host) and has been found in more exotic species including elephants and big cats.[22] The exclusive attachment of variola virus to man was ultimately its undoing. As there was no possible reservoir of infection in other species, it could be exterminated by a vaccination strategy that eventually starved it of susceptible human hosts. Other poxviruses could not be eradicated in this way – unless all possible reservoirs in rodents and man were somehow emptied by mass vaccination or extermination.

As everyone knows, cowpox figures prominently in the story of vaccination, together with Blossom (a cow), Sarah Nelmes (a milkmaid), James Phipps (a human guinea-pig) and Dr. Edward Jenner (the 'father of vaccination'). Cowpox produces firm ulcers on cows' skin, commonly around the udder. It can also infect humans when the virus-laden ulcer fluid enters the body through a scratch in the skin.[23] Cowpox in humans can be extremely unpleasant, with nasty ulcers at the infection site (usually the hand), swollen and tender glands in the armpit and often a fever. However, this suffering was worthwhile, as an attack of cowpox left the subject immune to smallpox for several years and sometimes for life – hence the rationale for vaccination, the deliberate infection of a healthy subject with cowpox to 'immunise' against smallpox. Cowpox was uncommon, even in Jenner's time, and is now very rare. Human cases are more likely to be acquired from domestic cats (another host species) than from cows.

The vaccinia virus is distinct from all other poxviruses, including cowpox. This is baffling, as vaccinia was isolated as the active ingredient from smallpox vaccine, of which cowpox was supposed to be the original source. Various theories have tried to explain where vaccinia could have come from, including hybridisation between cowpox and smallpox and (perhaps the most likely) mutation from the now-extinct horsepox.[24] The vaccinia mystery will be picked up again in Chapter 13.

Finally, that peculiar taterapoxvirus, named for the wild gerbil (*Tatera kempii*) from which it was isolated in 1975.[25] This now appears to be the ancestor from which smallpox was descended. Its DNA sequence is the nearest to variola's and therefore would need the smallest tweak by a cosmic ray or some other mutagenic force to turn into something like smallpox.

Molecular studies of the poxviruses' DNA sequences have recently been tied in with historical records in a stunning piece of genetic archaeology that shows how smallpox probably evolved.[26] It seems that taterapox underwent an initial mutation (A in Fig. 1.3) some time between 16,000 and 68,000 years ago (the limits are wide because it is difficult to calculate the interval between each tick of the mutational clock). This modified taterapox then mutated again (B) to spin off *Variola minor* between 1,400 and 6,300 years ago, coincidentally in the part of Africa where Shapona originated. Later, a separate mutation (C) gave rise to *Variola major*, probably in Asia between 400 and 1,600 years ago. This scenario makes sense of the gaps in the historical records – quite simply, full-blown smallpox had not yet arrived in classical and biblical times. It also suggests that Ramses V had *Variola minor* (and perhaps died of some other acute illness) and may explain why Rhazes thought that measles was worse than smallpox (which would also have been *Variola minor*).

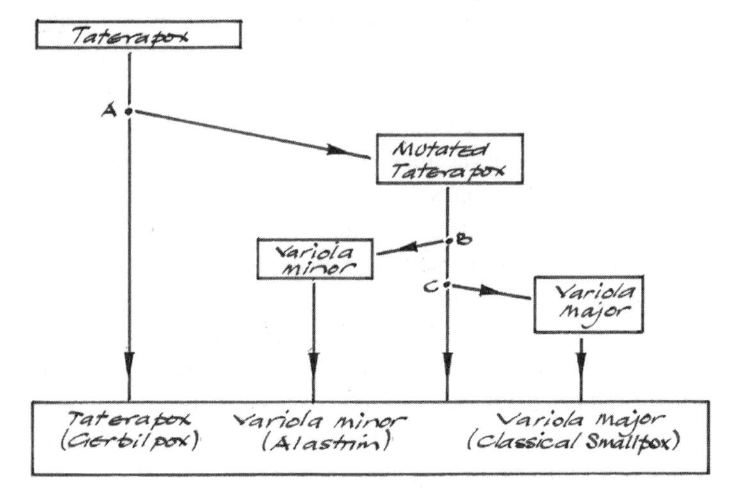

Figure 1.3 Evolution of the smallpox (variola) viruses from their precursor poxvirus.

These mutations changed the virus's preferences for ever and in two crucial ways. They enabled the new viruses to jump the species gap into man but also severed all connections with the gerbil and all other animals. The mutations therefore brought us the curse of smallpox but also enabled it to be eradicated by cutting it off from hiding in other species.

So much for the variola virus in its own right: a protein-wrapped block of DNA with an intriguing past. It is now time to add the human body, the ingredient that brought it to 'life', and see it in action.

Danger: Smallpox at work

"Lord Dalkeith is dead of the small-pox three days. It is so dreadful in his family that, besides several uncles and aunts, his eldest boy died of it last year and his only brother, who was ill but two days, putrefied so fast that his limbs fell off as they lifted him into the coffin."

Horace Walpole Letter, 2 April 1750

All infections need a portal of entry into the body. For the smallpox virus, this was nearly always the lung and the source was generally the air around an infectious patient, which was loaded with free-floating, highly contagious virus in tiny saliva droplets and particles of dust from dried blister fluid or scabs. This invisible cocoon extended for perhaps two metres around the patient and could rapidly contaminate a small room. With the help of defective air-conditioning, smallpox could spread further, including distant wards in a hospital, and it was even suggested that windborne virus could reach the shore from an isolation hospital boat moored in the Thames Estuary.[27] In general, though, there was safety in distance and effective isolation could clamp down on the infection. The eighteenth-century practice in Eastern Africa of dragging away patients to die outside the city walls probably worked, as long as the dogs which waited for them were prevented from bringing scraps back inside.[28]

The 1962 directive from the British Department of Health that mourners should be kept at least 20 feet from a victim's coffin was probably over-zealous; the virus would have found it hard to escape from the putty-sealed, disinfectant-filled coffin.[29]

The infectious phase usually began when blisters opened up in the throat (often two days before the rash appeared) and lasted until the rash had dried up and the last scab had fallen away – a total of four weeks and sometimes longer. All the lesions in the throat and skin, including the scabs, teemed with virus but the infection was usually transmitted by droplets of saliva spread by breathing and coughing. The patient's internal organs and fluids were also saturated with the virus and several twentieth-century doctors who had not kept up their vaccinations died after performing a post-mortem on a victim of undiagnosed smallpox.

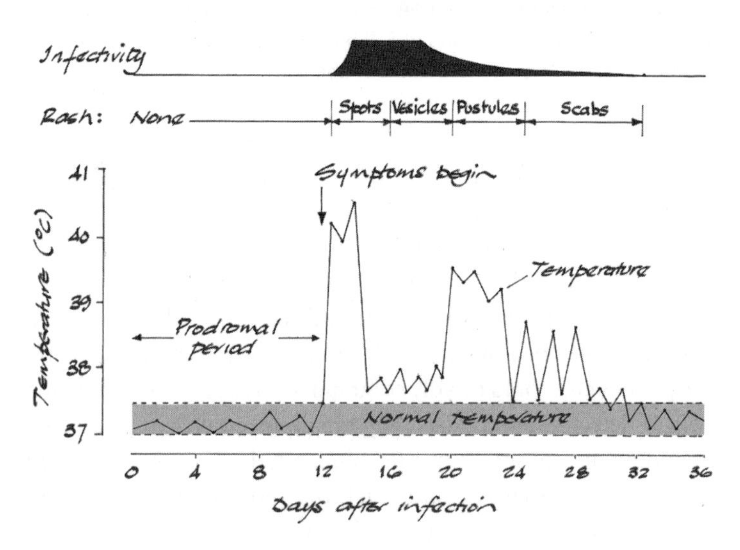

Figure 1.4 Clinical course of smallpox.

Bedding which mopped up pus from bursting blisters could remain infectious for several days. This fact was exploited by English settlers in North America in an early attempt at germ warfare and 'ethnic cleansing' by giving carefully contaminated blankets to unsuspecting Native American Indians (see Chapter 2). Bedbugs, mostly unnoticed against the misery of smallpox, could become infected by sucking up virus-laden blood but, because they were also confined to bed, were unlikely to spread the disease to others.

Outside its host, the variola virus was frighteningly robust. Dried blister fluid could reignite infection after three months while dried scabs remained viable for even longer.[30] This was well known to Chinese physicians who, for reasons that will be explained, stored smallpox scabs in stoppered porcelain flasks for many months. Bales of cotton contaminated during an epidemic in Alexandria caused an outbreak when they were opened in a mill near Manchester several months later. Under exceptional conditions, smallpox material has remained infectious for over two years. However, if the Curse of the Mummy had killed off those who desecrated the tomb of Ramses V, smallpox would not have been to blame.

First ports of call

Once inhaled, the viruses were carried deep into the lung to the cells that make up the walls of the air sacs (alveoli). The viruses slipped inside, shed their protein envelope and released the DNA core which immediately started up the programme to produce hundreds of copies of itself. DNA is normally confined to the cell's nucleus but this replication process took place in the cytoplasm, forming the densely-packed 'virus factories' which Guarnieri saw as particles down his microscope.

The stretch of DNA (genome) inside the smallpox virus is exceptionally long for a virus and contains 187 genes (about 1 per cent of the number in humans).[31] Its genes encode enzymes to replicate the viral genome, various proteins that wrapped up each DNA copy to make a complete new virus particle and other proteins that interfered with the host's immune system and blocked its ability to kill viruses. Smallpox destroyed the cells which it infected by starving them from within (scooping up essential nutrients to fuel viral replication) and finally stretching the moribund, virus-stuffed bag beyond bursting point.

When the alveolar cells ruptured, the viruses flooded into the tissue fluid (lymph) and were carried to the lymph nodes at the root of each lung.[32] The nodes, normally an important checkpoint for spotting and stopping infection, fell victim to the viruses and the same replication sequence then repeated itself – but on a vastly larger scale, thanks to the hundreds of copies churned out by each of the original invaders. Following this frenzy of replication, new viruses poured into the bloodstream (viraemia) and were then carried to every part of the body.

Until then, all this activity had taken place behind the scenes and passed unnoticed by the victim.[33] This incubation period typically lasted twelve days. Now, with the bloodstream suddenly awash with the toxic waste of dead cells, the first symptoms appeared: a fever up to 40 °C (104 °F) and a splitting headache, often with crippling pains in the back. This prodromal phase of symptoms generally lasted a couple of days. The victim felt and looked ill but did not yet have the diagnostic rash. Smallpox could not be confidently diagnosed and was often missed, especially in sporadic cases with no ongoing outbreak to raise suspicions. Unfortunately, the virus-laden blisters in the mouth and throat often made some patients highly contagious at this stage. Unsuspected cases commonly passed the infection on and triggered outbreaks before the diagnosis was made.[34]

I've got you under your skin

During the prodromal period the viruses began to replicate throughout the body. The most obvious target of variola was the skin, but post-mortem examination of fatal cases showed that it hit virtually every organ in the body to some degree.

In the skin, variola settled preferentially in the deeper layers and particularly in regions rich in the sebaceous glands that keep the skin oily. The pattern was centrifugal, affecting the head, hands and feet more than the trunk and, thanks to the distribution of sebaceous glands, was often worst on the face.[35]

The rash began as a flat red spots (macules), overlying centres of viral invasion. These developed over a couple of days into slightly raised papules.[36] As the virus-laden cells underneath swelled and burst, the papules rose into the typical blisters. As these lay deep in the skin, they initially felt tense – classically described as feeling like lead shot to the fingertips. As the eruption matured, the blisters enlarged and their roofs became thin enough to show the fluid inside. Early blisters (vesicles) contained clear fluid which after a couple of days turned milky and finally became thick yellow pus (pustules). The vesicle fluid was coloured by the body's immune cells that poured into the

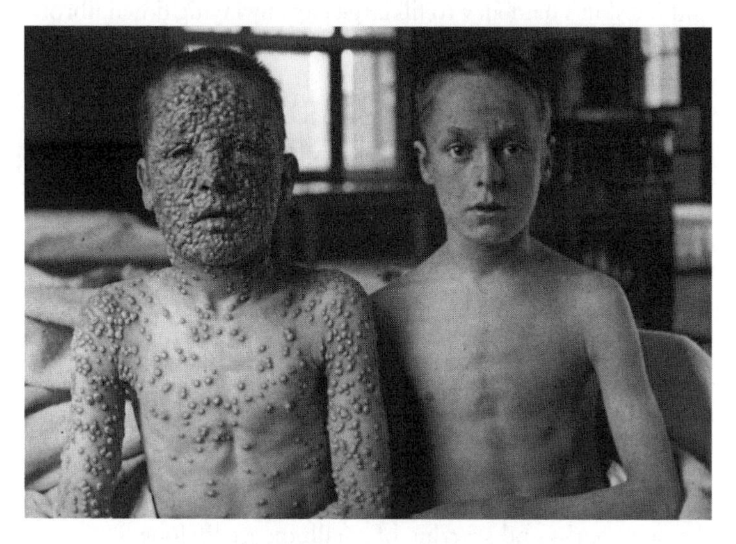

Figure 1.5 Smallpox rash in two boys exposed to the infection on the same day during an outbreak in Leicester, 1892. The boy on the right had previously been vaccinated. Reproduced by kind permission of the Edward Jenner Museum, Berkeley, Gloucestershire.

sites of invasion to try to kill off the virus and the cells which were breeding them.

If the patient survived, the blisters remained for a week or so and then began to collapse as the fluid inside melted away. Shortly after, the pustule's roof dried out to form a scab which fell off a few days later. With the loss of the last scab – sometimes four weeks after the rash began – the patient was no longer infectious. It was only then that the full extent of the damage to the face could be seen clearly.

Mementos

Many survivors of smallpox had to live with permanent reminders of their brush with the Angel of Death, especially the characteristic pockmarks. These pits are wider and deeper than those of acne. Each deep-seated blister left behind a large cavity, and the skin's tendency to fill large gaps by laying down fibrous tissue made things worse because this scar tissue contracted, holding the pit wide open.[37]

The virus's predilection for sebaceous glands meant that the face was often badly affected. A few facial pocks could be easily covered up but severely affected cases could be mutilated beyond recognition. Picture a bad case of acne with perhaps 30 nasty pimples on the face. Then think big: the deep, open pits of smallpox, with up to 250 pocks on the face in 'moderate' involvement and 500 in a confluent case. Beauty is in the eye of the beholder but, by medical criteria, about one-third of those who suffered an attack of variola major were left with 'significant' scarring, while 10 per cent were 'disfigured'.[38] As with severe burns, the skin runs out of ideas for repairing itself once it has resorted to patching up damage with fibrous tissue; the pockmarks and scarring of smallpox are lifelong.

Some unlucky cases were also left with permanent damage to the eyes. During a severe attack, many patients were spared the agony of seeing what was happening to them because their eyelids

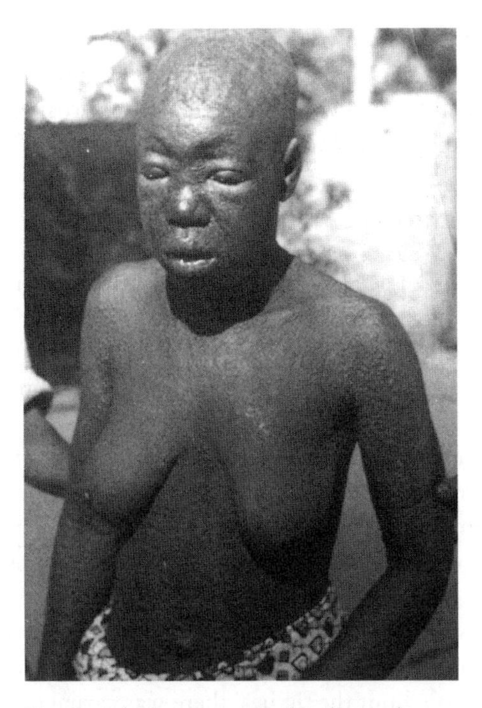

Figure 1.6 Scarring and blindness due to smallpox, in a young woman in Zaire, 1965. Reproduced by kind permission of the Edward Jenner Museum, Berkeley, Gloucestershire.

were pressed shut as the surrounding skin swelled and then glued together by pus as the blisters burst. Luckily, most were left with clear vision when the eruption settled but some also suffered inflammation of the cornea (keratitis) due to direct invasion by the virus. As in the skin, this left behind permanent and dense fibrous tissue – which unfortunately does not transmit light.

During the eighteenth century, smallpox was one of the commonest causes of blindness, responsible for over a third of cases in Europeans.[39] Those blinded by smallpox in childhood included the Syrian poet Abu'Ala al-Ma'ari and the Irish harpist and composer Carolan. Their loss of vision may have been mankind's gain because it brought out their particular creativity, but this was no comfort to the millions who lost their independence, livelihood or marriage prospects through being blinded by smallpox.

The end of the road

The mortality of *Variola major* ranged from 10 per cent to 50 per cent and averaged 20 per cent. The risk of death depended on the virulence of the smallpox strain (some were particularly lethal) and especially on the state of the patient's immune system and general health.[40]

The Angel of Death claimed its victims in various ways. Most died from toxaemia, with circulatory collapse and multiple organ failure from damage to the microcirculation, the dense network of tiny blood vessels that supplies the tissues. This could happen shockingly fast: patients could be dead within hours of starting to feel unwell and before any rash had appeared. Externally, the only abnormality might be some small haemorrhages in the skin; throughout the tissues, there was carnage. Severe lung involvement was also usually fatal.

The skin eruption itself could be life-threatening. Bursting of the blisters, helped by scratching or friction against bedding, exposed raw tissue like a third-degree burn that continually oozed plasma. Patients with an extensive rash could lose enough plasma to slide into cardiovascular collapse, when the circulating blood volume falls below the critical level needed to keep the vital organs perfused and functioning. In these cases, death usually occurred around the unlucky 13th day of the rash.

Finally, smallpox killed some of those who had survived the acute attack. For girls and women in particular, severe scarring and disfigurement meant the end of life as they were able to accept it. Some withdrew into their own world and wasted away through depression and self-neglect, while others took more direct action and killed themselves.[41]

From major to minor

"The distinct kind a nurse cannot kill; the confluent a physician can never cure."

William Wagstaffe 'Letter to Dr. Freind', 1722

It was long recognised that the rash of smallpox took different forms and had different outcomes. Patients with a discrete (or distinct) rash of relatively sparse lesions could expect a milder course and to survive, usually with minor scarring. This was the commonest variety and its mortality rate was generally 20 per cent or less. A confluent rash, with densely packed blisters that ran into each other and crowded out normal skin, carried a much higher mortality rate of over 60 per cent. Most survivors of confluent smallpox were severely scarred.[42]

Two rarer clinical presentations, together accounting for fewer than 10 per cent of all cases, were almost always fatal.[43] Malignant smallpox, characterised by dark flat lesions rather than blisters, killed over 90 per cent of its victims, while haemorrhagic smallpox with multiple bleeds into the skin, either with or without the usual rash, was almost always lethal.

By comparison, *Variola minor* (also known as alastrim) was a gift from heaven: a benign clinical course, with the risk of death lower than in measles and chickenpox, and the near-certainty of escape with little scarring.[44] The mortality rate in a large British series of cases was only 0.13 per cent, when coincidental causes of death were excluded. Even though their biological effects were dramatically different, the smallpox virus strains that caused *Variola major* and *minor* were structurally so similar that the immune response raised by infection with *Variola minor* also protected against *major*. *Variola minor* was so mild that most sufferers never bothered to consult a doctor or to isolate themselves, which favoured its rapid spread. By conferring immunity against *major*, it steadily crowded out its homicidal twin, especially in North America in the early twentieth century (see Chapter 15). Elsewhere, *Variola major* remained the dominant threat and continued to kill 20 per cent of its victims, as it had done one or two centuries earlier.

A bad one to get wrong

"Clinical observation can receive powerful aid from the virologist."

A.M. Ramsay and R.T.D. Emond *Infectious Diseases*, Chapter 14, 1977

A classical case of smallpox was unmissable and unforgettable but unfortunately smallpox sometimes ignored the rules. It could be mistaken for chickenpox, measles, syphilis (nicknamed the 'Great Deceiver' because it mimicked so many other skin diseases) and other rarer eruptions. Even experts missed the diagnosis from time to time, including the legendary Ricketts, author of an authoritative early twentieth-century textbook on smallpox.[45] Some of these missed cases were not isolated quickly enough and went on to cause fatal outbreaks.

The commonest cause of confusion was chickenpox.[46] Classically, smallpox was least dense on the trunk, favoured the backs rather than the fronts of the hands and arms, went through the same stage simultaneously across the whole body and was accompanied by severe general symptoms. According to the books, chickenpox did none of these things, but in real life, atypical cases of both diseases were not unusual.

Various laboratory tests were developed to diagnose smallpox.[47] The virus could be grown from samples of blister fluid and identified by electron microscopy (which could distinguish variola from chickenpox) or by inoculating live hens' eggs, incubating them at various temperatures for three days and then opening them to see how the virus had grown on the chick embryo's membranes. Characteristically, smallpox produced small white pocks and could not survive above 38.5 °C.

Particular proteins in the virus's outer coat could be detected in samples of vesicle fluid or scabs using 'reference' antibodies supplied to diagnostic laboratories. This specific-antigen test was favoured as it was extremely sensitive, reliable and quick to return a result. Alternatively, the antibodies which the patient made against the variola virus could be detected in blood samples. However, this test took seven to ten days to become positive, remained negative in some patients with overwhelming smallpox and could be falsely positive in subjects suffering from something else but who had been vaccinated against smallpox.

Diagnostic tests were most important for limiting the spread of infection by identifying patients who needed to be isolated; as there was no specific treatment for smallpox, there was little direct benefit to the patient. None of these tests was totally trustworthy and all took time: results sometimes returned after the patient's death. Performing the tests was also hazardous and occasionally fatal for the medical, nursing and laboratory staff who had to handle the highly infectious samples.

Today, the spectre of smallpox has faded from medical consciousness and the laboratory tests used to identify smallpox have long since fallen into obsolescence. If a case somehow appeared now, the diagnosis would probably be missed.

And in no time, the Angel of Death would again be stalking the streets.

Of Man and Angel

The Greatest Killer: smallpox in history.
Donald R. Hopkins Book title, 2002

The brief of the Angel of Death was even wider than the name implies. Smallpox also brought mutilation and blindness, dread and loathing and drove people to inflict hideous cruelty on others. Its impact on mankind was vast, wrecking societies and entire civilisations and ultimately shaping history.

This chapter picks out some highlights of our complicated relationship with smallpox. The focus is on the period of the Angel's greatest triumph which lasted from the Middle Ages until the start of the nineteenth century. As far as Western medicine was concerned, this era could also be called the Age of Passivity. There were many treatments for smallpox, and doctors and their patients believed in them, but unfortunately all were useless. It was only during the first quarter of the eighteenth century that Western doctors first took any direct action against the disease by adopting a peculiar and risky measure that, according to Asian and Arabic medical traditions, could protect against it.

Navigating a sensible path through this part of the story is a challenge. Donald Hopkins' *The Greatest Killer* gives exhaustive accounts of scores of outbreaks which would seem to justify the superlative in his title. However, mass murder on this scale soon loses all human relevance because the numbers of deaths are are too vast to comprehend. Instead, I have pulled out some headline statistics to illustrate the killing power of smallpox and have concentrated on what the disease meant to individuals.

This chapter also introduces some of the treatments touted for smallpox. These deserve mention, not because they had any useful impact on the disease, but because they were woven inextricably into the experience of smallpox. Like smallpox itself, some were a ritual that had to be endured; many enhanced the horror of the experience and some undoubtedly killed rather than cured.

A way of life, a way of death

"Infinitely more destructive than the plague itself."
John Baron *The Life of Edward Jenner*, vol. 1, p. 261, 1827

On breaking new ground, smallpox gradually settled into an unstable relationship with its host which was influenced by the size and growth rate of the population, its mobility and how effectively it isolated itself from renewable sources of infection. The variable interplay between these factors produced distinct patterns of smallpox infection.

Seventeenth-century London was a chaotic expanding sprawl with huge volumes of trade with all corners of the globe. Smallpox was always present but mostly smouldered away at a relatively low background rate before flaring up every few years into minor or major outbreaks that could kill up to 3,000 of its inhabitants. The year of the Great Plague (1665) was unusually quiet for smallpox, with just eleven deaths – half the recorded death-toll from 'worms' and twice the number who reportedly died of fright or grief. Evidently the plague, which that year claimed 4,237 lives, carried off most of those who were otherwise destined to catch smallpox.[1]

The picture was different in Boston which was relatively self-contained, had more restrained population growth and was partially protected by quarantine regulations that blocked some

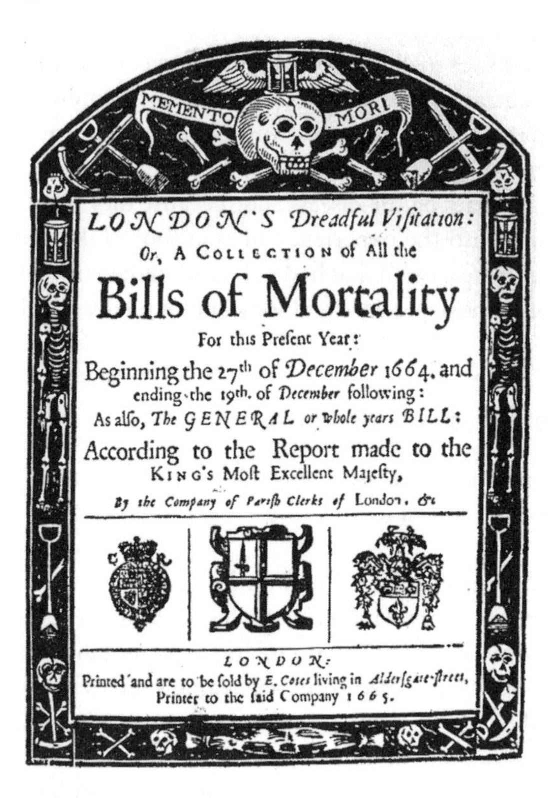

MEMENTO MORI

LONDON'S Dreadful Viſitation:

Or, A COLLECTION of All the

Bills of Mortality

For this Preſent Year:

Beginning the 27th of *December* 1664. and
ending the 19th. of *December* following:

As alſo, *The GENERAL or whole years BILL:*

According to the Report made to the
KING's Moſt Excellent Majeſty,

By the Company of Pariſh Clerks of London, &c.

LONDON:

Printed and are to be ſold by *E. Cotes* living in *Alderſgate-ſtreet*,
Printer to the ſaid Company 1665.

Figure 2.1 Front sheet of the London Bills of Mortality for 1664–65, during the Great Plague. The death rate from smallpox was unusually low during this period.

imports of smallpox through its busy harbour. Also, from the late seventeenth century, people with contagious diseases were forcibly isolated, together with their immediate contacts. The population of susceptible subjects – mostly children who had not yet met smallpox and had no immunity against it – grew back steadily after each outbreak, like underbrush after a wildfire. As a result, Boston had few or even no cases in most years, but when the susceptible population had built up to a critical level the balance would tip in favour of smallpox and the arrival of just one or two infected cases could trigger a devastating outbreak. This happened every twelve years or so.[2]

Elsewhere, for example in Mexico and many Caribbean Islands, smallpox burned constantly rather than smouldering and flaring up. Sustaining itself on the constantly renewed population of infants and children, it killed perhaps one in two of its victims and allowed the survivors to join the ranks of the immune. These countries had originally been infected from Europe and now regularly returned the compliment. The Caribbean Islands fed into busy sea-trade routes that converged on the ports of the eastern American seaboard and were just far enough away for new cases to reach ports such as Boston: well enough to go ashore and not yet showing a diagnostic rash but already highly infectious. The outcome was entirely predictable and the savage outbreak that hit Boston in 1721 figures in Chapter 5.

Throughout its dominion, smallpox could drop without warning on to anyone, irrespective of background, wealth or health. Smallpox was entirely democratic in its selection of victims. Kings and queens made news when they were struck down but they faced the same ordeals and uncertainties as the poorest of their subjects, whose battles for survival went unnoticed outside the illusory protection of the castle walls.

Smallpox became a fact of life and a rite of passage that nobody could ignore. As the French mathematician Charles-Marie de la Condamine put it, "a river that all must cross".[3] Surviving an attack was the gateway to a life that could still be cut short by infections such as tuberculosis and cholera but at least was secure against a return visit from the Angel.

Life during this period was a lottery that present-day citizens of the developed world can barely imagine. Life expectancy in most European countries averaged 35 years and was even lower among the poor, most of whom would have been presumptuous to look beyond their thirtieth birthday.[4] It is fortunate that children were overproduced, because only three in five would survive infancy and just two of them would reach reproductive age – and of course, childhood mortality rates were much higher among the poor.

During the Age of Passivity, smallpox was one of the major players that presided over this lottery, as calculated in the late eighteenth century by mathematicians such as James Jurin in London and Daniel Bernouilli in Basel.[5] Their conclusions added terrifying scientific credibility to the conviction that the Angel was out to get you. In England and other European countries, the chances of catching smallpox during a lifetime were almost one in three. For those who caught it, the chances of dying from it were around one in five. Overall, smallpox killed about one in twelve of all mankind. Each year, smallpox claimed over 200,000 lives in Europe and 600,000 worldwide. During one 25-year period in the eighteenth century, 15 million people died from smallpox.

Until vaccination took hold in the early 1800s, children were the most likely to die from smallpox. During the eighteenth century, the under-tens accounted for between 80 per cent and 98 per cent of all smallpox deaths in the countries of Europe. In late-nineteenth-century China, smallpox served as a retrospective birth-control measure, according to Dr. G.E. Morrison in his travelogue, *An Australian in China*:[6]

> Infanticide is hardly known in that section of Yunnan of which Tali is considered the capital. Small-pox kills the children. There is no need for a mother to sacrifice her superfluous children, for she has none.

During the seventeenth and eighteenth centuries, smallpox was not the only contender for the title of 'Greatest Killer'. The lingering curse of tuberculosis killed one in five English people during the eighteenth century, a higher proportion than smallpox.[7] In London, deaths from 'fever' regularly outstripped those from smallpox between outbreaks, although this ragbag heading covered many different illnesses. Measles was another competent killer which, like smallpox, targeted children.[8] The annual returns of deaths in the London Bills of Mortality did not distinguish between the two until 1700, but when they were counted separately it became clear that measles often carried off as many victims as in smallpox's quieter years. Also, measles was almost as destructive as smallpox when it burst into previously untouched territory such as Mexico, but overall never quite matched the global penetration of smallpox.

An obvious competitor was the 'Scourging Angel' of bubonic plague.[9] As the Black Death in the mid-fourteenth century, it had ploughed through Europe and annihilated 20 million people in just three years. It continued to terrorise the world for another five centuries, its highlights including the Great Plague of London in 1665. Mortality from bubonic plague was up to 80 per cent but, unless trapped in an outbreak, an individual's chances of catching it were very low. Macaulay, in his *History of England*,[10] captured the difference between the two marauding Angels:

> Smallpox ... was the most terrible of all the Ministers of Death. The havoc of the plague had been far more rapid: but the plague had visited our shores only once or twice within living memory; and the smallpox was always present, filling the churchyards with corpses ... .

Aftermath

Smallpox did more than just kill. It often made a savage mockery of the saying that "beauty is just skin deep". The human face is much more than an image in a mirror; it is the essence of our personality and identity and it conditions how we and others think of us. It is difficult to imagine the shock of looking into a mirror and seeing the face of a stranger. Unlucky survivors of smallpox had to confront the horror of a rough pitted mask with their eyelashes stripped out – a face that, a month earlier, would have made them turn away in disgust – and the realisation that this was how they and the world would see them for as long as they lived. The full severity of disfigurement is illustrated by Sister Isla Stewart's dispassionate account of a smallpox outbreak in London in 1884–5:[11]

> I have known two sisters live in the same camp, eat at the same table, and see each other daily for a week before they recognised each other.

Nowadays, seeing images of what smallpox did to people's faces can still bring some understanding of why mirrors were hidden until disfigured patients were judged strong enough to cope with what they would see – even during the 1950s[12] – and also why the Angel of Death claimed some of its near-escapes afterwards, through suicide.[13]

For society at large, smallpox scarring was taken for granted. Many people sported at least a few pocks. Sensitive women could cover up a few marks with the false moles known as 'beauty spots', or more extensive scarring with a plaster-skim of heavy cosmetics or a veil. A completely pock-free face stood out – as in the legendarily unmarked milkmaids of Gloucestershire – and was enough of a distinguishing feature among the criminal classes to be mentioned specifically on 'wanted' posters. Mozart had himself painted as he was, pocks and all, but George Washington had his skin smoothed by the artist (except for one true-life portrait, now hanging in a medical society's meeting room in North Carolina).[14] Stalin similarly had his heavy pockmarking airbrushed out of official photographs.

Smallpox scars ruined life's prospects for some but were a passport to employment for others. During the eighteenth century, household servants and nurses had to have pockmarks as hard evidence that they had already met smallpox and were therefore immune – and so could be trusted not to bring in the disease and to continue working if someone else did. In the words of a newspaper advertisement of 1774:[15]

> Wanted, a man between 20 and 30 years of age, to be a footman and under-butler in a great family; he must be of the Church of England and have had the smallpox in the natural way.

The therapeutics of despair

"Consult an honest physician."

John Wesley *Primitive Physick: or, an Easy and Natural Method of Curing Most Diseases*, 1747

Even in the 1970s, with the Angel of Death about to be banished from earth, smallpox was untreatable. There was never any drug that could rescue its victims in the way that antibiotics can cure bacterial infections. This did not prevent smallpox sufferers from being treated, often intensively, throughout history. More accurately, they were subjected to various rituals that were a valuable earner for doctors but were uncluttered by any evidence that they worked. Smallpox was by no means unusual in this respect. Up to the twentieth century and the advent of formal clinical trials, medicine was plagued with treatments that were spurious but so deeply embedded in tradition that nobody dared question whether they were useful or could be dangerous.

Many of the pseudo-treatments for smallpox can be blamed on the great fathers of European medicine. A fine example was the Hippocratic notion, later embroidered by Rhazes, that fever was caused by an excess of blood, the 'heating humour'.[16] This could be relieved by making the patient sweat, which also dragged the contagion of smallpox to the skin where it could be safely expelled. So the fire was stoked up, the windows and doors shut and the victim tightly wrapped in blankets – even if this tore open the fragile pustules. Some experts insisted that the bedding must not be changed until recovery; it must have been an additional bonus to be released from a stinking cocoon that was stiff with a three-week concretion of pus and tissue fluid. If sweating was inadequate, it could be forced with various herbs and toxic metal salts. 'Heat therapy' ruled until the mid-seventeenth century, when favour swung to the opposite extreme of the 'cold therapy' introduced by Thomas Sydenham.[17]

The colour red was also claimed to assist healing: red bedclothes and bedding warmed the skin, drawing out both sweat and contagion. The use of red cloth (especially flannel) to ward off or treat smallpox can be traced across Europe, Asia and America from thirteenth-century England and Spain to Boston in the 1760s and even to the red clothes, toys and sweets given to infected children in Japan in the 1830s.[18] Red and heat therapies could be combined, as in the twenty yards of red English flannel in which Joseph I, Emperor of Austria, was enveloped to await death from smallpox in April 1711.[19] The ultimate extrapolation of red therapy was the 'red light' treatment developed in the early 1900s and which remained fashionable for a quarter-century after it was shown to be useless (see Chapter 14).

Bleeding was the most dramatic legacy of the theory that too much blood caused fever.[20] This was one of the doctors' few stock tricks to prove their mastery over the body (others included the induction of diarrhoea with purgatives or vomiting with emetics). The patient was bled either by cutting into a distended vein with a lancet or sticking on a couple of dozen well-starved leeches. Typically, several ounces (half a wine bottle) of blood would be removed at one sitting, although enthusiasts aimed for a whole pint. This was repeated as necessary, sometimes until the patient fainted or could not stand. For the smallpox sufferer, the first prodromal symptoms of fever, headache and vomiting would summon the doctor with his lancet and/or jar of hungry leeches. Once a bad eruption had blown up, leeches were harder to use because there was so little normal skin for them to get their teeth into. One expert ingeniously exploited a loophole in the coverage of the rash, by recommending the application of leeches to the anus.[21]

Bleeding was bizarre, pointless and potentially dangerous but trusted by patients and doctors alike. Even Edward Jenner was an enthusiastic bleeder, to the extent of being able to conduct systematic experiments on the growth-promoting effects of

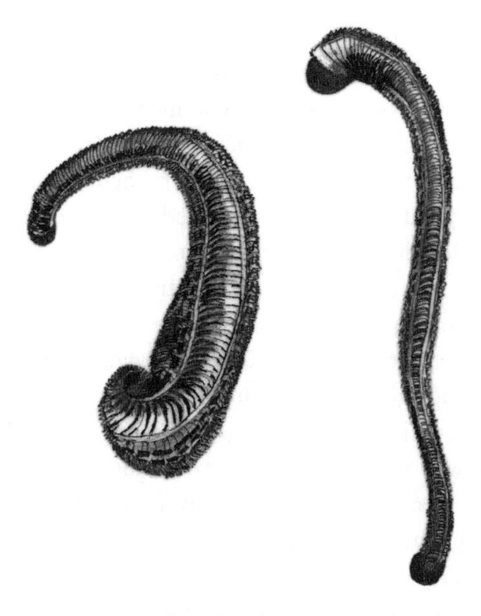

Figure 2.2 Medicinal leech, used for therapeutic bleeding and the treatment of smallpox up to the end of the nineteenth century.

human blood on plants (see Chapter 9). Astonishingly, bleeding was still practised until the end of the nineteenth century. The expert who recommended anal leech therapy in smallpox was Dr. Hubert Boëns of the Belgian Academy of Medicine. A staunch opponent of Pasteur, Jenner and vaccination, Boëns wrote in the security of knowing that his views would be widely and warmly received; his paper was published in 1884.[22]

More therapeutic suffering piled up for the Angel's victims. Hippocrates *et al.* had decreed that fevers must be treated by starvation, so they were starved.[23] Some doctors also withheld all fluids. Some patients might have preferred a raging thirst to the agony of trying to swallow with a badly ulcerated throat, but dehydration could rapidly endanger those already compromised by massive fluid losses across raw skin. Eventually, in the mid-seventeenth century, Thomas Sydenham managed to swing fashion with the sensible measure of encouraging drinking.[24] Plain water was too simple, so recommended fluids ranged from the bright red (pomegranate juice or mulberry wine gargles) to beer given an extra tang with sulphuric acid. The latter might not have slipped easily down ulcerated throats, especially at the recommended dose of twelve bottles a day.[25]

Vomiting and diarrhoea were additional miseries commonly inflicted by smallpox. Doctors saw these as laudable evacuations and evidence that the body was fighting back, and often helped Nature along with emetics and laxatives.[26] The nastiness of the outcome is hard to picture: nausea or griping pains in the belly, followed by urgent retching or defaecation, especially in patients so weak that they could not even lift their head from the already soiled bed. Yet the doctors persevered: after all, these were among the few treatments that produced any effects obvious to the patient and onlookers.

The doctor's bag was also plundered for various herbal and chemical treatments. During the Age of Passivity, several genuinely active drugs were in use, although the process of applying them to their target diseases had been a random journey and there was no inkling of how they worked.[27] Quinine-rich cinchona bark from Peru could treat malaria, autumn crocus (active ingredient, colchicine) was useful in gout, while opium from the sleep-giving poppy killed pain; the 1760s brought extracts of willow bark (later shown to be a source of aspirin) and foxglove (digitalis) to treat fevers and heart failure respectively.

Many extravagant claims were made for remedies for smallpox, which ranged from the banal to the top secret. Lentils and rape seed were a "certain cure", according to John Wesley in his encyclopaedic and monumental herbal, *Primitive Physick*, of 1747.[28] At least these will have obeyed the medical dictum *Primum non nocere* ('first do no harm'), unlike the patent 'powders' which turned out to be a toxic slurry of antimony and mercury salts (see Chapter 6). In real terms, this was no advance on the Macbethian mixture of powdered crabs' eyes and flour which King William III of England had applied daily a century earlier to ward off the smallpox which had killed his wife.[29] The skin eruption sometimes received special medical attention.[30] Bursting each pustule, ideally with a silver or gold needle, was advised to speed recovery and reduce scarring. Blisters were also painted with carmine, applying the magical red directly to the contagion.

What about the doctors themselves? They were the only consistent beneficiaries of the Angel's visitations, as smallpox created a steady income for them, especially during outbreaks. Many made their reputation by adding their own dash to the mystical mess of smallpox treatment. The total absence of any supporting evidence never prevented them from defending their position to the hilt – literally, in the cases of Richard Mead and John Woodward. In 1719, these illustrious physicians and Fellows of the Royal Society fought a duel with swords over which end of the smallpox patient's bowel should be attacked with drugs (Mead favoured purges and Woodward emetics).[31] Both survived and continued to inflict their own brands of intestinal quackery on their victims (and both reappear, better behaved, later in this book).

By default, this was the age of the art of medicine. Even if the physician-scientist had nothing to offer, the doctor-comforter could bring reassurance – after all, some of his previous patients had survived. Payment for his services also helped to bolster the illusion that something active was being done. Smallpox was always a powerful magnet for quacks but most doctors probably believed that they were doing good, and it was in their patients' best interests for them to cling to that belief too.

Doctors were good at judging when things were going badly, even though the classical 'Hippocratic facies' (a leaden complexion and sunken features) of those about to die was hard to detect beneath a confluent rash. Other 'vital signs' were not sought. The pulse was not checked systematically until the nineteenth century (when its interpretation became an art-form) and blood pressure could not be measured at the bedside until the early twentieth century. Temperature was judged with the hand, but probably not in a bad case of smallpox. For clinicians, the thermometer remained a quaint curiosity until the late nineteenth century; Jenner used a prototype in 1788, but only to measure the internal temperature of hibernating hedgehogs (Chapter 9).

On balance, doctors were a liability whose treatments helped smallpox to kill its victims. This was pointed out by Sydenham, who noted that mortality rates were higher among the wealthy than among the poor who could not afford medical help.[32] Some explanations seem obvious. Patients teetering on the brink of circulatory collapse could be pushed over the edge by treatments that further reduced the blood volume – bleeding, massive fluid losses across skin stripped raw by friction with heavy bedding and the effects of purgatives and emetics – and not being allowed to drink. 'Cures' based on toxic metals were poisonous enough in health, let alone when the victim was locked in mortal combat with smallpox.

Many patients might have fared better under some traditional medical regimes of 'uncivilised' peoples than with the state-of-the-art 'primitive physick' of European medicine. In Sudan, smallpox sufferers were smeared with butter from head to foot, probably a kinder experience than being trussed up in a shroud of heavy flannel.[33] In Buru, East India, they were tapped with a bundle of branches which were then thrown into a boat that was rowed out to sea and abandoned with an appeal to the local incarnation of the Angel: "Grandfather Smallpox, go willingly away."[34] Even if Grandfather Smallpox had clung on, this might have been a kinder end than the misery inflicted by the clever doctors in London or Paris.

Journeys with the Angel

"There is no disease that the medical writer has to describe, which presents a more melancholy scene than the natural smallpox."

Dr. John Thornton *Philosophical Magazine*, vol. 20, p. 146, 1805

When people met smallpox, they had no idea how it would end: a lucky escape with a few pockmarks, the living hell of mutilating scarring and blindness, or death. And while the victim fought his or her solitary battle with the Angel of Death, their family and friends would be suffering their own agony, waiting in dread to see what would emerge from the sickroom.

The following accounts give some insights into what it was like to have smallpox.

Extracts from Macaulay's History of England,[35] *describing the final illness of Queen Mary II, wife of King William III of England. She died aged 32, just before the New Year of 1695.*

> At length the infection spread to the palace, and reached the young and blooming Queen. She received intimation of her danger with true greatness of soul. She gave orders that every lady of her bedchamber, every maid of honour, every menial servant who had not had the smallpox should instantly leave Kensington House. She locked herself up during a short time in her closet [and] calmly awaited her fate.
>
> During two or three days there were many alterations of hope and fear. The physicians contradicted each other and themselves in a way which sufficiently indicates the state of medical science in that age ... At length all doubt was over [and] it was plain that the Queen was sinking under smallpox of the most malignant type ...
>
> Tenison [the Archbishop of Canterbury] undertook to tell her that she was dying. He was afraid that such a communication abruptly made might agitate her violently, and began with much management. But she soon caught his meaning, and with that meek womanly courage which so often puts our bravery to shame submitted herself to the will of God. She called for a small cabinet in which her most important papers were locked up, gave orders that, as soon as she was no more, it should be

delivered to the King, and then dismissed worldly cares from her mind. She received the Eucharist, and repeated her part of the office with unimpaired memory and intelligence, though in a feeble voice ...

After she had received the Sacrament she sank rapidly, and uttered only a few broken words.

From an account by missionaries of the outbreak of smallpox in Greenland in 1733.[36]

In one island the only living creatures they found were a little girl covered with the smallpox, and her three younger brothers. The father, having buried the rest of the inhabitants, had laid himself and his youngest child in a grave of stones, bidding the girl to cover him with skins; after which she and her brothers were to live upon a couple of seals and some dried herrings til they could get to the Europeans.

Extracts from the reminiscences of Emily Spence, born in Yorkshire in 1827.[37]

I don't know how old my Father was when he took the smallpox. He was eight years younger than his brothers and sisters and a great pet with them all. He was a very handsome boy, his head covered with clustering curls which remained with him when his head was white as snow; he never became bald though he lived to be eighty-one.

He took the smallpox badly, vaccination not being know then. One day, when they were dining with a large company of Friends during the Quarterly Meeting, one of the Friends said, 'Anne, where is my beautiful boy?' and poor Grandmother with tears in her eyes pointed to the dear little disfigured boy at her side.

Notes to nurses about the care of smallpox patients, by Sister Isla Stewart of the London Metropolitan Asylum Board isolation unit, in 1888.[38] *The unit consisted of three ships, the* Atlas, Endymion *and* Castalia, *moored in the Thames Estuary at Long Reach, near Dartford, with a convalescent camp on shore at Darenth.*

The patient should have as much air as possible, and his room should be kept at uniform temperature. The delirium is sometimes violent, but as little restraint as possible should be used. He should, however, never be lost sight of.

High temperature may be relieved by tepid sponging. The eyes should be carefully watched and kept clean, and the doctor's attention *at once* called to any soreness or swelling. Itching may be relieved by the application of Vaseline. The patient should be nursed on a water bed, and great care should be taken to keep his mouth as clean as possible. The bowels should be kept open and the condition of the urine observed, as haematuria [blood in the urine] is a symptom of grave import.

In nursing small-pox in hospitals the great question of cleanliness rises. If the case is confluent, the scalp is usually too sore to comb, and pediculi [head lice] congregate and multiply with great rapidity below the scabs. All that can be done is to cut the hair off as closely as possible and apply carbolic oil as freely as possible. A piece of lint cut in the shape of a mask, with holes for the eyes, nose and mouth, smeared with Vaseline and applied to the face, facilitates scabbing.

Such is the nursing needed for confluent cases. In discrete, little is needed, and in haemorrhagic it is of little avail except to soothe the last days of the unfortunate patient.

From the medical records of George Pinch, aged eleven.[39] *He was diagnosed with smallpox in London and transferred to the hospital ship* Endymion *in late April 1891.*

April 29th.	Diet low. Gargle potassium permanganate.
May 9th.	Abdomen very tender, legs drawn up.
May 17th.	Small bit of fish given.
May 25th.	Poultice to eye. Received visitor. Vomiting twice during the night. Soap and water enema.
June 3rd.	Complains of pain at base of neck and keeps crying out. Temperature 101.6.
June 4th.	Strabismus [squint], unconscious, left side of face red and swollen. Temperature at 9.45 p.m. 104.0.
June 5th.	Death at 2.30 a.m.

A personal account of an attack of confluent smallpox by Sister Margaret Magrath.[40] *She was stationed at the French military base in Langhouat, Algeria in spring 1945 and was then 36 years old. In her childhood, the local doctor had agreed to her father's request to 'forget to vaccinate her' because her elder sister had reacted badly to vaccination.*

The isolation unit was on the other side of the oasis, some two or three miles away. There was no means of transport so it was arranged to get an ambulance sent from Algiers, some 400 miles away, to carry me the short distance. By the time I was transported, the top half of my body was already pustulating badly and no one dared touch me. I got out of bed by myself and lay on a stretcher, then the doctor and one of my own nursing sisters carried me to the ambulance. I remember the doctor sitting beside me and continually taking my pulse. At the end of that short journey, he threw my arm down and saying "You'll never die." I learned later that no one had expected me to survive the journey.

Day and night, someone continually bathed my eyes which eventually stuck with with discharge. One morning, the nurse said to me that she had expected to find me blind – but I could see all right. I had had my hair cut off when the pimples first showed and later my head was one mass of suppuration. I remember shouting out with the pain.

I could not bear light or noise. All the openings of my body were filled with pus. My throat was especially painful and I could only whisper.

I don't remember other details except that the eruption on my body gradually extended to my feet. By that time the head, face and chest were developing a thick crust which was very itchy and difficult not to scratch. Gradually all large areas of crust fell off and these scabs were all burned. I remained conscious throughout all these weeks and knew and understood all that went on around me.

I fell ill in January 1945 and it was May before the last scabs left me, leaving me a bright, raw red. The doctor later told me that when the news got around that a European was dying with the disease, the Arabs – both sedentary and nomads – asked to be vaccinated. He said, "You have done a lot of good!"

He had reported me dead to the Pasteur Institute in Algiers as he believed it impossible that I should survive such a severe confluent attack.

Force majeure

According to chaos theory, the flap of a butterfly's wings in Brazil could set off a tornado in Texas. The wingbeats of the Angel of Death had more obvious consequences and triggered many events that changed the course of history. This Angel had global ambitions which affected every nation to some degree. Sometimes smallpox was a seismic force that annihilated civilisations; elsewhere, it simply gave a subtle nudge at decisive moments. Strangely, smallpox sometimes assisted Man's own intentions and even more bizarrely, Man sometimes gave the Angel a helping hand.

El Dorado

The conquest of South America by the Spanish and Portuguese owed much to smallpox. It was an integral part of the offensive and often spearheaded the campaign, ploughing relentlessly through countries and civilisations.

By the time Hernando Cortéz left Cuba to conquer the Aztecs of Mexico in 1519, smallpox already had a foothold on that island and two other Spanish possessions in the Caribbean, Hispaniola and Costa Rica.[41] Imported through the slave trade from Africa, it had already killed over a third of the native inhabitants of these islands and was poised for an assault on the virgin territory of the mainland. It was a single infected African slave with the rival force led by Pánfilo de Narvaez (sent to prevent Cortéz from establishing a monopoly) who carried the infection across.

The Spanish soldiers might have stood amazed at the wonders of the Aztecs but this did not stop them from destroying all that they had created. Even more ruthless was the new disease that struck down the Aztecs in their millions and which probably wiped out over half the population in just two years. The Spanish, being mostly immune following encounters with smallpox at home, were apparently untouched and this added to the supernatural reputation of the invisible curse. A Spanish monk who was there wrote:[42]

> They died in heaps, like bedbugs ... As it was impossible to bury the great numbers of dead, they pulled down the houses over them in order to choke the stench that rose from the dead bodies, so that their homes became their tombs.

Smallpox killed off the Aztecs' leaders and their great, highly ordered civilisation collapsed in ruins. The Angel swept the board and all Cortéz had to do was clear up afterwards – with grateful thanks to God for having sent the pestilence to ensure that another country was claimed for Spain and Christianity.

To the south was another magnificent civilisation and another vast susceptible population: the Incas of Peru, and the target of Francisco Pizarro some 20 years later.[43] It was even easier for him because the trade routes with Mexico had already carried

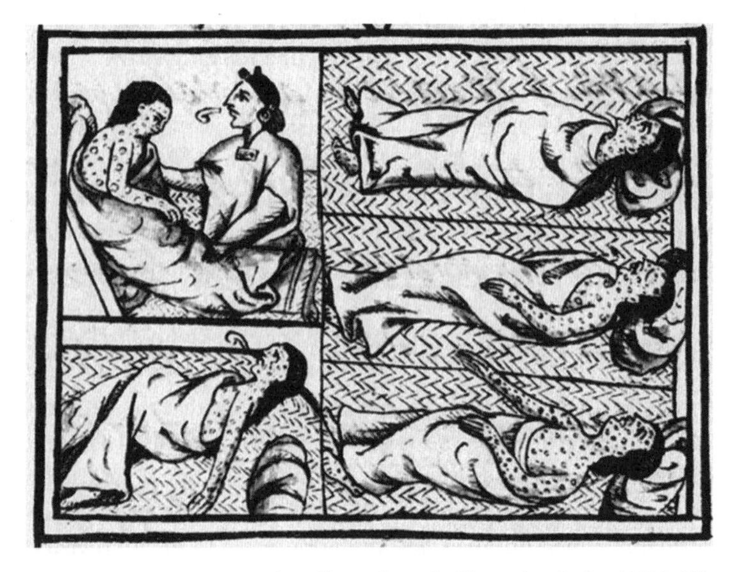

Figure 2.3 Aztec victims of smallpox, from the Florentine Codex (1540–85).

in the infection. When Pizarro arrived in 1532, the Incas had been thoroughly softened up. Much of the population had died, either directly from smallpox or the civil war that had broken out in the power vacuum which the Angel created by picking off various high-ranking commanders.

Both Cortéz and Pizarro were courageous, resourceful and determined, but each had a ridiculously small force – just 900 men for Cortéz, each one outnumbered at the start of the invasion by perhaps 5,000 Aztec men of fighting age. Hence the argument, which seems reasonable, that neither could have succeeded without the Angel's help.

The Angel's trail of devastation continued southwards, reaching Chile (1554) and Brazil (1555), following incursions by the Spanish and French.[44] In Brazil, God was again implicated in countless deaths, thanks to a Jesuit campaign to convert the natives: mass baptisms created ideal conditions for spreading smallpox and were soon followed by mass burials. Here, as in Mexico and Peru, the virus appears to have been a particularly malignant strain, causing haemorrhagic smallpox and a very high mortality rate.

South America remained in the grip of smallpox until 1971, with the last case reported in Brazil (see Chapter 15). The subcontinent was swept by numerous epidemics but none was as devastating as those which wiped out two major civilisations and handed possession of vast tracts of land to a foreign aggressor.

The Land of the Free

North America also contained many untouched and susceptible populations in the tribes of Native American Indians, mostly concentrated along the coastal strip of what became Massachusetts. They enjoyed a century of respite after the first epidemics hit South America but rapidly fell victim to smallpox when European settlers – English, French and Dutch – established permanent bases on the eastern seaboard during the first quarter of the seventeenth century.

Smallpox, possibly spreading down from French settlers in Nova Scotia, was probably the pestilence that swept through in 1616 and killed 90 per cent of the Algonquins and Narragansetts.[45] Sixteen years later, a smallpox epidemic killed off Indians around New Plymouth, where those on the *Mayflower* had built their colony. The Christian settlers viewed their destruction as a just and appropriate intervention by God in their squabble with the Indians over territorial boundaries.

From there, the Angel of Death set out on a steady march into the interior, guided by the new trade routes that were opening up for fur and other commodities. The communities were more thinly scattered than on the coast but the virus leapfrogged from one to another, and tribe after tribe fell during the three grim decades between 1635 and 1665.[46] The Hurons suffered the most: their numbers were halved between 1636 and 1640 and they were extinct by 1650. The Iroquois and Mohegans (Mohicans) also sustained devastating losses.

A consistent feature of the advance of smallpox through North America was the callousness with which the settlers regarded the plight of the Native Americans. This was genocide, humanitarian disaster and the destruction of centuries of tradition, but to the settlers the Indians were just another obstacle to be cleared to make their New World habitable. Smallpox, clearly sent by God, simply furthered that cause: "Towns of them were swept away, in some not so much as one Soul escaping destruction."[47] The author of these words was clearly not unhappy with the outcome; these were expendable Souls. He was a Puritan preacher from Boston, the Reverend Increase Mather. His son Cotton Mather followed his father into the Church and shared his views about the "pernicious creatures" who were their countrymen. Both Mathers crop up again later in this book (Chapter 5). Their sentiments survived until at least 1932, when the American historian Woodward wrote, "smallpox was the blessing in disguise that gave our emigrant ancestors an opportunity to found the state".[48]

Some American settlers went further and gave God a helping hand. Smallpox was not the first disease to be used deliberately to kill enemies en masse. Christians besieged in Kaffa in 1346 had to contend with the corpses of plague victims crashing to the ground among them, having been lobbed over the walls from catapults by the surrounding Tatars.[49] For Europeans wishing to clear an inconvenient Native American village, the durability of the smallpox virus and the susceptibility of the Indians provided an ideal opportunity.[50] The targets were presented with Trojan-horse gifts of blankets, hankerchiefs and even a keg of rum wrapped in a flag, all carefully contaminated with smallpox pus. And it worked: those who did not die generally fled in panic. This ploy appears to have been used several times and possibly by British commanders during the war with the Indians, led by Pontiac, during the late eighteenth century.

And back to Europe

Smallpox made many interventions in Europe, killing royalty, ending dynasties and pulling down governments. Queen Mary of England, whose dignified death at the end of 1694 has already been described, fell in the middle of her dynasty and of a three-generation curse on the House of Stuart. Both her parents-in-law (William II and her namesake, Mary) had died of smallpox ten years apart, as had her own baby brother. Mary had no children; her sister Anne had 18 pregnancies but only five live births, and three of these children were killed in infancy by smallpox. It was the death of the last, Prince William, that ended the Stuarts' reign and ushered in the House of Hanover.[51]

In the eighteenth century, numerous European Royals also succumbed to smallpox, including kings (Luis I of Spain, 1724 and Louis XV of France, 1774), queens (Ulricka Eleonora of Sweden, 1741), emperors (Joseph I of Austria, 1711), a tsar

(Peter II of Russia, 1730) and heirs to thrones (Louis of France, 1711 and José of Portugal, 1788). Repercussions included the reshuffling of various successions, including the complicated sequence which gave Austrians their successful and much-loved Empress Maria-Theresa.

Another quirk of smallpox that may have pushed European history down a different route took place in the summer of 1779, with France at war with England and the might of a combined French and Spanish fleet, commanded by Admiral d'Orvilliers, assembling in the English Channel.[52] All that d'Orvilliers needed to launch a decisive final assault was command of the Channel for a few days – an opportunity that was torn out of his hands when smallpox broke out on his ships and rapidly incapacitated half of his men. As the disease took grip, bodies had to be thrown into the sea in such numbers that the waters of the Channel were polluted as far as Plymouth (where the inhabitants ate no fish for a month). The intervention of smallpox on that occasion probably saved England from invasion. This was a lesson that had unexpected consequences for French ambitions to conquer England. A quarter of a century later, even though France was back at war with England, Napoleon would declare himself an admirer of Jenner – and, unable to refuse a request from him, agreed to release two English doctors who were being held prisoners-of-war in Paris.[53]

The Angel appraised

Smallpox was a vile disease. It probably does not merit the title of 'Greatest Killer', at least on numerical grounds. Its worst excesses are dwarfed by the 50 million people killed during the Great Influenza Pandemic of 1918 and by the continuing menace of malaria, with 300 million cases at any time and 3 million deaths annually; for comparison, the atomic bomb at Hiroshima killed about 110,000 people. But if you wanted to invent a curse to terrorise your worst enemy, the horrors in the Angel's repertoire would be hard to beat.

Because of its global reach and impact, smallpox is a fascinating subject for playing 'what if'. What would have happened if smallpox had spared Queen Mary, the Tsarevich Peter or d'Orvilliers' armada massing in the Channel? And what if some of the near-misses had been direct hits? We would have had a Russia without Stalin, an America that had to find its feet without George Washington or Abraham Lincoln – and a world without the music of Mozart.

Returning to the plot, we have now reached the early eighteenth century and the end of the Age of Passivity is approaching. Western medicine is about to discover some curious customs built around the improbable claim that these protected against smallpox. And the reason that Chinese physicians carried smallpox scabs about with them is shortly to be revealed.

3 First Steps in the Right Direction

"When the Patient has once had the Disorder in this way, though ever so mild, we are assured by Experience that they never have it again; and therefore the Opinion of those seems to be well grounded who think the Propagation of the Small Pox by Inoculation might be of general Use and Benefit to Mankind, in preserving the lives of some, and the most important Members of others, as the Face, Eyes, Hearing, Viscera &c."

Lorenz Heister *Chirurgie*, Chapter XV, 1740

Pembrokeshire forms the south-western corner of Wales and has always been out on a limb. In the Middle Ages, two pilgrimages to St. David's Cathedral on its western extremity counted as one to Rome. Its people have long had a reputation to match: keeping themselves to themselves and stuck in tradition and the past.

The village of Marloes lies just south of the Landsker, the line drawn soon after the Norman invasion of 1066 to separate the inhabitants of 'Little England beyond Wales' from the Welsh-speaking wilds to the North. Marloes sits on the base of a headland pinched off the coast where the Bristol Channel opens into the Irish Sea. Its local resources include good agricultural land and plentiful crabs and lobsters in the coastal waters but many of its inhabitants did not enjoy health or wealth. During the 1700s, smallpox was rife and visitors noted with disdain the ragged, barefoot children, the adults' dedicated pursuit of drunkenness and a high prevalence of insanity.[1]

Despite its obscurity, Marloes deserves two separate entries in the history of medicine. During the eighteenth century, it was internationally renowned for the high-quality leeches which flourished in Marlais Mere, now drained but originally covering some 60 acres. The reputation of Marloes leeches went before them and, as well as serving the Welsh and English medical markets, they were exported in bulk to France through the nearby port of Haverfordwest.

The village's second claim to fame comes from a curious custom which, the locals claimed, protected against smallpox. This defied common sense and any notion of hygiene. Pus or scabs were collected from the skin of a smallpox victim and used deliberately to infect a healthy person who had not yet met the disease. Most people favoured pus. Several blisters would be punctured and the fluid squeezed out and rubbed into the recipient's skin (usually on the hand), which had been prepared by scratching with a needle or blade. Alternatively, when the rash had passed through the pustular stage and was drying out, a couple of dozen scabs could be picked off and kept tightly gripped in the recipient's hand for a day or more.

The practice was known locally as 'buying the smallpox'.[2] Money did change hands: the going rate in 1700 was three pence for the juice from twelve pustules.[3] According to tradition, the purchaser would get away with a mild dose of smallpox that left few scars afterwards and would then enjoy freedom from smallpox for the rest of their life. If true, this was a bargain – especially as the donor faced all the usual risks of smallpox (and some did not live to spend their earnings).

Two doctors from Haverfordwest first drew attention to 'buying the smallpox' in the early 1720s. Their reports were anecdotal but both seemed to confirm the far-fetched claim that this could protect against the worst horrors of smallpox.

Dr. Perrott Williams described the custom in a letter he wrote on the 28th September 1722 to a colleague, Dr. Samuel Brady, who was physician to the Garrison at Portsmouth.[4] Williams had been aware of the practice for some time and that it was established in Pembrokeshire since "time out of mind." Several acquaintances had tried it, including a neighbour who had treated her daughter just 18 months earlier. As promised, the girl had only a mild attack and seemed to be protected as she had remained healthy despite close contact with active cases during a smallpox outbreak the following summer.

Williams provided some technical details. One "Learned and very Ingenious Gentleman of this Country" claimed that at school in around 1700, he and several other boys had scraped the skin off the back of their hands with penknives and then rubbed in smallpox blister fluid bought from a young woman. The Gentleman had a worse attack than he had bargained for but, like all the others, recovered uneventfully. He had never had a second attack of smallpox even though he "frequently conversed with such as were sick of that Distemper." This respectable witness – later revealed to be George Owen, a lawyer and the son of an archdeacon at St. David's Cathedral – was left with a "very visible" scar on his left hand, in case any doubts lingered about his credibility. Williams interviewed others who had "made the like Experiments on themselves" and found the same. Remarkably, wrote Williams, he had not heard of any cases that later caught smallpox, even up to 20 years afterwards.

He unearthed some additional snippets and included them in a further letter on 2 February 1723, but sent this time to James Jurin, Secretary of the Royal Society in London.[5] As Secretary, Jurin was the addressee for communications to the Society. He was also a renowned statistician with a particular interest in smallpox and its mortality rates. Williams explained again that 'buying the smallpox' had been widespread and an "immemorial Custom in these Parts" and that it was used by well-informed adults as well as schoolboys. He gave no data but was confident that the method was safe and conferred complete and lasting protection. By contrast, he pointed out that the smallpox outbreak which had flared up around Haverfordwest had already killed 52 of the 227 people infected – a 25 per cent mortality rate.

Towards the end of this second letter, Williams dropped in a first-hand observation that powerfully supports his contention:

> To make it appear that Inoculation is a sufficient Preservative against receiving the Small Pox a second time; about six Weeks ago, I caused my two Boys, who had been inoculated this last Summer, not only to see, but even to handle a child, dying of a most malignant sort of Small Pox; who notwithstanding, I thank God, continue in perfect Health.

It is surprising that he omitted to mention this before; perhaps he was worried about how it would be received by the Royal Society or his fellow doctors. We do not know how old his sons were and can only speculate about how much coercion was involved in 'causing' them to handle the moribund child about to lose its battle with smallpox. We also presume that he was courageous rather than foolhardy. Being in the thick of an outbreak, he would have known all too well the terrifyingly high stakes that he was playing for – but he must have convinced himself that enough guinea-pigs had already put the old-wives' tale to the test for it to be safe for him to gamble with his sons' lives. Nonetheless, the relief in his "I thank God" is nearly palpable.

Williams' accounts were supported by Richard Wright, a surgical colleague in Haverfordwest. Wright wrote to an apothecary friend in London in February 1723, just after Williams sent his letter to Jurin.[6] All these letters were printed together in the March 1723 issue of the Royal Society's *Philosophical Transactions*, the world's first and, at the time, its most prestigious scientific journal. Wright agreed that this was "a very ancient and frequent Practice, among the common People", notably in Marloes and the nearby village of St. Ishmael's and that it was still in use. From the testimony of 90-year old William Allen of St. Ishmael's, given to "Persons of good reason and Integrity", it had been widespread in his mother's time (and so presumably dated back to c. 1615 or earlier). Protection against smallpox lasted over half a century, at least in the case of one woman who had been sold smallpox 54 years earlier by Joan Jones, now a 70-year-old midwife "of good Credit, and perfect Memory".

There was, however, a potentially nasty fly lurking in this otherwise miraculous ointment. Two years previously, a 20-year-old woman had "procured the Distemper from a Man then dying of a very malignant Small Pox". Even at the time, with an outbreak looming, it should have been obvious that this was a transaction to avoid. Sadly, the woman proved the principle of *Caveat emptor* – Buyer, beware! – and died from her purchase.

Nevertheless, the bottom line of Wright's letter is strongly positive about the results of the Pembrokeshire experiment: "That hundreds in this Country have had the Small Pox this way is certain; and it cannot produce one single Instance of their ever having them a second time."

It might seem odd that two country doctors from the depths of rural Wales should be corresponding with the Royal Society about an odd piece of peasant folklore that had been around since the early 1600s. Even odder is the fact that the Royal Society was already aware of the custom of infecting healthy subjects with smallpox to protect them against getting a full-blown attack. At the time, with smallpox killing nearly one person in ten, the discovery of something that shielded people against the disease should have been one of the greatest possible gifts to medicine. Yet the Royal Society, that powerhouse of scientific and medical thought, had been strangely slow to act. In fact, they did nothing obvious for over a decade after the first reports reached them and it took another decade for the medical establishment to do the obvious experiment to see whether 'buying the smallpox' actually worked. Even then, they were less decisive than Dr. Perrot Williams of Haverfordwest had been.

During this time, people in several parts of the world – China, Africa, Arabia and Pembrokeshire, barely 200 miles from London – were continuing to protect themselves against smallpox, as they had done for decades if not centuries. Meanwhile, English medicine continued to treat smallpox to the best of its ability – with a well-stoked fire, leeches and red flannel.

A favourable smallpox

The Welsh custom of scratching smallpox blister fluid into the skin was just one example of what became known as 'variolation' (the term 'inoculation' was used interchangeably during the seventeenth and eighteenth centuries). Variolation was the deliberate infection of a healthy subject with smallpox in the hope of preventing natural attacks of the disease. This aimed to improve on nature, building on the observation that people who had survived smallpox never caught it again. Variolation was essentially an extension of the practice of deliberately exposing healthy children to infections in the hope that they would escape with a mild attack which would still grant them lasting immunity. This has long been done in many cultures for infections that are seen as trivial in childhood but serious in adult life, such as chickenpox and mumps; the custom lives on as the 'chickenpox parties' of the present day. With something as dangerous as smallpox, however, this would be playing with fire. Ideally, variolation induced smallpox when the subject was in good condition, hopefully producing a mild 'artificial' infection that would not kill or scar, yet would provide lifelong immunity against natural attacks. Given the unpredictable and often lethal behaviour of the virus, this might seem a tall order. Results were variable, but at its best variolation did what it was claimed to do.

Variolation was the first active step against smallpox, as opposed to passive measures such as quarantine and the isolation of cases. It was also the first successful attempt to manipulate the body's immune system so as to throw up defences against a specific infection. In turn, variolation was the conceptual and technical forerunner of vaccination, which is the administration of the cowpox virus to a healthy subject to stimulate immunity against smallpox. The cowpox virus lacks the homicidal tendencies of the smallpox virus itself but is closely enough related to smallpox that it provokes the immune system into mounting defences that will also stop smallpox in its tracks.

Assuming that it works, vaccination has obvious advantages over variolation and Jenner's discovery should have rapidly swept variolation into the dustbin of medical practices that have outlived their usefulness. However, for reasons of tradition as well as fears over vaccination, variolation clung on for decades in England. Astonishingly, it was still being practised in Afghanistan and Ethiopia right up to the extinction of smallpox during the late 1970s (see Chapter 15).

The birth of a ridiculous idea

"That although at first the more prudent were very cautious in the use of this Practice; yet the happy Success it has been found to have in thousands of Subjects for these eight Years past, has now put it out of all suspicion and doubt."

Emanuel Timoni, Letter to the Royal Society, 1713

Variolation is such a huge affront to common sense that it is difficult to think of anyone coming up with the idea, let alone being foolish enough to try it out. Smallpox was feared and loathed in every culture because what it did to people was cruel and revolting. This makes it all the more incredible that anyone would deliberately collect pus or scabs from a smallpox victim and use this material to infect someone who had been lucky enough to escape the attention of the Angel of Death.

Every stage in the process of developing variolation demanded a step-up in understanding, even if subconscious at the time. Simply working out that those who survived smallpox did not get it again was a landmark of clinical observation – especially in times when life was short and under constant threat from numerous infections, some of which (such as chickenpox and measles) were easily confused with smallpox. Thinking then had to move up a gear for this observation to lead to its correct interpretation, namely that the protection against smallpox was actually conferred by the first attack itself. In many cultures, it would have been automatically assumed (and heretical to presume otherwise) that survival and freedom from subsequent attacks were due to some divine favour. Finally, a breathtaking leap in logic is demanded to dream up the practicalities of how to collect infected material and administer it to the recipient. The skin, as the obvious site of infection, would be an intuitive choice for both donating and receiving smallpox material but how the Chinese method described below might have originated is anyone's guess.

Smallpox would have been a good subject for early experiments in preventative medicine. As it was ever-present, variolated subjects would not have to wait long to meet the disease again to demonstrate that they were protected. It would have been much harder, for example, to follow any attempt to prevent plague, which only broke out every 20 years or so. The clinical end-point of smallpox infection was also obvious – as was the continuing good health of the variolated subject, especially during an outbreak.

The significance of variolation as a medical discovery in its own right must not be underestimated. The people who first thought of this and tried it out were as visionary and bold as Jenner himself, and it was a leap in the dark for them. When Jenner's time came to experiment with vaccination, he at least had the framework of variolation on which to hang his ideas.

Time out of mind

"It is withall so ancient in the Kingdoms of Tripoly, Tunis and Algier, that no body remembers its first rise, and it is generally practised not only by the inhabitants of the Towns but also by the wild Arabs."

Cassam Aga, Ambassador of Tripoli, Report read to the Royal Society, 1729

Variolation can be traced back over many centuries before it filtered through to the consciousness of doctors and scientists in Western Europe and North America in the first quarter of the eighteenth century. Two main traditions were identified, one Chinese and the other Arabic and African. These differed fundamentally in their ways of introducing the infection and probably sprang up independently.

Ancient Chinese medicine can claim the longest documented use of variolation, with accounts going back to 1550 or earlier of the bizarre method of 'nasal insufflation', or the blowing of powdered smallpox material (usually scabs) up the nostrils.[7] Various insufflation techniques were described in detail during the seventeenth and eighteenth centuries by Chinese physicians such as Dai Manquang (c. 1650), who spread the knowledge to Japan and Russia.[8] Mild smallpox cases were carefully selected as donors, in the hope of sparing the recipient a serious attack. Most operators favoured scabs that had been left to dry out for some time; fresh scabs were too likely to produce a severe infection. Three or four dried scabs were either ground into a powder, or mixed with a grain of musk and bound in cotton. Sometimes blister fluid was used, drawn up into a cotton wick. The infected material was loaded into a blowpipe and puffed up the recipient's nostril. The procedure was ritualised; the blowpipe was often made of silver, while the right nostril was used for boys and the left for girls. Variolated subjects were treated as infectious, as with naturally acquired smallpox, and kept apart from other children until the rash had cleared.

By contrast, the Arabs and Africans used inoculation of smallpox material through the recipient's skin, as in Pembrokeshire.[9] There were various regional and cultural variations on this basic theme. Two traditions were described in the Sudan in the late eighteenth and early nineteenth centuries. Both were long-established and apparently stemmed from Arabic practice. *Tishteree el Jidderi* (literally, 'buying the smallpox'), was described by Bruce in 1790 as practised "since time immemorial" by the women of Sennar in Central Sudan.[10] On hearing of a newly-infected child, the mother of an unprotected child would visit the house and tie a cotton cloth around the victim's arm, leaving it there while she haggled with the child's mother over the price (usually one piece of silver) and the number of pustules that would be passed on to her own child. When the bargain was struck, the woman would return home to tie the cloth around her own child's arm. According to tradition, the recipient respected the bargain and developed no more than the agreed number of pustules. Variations included the recipient bringing gifts of confectionery or fruit to the donor.

An alternative method – *Dak el Jedri* ('hitting the smallpox') – was closer to that used by the Turks and eventually imported into England in the early 1720s.[11] Fluid collected from a smallpox pustule was rubbed into a cut made into a recipient's skin, usually on the leg. This technique spread more widely throughout Africa and was, for example, well known in the area around Tripoli – from where a young man abducted into slavery at the start of the eighteenth century would later bring the first news of the technique to New England. The tradition may also have tracked with merchants and pilgrims along the Middle Eastern caravan routes and into Turkey and Greece.

By the mid-seventeenth century, variolation was commonplace in China and parts of Africa, and cutaneous inoculation was also being practised in isolated foci in various European countries as well as in Pembrokeshire. However, all this was out of sight and out of mind as far as Western European doctors were concerned. The last quarter of the seventeenth century saw some attempts to bring variolation to the notice of the medical establishment. A few scattered reports of cutaneous inoculation were published in Denmark in 1666 and again in 1673, Switzerland in 1683, and Poland in 1674.[12] These described the practice and its supposed purpose, but did not attempt to test its claims. Possibly these accounts came across as oddities from folk traditions that lacked any medical credibility. In any event, they failed to generate interest in their native countries and if the news did percolate as far as England, it was either ignored or not understood.

England became the setting for the next few attempts to raise awareness of variolation. The first was an eye-witness account of the Chinese technique from Joseph Lister, a trader based in Amoy and working with the East India Company.[13] In January 1700, he wrote to Dr. Martin Lister, then Secretary of the Royal Society in London, and reported that he had watched physicians "opening the pustules of one who has the Small Pox ripe upon them and drying up the Matter with a little Cotton". The fluid was kept in a closed box until use. Joseph Lister's letter was followed a month later by another description of the Chinese method, by Dr. Clopton Havers, also sent to the Secretary of the Royal Society.[14] Neither of these communications saw the light of day in the pages of the Royal Society's journal, the *Philosophical Transactions*. Perhaps the Fellows of the Society had some misgivings about them – as did their successors some 80 years later, when they rejected Jenner's first paper on vaccination.

The next landmark was a celebrated pair of reports of cutaneous inoculation, both from doctors working in Turkey. These were published in the *Philosophical Transactions* but the first did not appear until 1714, after a gap of 14 years. Then, after a further latent period of several years, there was the brief flurry of correspondence from Williams and Wright in Haverfordwest.

Looking back now, it seems incredible that the Royal Society and the medical establishment sat for so long on the initial reports of variolation. Fourteen years is a long time to do nothing, especially while smallpox was killing tens of thousands of people every year in England alone. The notion of squirting smallpox up the nose might have seemed outlandish, but with such huge stakes to play for, it was surely worth looking into.

In fact, there was some activity behind the scenes at the Royal Society. In 1712, Dr. Richard Waller, who succeeded Lister as Secretary, heard that Dr. Edward Tarry had returned to England with intriguing news from Pera, across the Golden Horn from Constantinople (present-day Istanbul). Tarry reported that the local Galatians had a tradition of scratching smallpox blister fluid into the skin to ward off a full-blown attack. He claimed to have seen over 4,000 people treated in this way and believed that it worked.[15]

This hooked Waller's attention. Seizing the initiative in a way that his predecessor had failed to do, he decided to follow up Tarry's lead and wrote to all the Fellows of the Royal Society, asking them for any information that they might have come across about methods that were supposed to protect against smallpox.[16]

Waller's circular letter was sent out in late 1712. It was a whole year before a substantive response came back. By then, Waller had been replaced as Secretary by Dr. John Woodward, the physician who would later fight a duel with Richard Mead in defence of emetics to move the bowels in the treatment of smallpox. It was therefore Woodward who went down in history as the promoter of the report, by Dr. Emanuel Timoni of Constantinople, which finally put variolation on the map.

An ingenious discourse

Timoni, also known by his Latinised name of Timonius, was an Italian who was born on the Greek island of Chios. Trained as a physician at Padua and Oxford, he had been elected to Fellowship of the Royal Society in 1703. During the previous eight years, Timoni had been working in Constantinople as the doctor attending the British Embassy, which was based in Pera. Soon after arriving there, he had been struck by the custom of "incision" or "inoculation", which he was told had been introduced into Constantinople some 40 years earlier by "Circassians, Georgians and other Asiaticks". His description of the technique, in a letter dated December 1713, was summarised by Woodward (the customary role of the Secretary at the time) and the resumé was published in the *Philosphical Transactions* in early 1714.[17]

(72)

V. *An Account, or Hiſtory, of the Procuring the* SMALL POX *by Inciſion, or Inoculation; as it has for ſome time been practiſed at* Conſtantinople.

Being the Extract of a Letter from Emanuel Timonius, Oxon. & Patav. M. D. S. R. S. *dated at* Conſtantinople, December, 1713.

Communicated to the Royal Society *by* John Woodward, M. D. Profeſ. Med. Greſh. *and* S. R. S.

THE Writer of this ingenious Diſcourſe obſerves, in the firſt place, that the *Circaſſians, Georgians,* and other *Aſiaticks,* have introduc'd this Practice of procuring the *Small-Pox* by a ſort of Inoculation, for about the ſpace of forty Years, among the *Turks* and others at *Conſtantinople.*

That altho' at firſt the more prudent were very cautious in the uſe of this Practice; yet the happy Succeſs it has been found to have in thouſands of Subjects for theſe eight Years paſt, has now put it out of all ſuſpicion and doubt; ſince the Operation having been perform'd on Perſons of all Ages, Sexes, and different Temperaments, and even in the worſt Conſtitution of the Air, yet none have been found to die of the *Small-Pox*; when at the ſame time it was very mortal when it ſeized the Patient the common way, of which half the affected dy'd. This he atteſts upon his own Obſervation.

Next he obſerves, they that have this Inoculation practiſed upon them, are ſubject to very ſlight Symptoms, ſome being ſcarce ſenſible they are ill or ſick; and what is

Figure 3.1 Front sheet of the paper by Emanuel Timoni in the *Philosophical Transactions* of the Royal Society, reporting variolation in Constantinople. Reproduced by kind permission of the Royal Society of London.

Timoni explained that the donor was typically a healthy lad with 'distinct' (mild) smallpox in its twelfth or thirteenth day. Blisters on the legs were pricked with a needle and the fluid squeezed out into a clean glass vessel, which was kept warm, and used as quickly as possible. The recipient's arm was cut with a surgeon's needle or a lancet to make "several little wounds" in one or more places, and a drop of blister fluid was rubbed into the the incisions using "a blunt Stile, or an Ear picker". A concave shield, such as a half-walnut shell, was tied over each inoculation site for a few hours to prevent the fluid being rubbed off by the clothes. No prior preparation was required, although the custom in Constantinople was to avoid meat and broth for 20–25 days afterwards. Inoculation was traditionally performed either at the start of winter or in the spring.

Timoni portrayed the procedure as innocuous – the subject had only "very slight Symptoms, some being barely sensible they are ill or sick". There was some continuing discharge from the incision sites with local redness ('efflorescence') after seven days or so, followed by the appearance of smallpox blisters elsewhere. Importantly, according to Timoni, these generalised pocks were relatively sparse – usually fewer than 20 and only rarely over 100. Unlike natural smallpox, they contained a thin fluid, not thick pus, and rapidly dried up. Crucially, and a point "which is valued by the Fair", the procedure "never leaves any Scars or Pits on the Face"; Timoni claimed that the only permanent reminder was a scar at the incision site. There were rare exceptions to this mild pattern. Timoni reported four cases who each developed a confluent rash but who probably were already incubating naturally-acquired smallpox, as their eruptions appeared unusually early. Otherwise, he was convinced about its safety:

> I have never observ'd any mischievous Accident from this Incision hitherto; and altho' such Reports have been sometimes spread among the Vulgar, yet having gone on purpose to the houses where such Rumors have arisen, I have found the whole to be absolutely false.

Timoni was equally positive that inoculation worked. He praised the "happy Success ... in thousands of subjects for these 8 years past", pointing out that these included "Persons of all Ages, Sexes, and different Tempraments" and "even in the Worst constitution of the Air". None had died of smallpox, at a time when the natural infection carried off half of its victims. Timoni ended his letter with five pages of 'aetiological' speculation about how inoculation might work, which Woodward left intact in its original Latin.

From Woodward's edited transcript, it is difficult to know how much of Timoni's account reflects his own experience or information that was at least second-hand. Woodward notes that Timoni "attests upon his own Observation" but given Timoni's duties at the Embassy, it is unlikely that he followed up closely the thousands of subjects who had been inoculated over the previous eight years. Timoni was clearly a convert to inoculation, which is not surprising if we are to believe the glowing statistics which he quotes. He must have realised that he had stumbled across something big; possibly he was taking his time to gather more follow-up evidence and was then prompted to publish by Waller's request for information about protective measures against smallpox.

Timoni's paper[18] appeared in the *Philosophical Transactions* alongside a mixed bag of articles, ranging from van Leeuwenhoeck's landmark description of the microscopic structure of mammalian muscle fibres, to banal accounts of some oddities of nature from North America. Timoni may have been surprised by the initial response to his letter, or rather the lack of it: no correspondence, no apparent interest from his fellow doctors. By providing a detailed, step-by-step account of how to inoculate, he might also have hoped that doctors in England could be excited into experimenting with variolation for themselves and so put this strange foreign custom to the test. Here, too, he would have been disappointed.

However, Timoni's flag-waving in the *Transactions* was spotted by others, with whom it rang various bells. On the far side of the Atlantic was a Puritan preacher and sometime witch-hunter in Boston, who had already received similar intelligence but had doubts about the reliability of his source. For him, Timoni's letter made everything fall into place, but the time was not right to try it out. He filed it away for future reference.

Closer to home, Timoni's letter also prompted the energetic Waller to chase up his own contacts in Turkey to see if the stories about inoculation could be corroborated. His source was Dr. William Sherrard, the British Consul in Smyrna (modern-day Izmir) who was also a noted botanist. Sherrard took his time to reply but eventually came up with the goods. In 1716, he sent Waller a copy of a pamphlet which had been published the previous year in Venice by Dr. Giacomo Pylarini (Jacobus Pylarinus).[19] Pylarini was another doctor with a diplomatic brief: trained as a physician in Venice, he had been posted to Smyrna as the Italian Consul but had now returned home. He also described inoculation as practised in Constantinople, where he had worked some years earlier. His account echoed Timoni's and reinforced the basic message that inoculation with smallpox was effective and safe, had been extensively tried and tested over many years and lived up to its far-fetched claim to protect against the disease.

This gave Waller the confirmation that he had been waiting for, and he acted quickly. He introduced Pylarini's pamphlet for discussion at a meeting of the Royal Society in May 1716, and it was published in edited form in the first issue for 1717 of the *Philosophical Transactions*.[20] But then, despite Waller's efforts, there followed another long period of inactivity. As far as the medical establishment in England was concerned, any spark of interest that might have been ignited by the reports of inoculation simply fizzled out.

Across the English Channel, Dr. J. N. B. Boyer of Montpellier had also been studying inoculation and wrote a thesis about it, which was published in 1717.[21] Boyer was destined for great things as Dean of the Medical Faculty in Paris, but this particular work stirred up little interest. Elsewhere, people as far apart as China, Pembrokeshire and Constantinople carried on as usual with their various traditions of variolation. In Constantinople, facing the prospect of a smallpox outbreak, Timoni demonstrated his trust in the wisdom of the Circassians by having his daughter inoculated in 1717. Mr. Hefferman, the Secretary to Sir Robert Sutton, the British Ambassador, followed suit with his two children. They were later repatriated to London, where their inoculation scars became medical exhibits, available to be inspected by any interested doctors. Curiously, few seem to have taken up the offer.[22]

Lost in transmission

At that point, with barely a handful of publications to sustain it, English interest in variolation could easily have foundered on the sands of indifference – a sad fate for something that promised so much and that was crying out to be put to the test. Something must have prevented the medical establishment from seeing what variolation had to offer and from taking it seriously.

Paradoxically, the *Philosophical Transactions* may have been partly to blame. This was the world's first scientific journal and it had managed – partly because there was little competition – to maintain its reputation as the best. However, a quick leaf through issues from the mid-1710s shows that its many nuggets of true genius, from the likes of Newton, van Leeuwenhoeck and Herschel, were intermingled with rubbish. The papers by Timoni and Pylarini appeared alongside a dubious recipe that claimed to cure distemper in cattle,[23] North American Indian herbal cures for gout, dropsy (fluid accumulation) and scrofula (tuberculosis of the lymph glands) – and the suggestion, made in all seriousness, that birds migrating from New England took shelter on a moon that had somehow managed to avoid detection.[24] The Secretaries – Lister, Waller and Woodward –were reponsible for pruning submissions into the journal's style but do not appear to have been so rigorous in vetting the quality of the science. Thus, the reports on variolation – even if flagged up as "an ingenious discourse" may not have appeared exciting or convincing against the other curiosities that found their way into the hallowed pages of the *Transactions*. Also, this journal was not regular reading for jobbing doctors. It would have been seen by the 'Great and the Good' of the Royal College of Physicians, many of whom were also Fellows of the Royal Society, but none of them was sufficiently moved to comment on this strange practice, let alone follow it up.

The miraculous claims made for variolation may also have prevented it from being taken seriously. According to Timoni, inoculation produced a trivial infection, always worked and never left scars on the face. This must have seemed too good to be true (as indeed it was). Everyone knew that smallpox varied hugely in its severity: what was so special about this procedure that could possibly guarantee a mild attack? And smallpox was highly contagious – yet Timoni never mentioned whether the thousands of those who had been inoculated had ever passed the infection on to others. It simply didn't add up. Even though doctors in the eighteenth century were happy to charge for miracle cures, they were also well used to them failing to live up to expectation. Variolation might have been dismissed as just another example of ever-hopeful hyperbole.

Baser human instincts undoubtedly played their parts. Nationalism and chauvinism would have dragged down anything coming from the "illiterate and unthinking" Turks[25] and, even further away, the Chinese. How could such uncivilised heathens possibly have any offering of value to lay before the sophisticates of London? When it later turned out that inoculation was not practised by Turkish doctors but by people with no medical understanding, this also undermined its credibility. Finally, as would soon be pointed out with typical bluntness by a forceful woman correspondent from Constantinople, the self-interest of doctors was a force to be reckoned with, and any new development that could threaten the income they made from smallpox could be guaranteed a rough ride.

In the end, the notion of inoculation was refloated and remained seaworthy for another 120 years in England. Its rescue owes little to the intellectual heavyweights of the Royal Society or the leaders of medical fashion at the Royal College of Physicians, but rather to the efforts of two colourful characters whose convictions and personalities more than made up for their lack of medical qualifications. One was the fearsome Bostonian preacher for whom Timoni's paper had made everything click into place. He was Cotton Mather, son of the Reverend Increase Mather whose views about the usefulness of smallpox in clearing away the original inhabitants of Massachusetts we have already heard.

The other was a gifted, beautiful and pushy English Lady – the genuine article, with a hereditary capital 'L' – and it was she who moved first. Lady Mary Wortley Montagu became an evangelist with a mission and two big scores to settle – the first with the Angel of Death who nearly killed her and left her face in ruins, and the second with the medical profession which had failed to rescue her from smallpox. Like Timoni, she found herself broadcasting the news of variolation from the social and intellectual exile of Turkey. Compared with him, she had the disadvantages of being a woman and lacking both medical training and any credibility with the Royal Society. However, she had a huge force of personality, great fluency and powers of persuasion, and – crucially – a ready-made network of personal contacts that went right to the top of the social tree. She kept these ticking over while in Constantinople through a constant stream of letters home, and was able to mobilise high society when she returned to London.

And she succeeded where Timoni and Pylarini had failed – in dragging variolation out into the open and into the suspicious gaze of doctors, scientists and the wider public.

The Ambassador's Ingenious Lady

"We travellers are in very hard circumstances. If we say nothing but has been said before us, we are *dull*, and we have observed *nothing*. If we tell any thing new, we are laughed at as *fabulous* and *romantic*."

Lady Mary Wortley Montagu, Letter to her sister, Lady Mar, April 1714

Thanks to both genes and environment, Lady Mary Wortley Montagu was always going to make her mark. Born in April 1689, Lady Mary Pierrepoint was the eldest child of Evelyn Pierrepont, a serial inheritor of hereditary titles who eventually became the Duke of Kingston. Wealth and social standing were therefore waiting for her at birth. Her own gifts turned an already excellent hand into a truly stunning one: she was beautiful, terrifyingly intelligent and a brilliant conversationalist and correspondent. She was also outspoken and fearless, in an age when these qualities were frowned on and actively suppressed in women. Indeed, before she met smallpox, Lady Mary had everything.

Lady Mary was succeeded by two sisters, Frances and Evelyn, and finally by William who was born into the title of Viscount Newark and the family inheritance. William's arrival in October 1692 was complicated and led inexorably to his mother's departure shortly afterwards. Lady Mary was three and a half years old when her mother died.

Being a girl, she was naturally denied any formal education. She made up for this with an insatiable drive to read and learn, which she indulged in her father's well-stocked library in the family home, Thoresby Hall, on the fringes of Sherwood Forest in Nottinghamshire.[1] While still a child, she taught herself French and Italian so that she could read the great romantic works in those languages – a gift that she would reawaken much later to learn Turkish on the hoof in Constantinople. Also at an early age, she discovered that she got a kick from being in the company of clever and powerful men. Far from being dazzled by their brilliance, she gave as good as she got and later found herself being taken seriously by some of the greatest names and brains of her day. An early experiment in manipulating men was the evening when her father led her into the alcohol-charged cavern of the Kit-Kat Club in London. There, she was declared the toast of the all-male assembly and had her name cut into the traditional glass with a diamond; petite and wide-eyed in her best gown and jewels, she was just ten years old.[2]

Mary Pierrepoint delighted in smashing moulds, beginning with the assumption that a society girl would do as her father bid her. They fell out frequently over many things, including her determination to write – even though he was happy to parade her and her talents to impress his influential and career-enhancing friends. In 1704–05, not yet 15, she produced a secret volume of *Poems Letters Novels Songs etc.*[3] This book had limited circulation, just her favourite sister Frances and her close friends, Sarah Chiswell and Philippa Mundy. In the introduction, she pleads for the reader's understanding, because of her extenuating circumstances:

> 1. I am a woman. 2. Without any advantage of Education. 3. All these was wrote by me at the age of 14.

Faults there are, but the collection – including 15 poems – already had originality and confidence.

Love was something else to hide at all costs from her father. In their late teens, she and her friends devised a secret code to describe the kinds of marriage that they might find themselves locked into: 'Paradise', 'Hell' and the connubial indifference of 'Limbo'.[4] They had already resigned themselves to the possibility that their fathers might push them into Hell or Limbo. In the event, none of them reached Paradise. Neither did William, married off by his father at the age of 20 to a wealthy but dim heiress.

Lady Mary and her father clashed over his belief that he had the right to select a husband for his daughter. At issue was one Clotworthy Skeffington, heir to Viscount Massereene in County Antrim, who unfortunately lived up to his improbable name. He was worthy (i.e. wealthy) but also, by comparison with his intended wife, a clot. For Mary, this was Hell at its grimmest: "I would rather give my hand to the Flames than to him."[5]

At the time, her options were limited. She had been in a clandestine, up/down relationship with Edward Wortley Montagu, a Whig activist who was heading for Parliament. Wortley fell far below the standards of her imagination, and their relationship – largely conducted through letters that swung between declarations of love and fault-finding – was currently down rather than up. However, the dire prospect of being banished to the wilds of Northern Ireland as Lady Mary Skeffington (and the prospect of making her father apoplectic with rage) won the day.

On Monday 18 August 1712, aged 24, Lady Mary sneaked away from home for the last time, leaving behind the dress recently bought for her wedding to Skeffington and any hope of reconciliation with her father.[6] At risk of his father's wrath, brother William went too, keeping her company on the long coach-journey to her assignation with Wortley at an inn somewhere on the road to Dover. Their elopement was more farce than passion. On arrival, she had a fever and immediately went to bed. Turning up later, Wortley prowled around and tried to find her, looking shifty enough for the innkeeper to take him for a criminal. The next morning, both were smitten with indecision rather than desire; Wortley even went back to London for a couple of days. In her poem, *The Lover*, she manages to pack humour, pragmatism and wistfulness into a single poignant line. This was written some years later, but may say it all about the occasion:[7]

> And we meet, with champagne and a chicken, at last.

Eventually, they married on 23 August and their first child, Edward, was born on 16 May 1713, just nine months later.

Paradise deferred

> "*Derdime derman bul:* Have pity on my passion.
> *Ver bize bir umut:* Give me some hope.
> *Bize bir dogru haber:* Send me an Answer."
>
> Extracts from a Turkish love-letter, quoted by Lady Mary Wortley Montagu, Letter to Lady —, 16 March 1718

Lady Mary settled into a routine of domestic Limbo, with Wortley often away on the Parliamentary campaign trail. Living in rented accommodation with a young baby must have been a challenge for the hereditary Lady, especially since her marriage to the commoner Wortley had deprived her of the automatic privilege of being a Lady-in-Waiting at the Court of the new King George I. If Wortley loved her, it is not obvious from their correspondence. She complained repeatedly that he never asked after her or their son, and when she developed a large facial abscess that had to be lanced, he wrote back eventually to say that he hoped it was not as bad as she had made out.

Worse was soon to come. Within a few weeks of her son's birth, her life was shattered when smallpox crossed her path for the first time. William, her closest confidant and ally during the wars at home, fell ill in mid-June 1713. Smallpox had broken out in London some weeks earlier and her hopes that it would turn out to be something else were dashed on 22 June, when the rash appeared. The night before, she had broken off another emotional letter to Wortley when the light faded. On the following day, her letter continued:[8]

> My first News this Morning is what I am very Sorry to hear. My Brother has the Small pox. I hope he will do well; I am sure we lose a Friend, if he does not.

After this brief matter-of-fact statement, she picked up where she had left off the previous evening:

> If you break your Word with me, and I have no letter, you do a very cruel thing and will make me more unhappy than you imagine.

William did not do well. Lady Mary was banned from visiting him and had to rely on second-hand information, which made the agony of waiting for news all the more unbearable. Dr. Samuel Garth, the family doctor and one of her father's friends from the Kit-Kat Club, was in constant attendance. On 25 June, she wrote again to her husband:[9]

> My Brother, they send me word, is as well as can be expected. But Dr. Garth says 'tis the worst sort … which I should think very forebodeing if I did not not know all Doctors (and particularly Garth) love to have their patients thought in Danger.

Unfortunately, Lady Mary was snatching at straws. William rapidly lost ground and died on 1 July 1713. His ordeal was terrible for a favourite sister to hear, even when watered down by distance, time and informants who did their best to protect her from the worst.

William's death smashed through the shield of security that her privileged upbringing had raised. For many months, her formidable confidence abandoned her, leaving her badly shaken, because she had met something that she couldn't outwit or charm her way round. Her letters show a withdrawn, suddenly paranoid woman, who had lost her sophistication and eloquence. Just before William's funeral, she wrote to Wortley:[10]

> I know not what to do, but I know I shall be unhappy till I see you again ... Your absence increases my Melancholy so much I fright my selfe with Imaginary terrors, and shall always be fancying Dangers for you while you are out of my sight ... I am afraid of everything ... since the loss of my poor unhappy Brother, I dread every Evil ... I have been today at Acton to see my poor Brother's melancholy Family. I cannot describe how much it has sunk my Spirits. – My Eyes are too sore to admit of a long letter.

From then on, smallpox hovered on the fringes of her life together with bereavement, marital disappointment and possible post-natal depression. Wortley simply made things worse: too busy with his career to come home, and indifferent to all her pleas, as her letters show. In January 1714: "You have forgot, I suppose, that you have a little Boy." In August 1714: "You made me cry 2 hours last night." And on 23 October 1714: "You seem not to have receiv'd my Letters, or not to have understood them."[11]

Finally, after two years of dedicated neglect, Wortley returned to his wife and baby son. In November 1714, he relocated his family to London, to be nearer the movers and shakers who could give him a leg-up onto his career ladder – which they did. Wortley became Member of Parliament for Westminster, elected unopposed in February 1715, and took up a post in the Treasury a couple of months later.

The move to the capital could have been smoother. Wortley found a house for them in Duke Street. She wanted reassurance that this was not the notorious house which had stood empty for over two years, following the death there of a prominent socialite (also called Montagu) and her young child from smallpox. Lady Mary was terrified that she and her son would fall victim to the same curse: "infection may lodge in Blankets &c. ... I should be very much afraid of coming into a house from whence any body dy'd of that Distemper, especially if I bring up your Son."[12] Luckily, it turned out to be a different house. He could easily have put her mind at rest but instead chose to keep her in suspense for a fortnight, even though it was obvious that she was working herself up into a state of high anxiety.

And so Lady Mary found herself and her young son right in the heart of London, the smallpox capital of England. Edward Wortley Montagu MP immediately struck off into the jungle of London politics, leaving Lady Mary to infiltrate the city's high society. She was soon bored by "the perpetual round of hearing the same scandal, and seeing the same follies acted over and over" and took up writing. The quality and wit of her poems and essays outshone any impediment of being "a woman without any advantage of education" and Lady Mary was soon drawn into the leading literary set of the day alongside John Gay, poet and playwright, and Alexander Pope, poet, satirist and master of the acid epigram. Gay, Pope and Lady Mary had complementary tastes and initially settled comfortably into a cosy, mutually congratulatory *ménage à trois*. Writing was the glue that held them together, although Pope later made it abundantly clear that he would have preferred the more conventional bonding of sex with Lady Mary. She wrote poetry including a series of poems that she called her *Town Eclogues*.[13] These covered perennial themes such as seduction, unrequited love and gambling, with one named for each day of the week. For some reason, she began with Monday, then jumped to Thursday and Wednesday. By the end of 1715, she had completed six of the seven *Eclogues*, leaving only Saturday to be written. This gap – an invitation for fate to step in with something truly memorable – was soon filled.

Sufficient reason to dread

"*Esking-ilen, oldum Ghira*: I burn, I burn, my flame consumes me. *Uzunu benden ayirma:* Don't turn away your face."

Extracts from a Turkish love-letter, quoted by Lady Mary Wortley Montagu, Letter to Lady —, 16 March 1718

Smallpox had never left London after the outbreak that killed William, and in mid-December 1715, Lady Mary's nightmare finally closed in on her.[14] She fell ill with a splitting headache, backache and fever – non-specific symptoms that can accompany many other illnesses, but typical of smallpox. The following days must have passed in an agony of dread, watching herself relive, one by one, the stages of her brother's downhill course. First, isolation, with no visitors, and her son sent away with his nurse. An increasingly grave Dr. Garth in attendance, then calling in two other colleagues – Dr. Richard Mead and the same Dr. John Woodward who had received Timoni's letter from Constantinople. All were esteemed physicians and all Fellows of the Royal Society, and none had anything useful to offer.

When her rash broke out it was confluent. With it came a high fever, sores that stripped off the lining of her mouth and throat, and a constant outpouring of saliva which was unbearably painful to swallow. All the mirrors in her room were removed even though she couldn't see; her face had swelled up so much that her eyelids were squeezed tight shut and glued together with pus.

Mortality from confluent smallpox was generally about 50 per cent. Between them, William and Lady Mary reinforced that statistic. She ran a course perilously close to her brother's – until the very end, when something that had flipped against him came down in her favour. After ten days of delirium and fever, with pus and blood weeping into the sheets, she edged past the crisis point. As the New Year of 1716 was rung in, she began the slow climb to recovery. The rash dried up over a few days and began to scab, and she could swallow again. The swelling of her face gradually subsided, allowing her eyelids to peel apart. Luckily, the smallpox had spared her eyes and when she could open them again her vision was clear. Three weeks after taking to her bed, she was finally strong enough to sit out and the London papers reported that she was recovering. She would have been able to see the damage that the smallpox had done to her limbs and trunk and would have been constantly reminded of the ordeal that still lay ahead by the spaces in her room where the mirrors had been. Finally she was judged ready to see her own face.

For Lady Mary, beauty had been an unquestioned gift in the bag of tricks that she exploited to carve her path through life. That first fearful glance into the unveiled mirror must have been devastating for her. The smallpox had torn through her skin, gouging out hundreds of pock-marks and leaving a leathery mask. Her eyelashes had been stripped out and never regrew, giving "a fierceness to her eyes" that robbed her features of their renowned humour and grace. In three weeks, she had been transformed into just another scarred ex-beauty, dumped in the wake of smallpox. She was not the only one to be wounded by the experience. While she was fighting for her life, her husband was said to be 'inconsolable' in case she lost her looks, thus damaging his standing at Court and his career prospects.[15]

But at least she was alive, and determined to stay that way. She had been to hell and back and lived through her worst nightmare. Her beauty might have been been scraped away, but everything else was intact and her personality, always forceful, grew to fill the gap.

She started writing again. 'Flavia', completed while still convalescing but not published until over 30 years later, became Saturday's entry in her *Town Eclogues*. Against its predecessors, 'Flavia' sticks out because of its anger, bitterness and sorrow.[16]

> The wretched FLAVIA on her couch reclin'd,
> Thus breath'd the anguish of a wounded mind;
> A glass reversed in her right hand she bore,
> For now she shunn'd the face she sought before.
> "How am I chang'd! alas! How am I grown
> A frightful spectre, to myself unknown.
> Where's my Complexion? Where my radiant Bloom?
> That promis'd happiness for years to come?"

There are also contemptuous stings in the tail for the clever doctors who had wrung their hands at her bedside but were impotent against the disease:

> MIRMILLO came my fortune to deplore,
> (A golden-headed cane well carv'd he bore)
> Cordials, he cried, my spirits must restore:
> Beauty is fled and and spirit is no more!
> GALEN the grave, officious SQUIRT, was there,
> With fruitless guile and unavailing care;
> MACHAON too, the great MACHAON, known
> By his red cloak and his superior frown;
> And why, he cry'd, this grief and this despair?
> You shall again be well, again be fair ...
> False was his oath; my beauty was no more.

Mirmillo (a Roman gladiator) was intended to be Mead, identified by his trademark gold-headed cane,[17] while Galen the Squirt (slang for doctor) was Woodward and Machaon was the red-cloaked Garth. Lady Mary's revenge was largely retrospective; by the time the poem was openly published, in 1747, only Mead was still alive.

Lady Mary's force of character returned with her physical strength, and pushed her firmly into the difficult business of picking up the strands of her previous life. Her son, now two years old, had been nursed well away from her and had luckily escaped smallpox. Her husband, shaken by nearly losing her, was attentive, for the time being. Initially, she wore a veil, but soon abandoned this and faced the world as she now was.

An uncommon voyager

"Your whole letter is full of mistakes from one end to the other. I see you have taken your ideas of Turkey from that worthy author Dumont, who has writ with equal ignorance and confidence."

Lady Mary Wortley Montagu, Letter to Lady —, 17 June 1716

In 1716, Edward Wortley Montagu was appointed British Ambassador to the Ottoman Empire (Turkey), with a demanding brief. Turkey had fought several large battles with Austria and was readying itself for all-out war, and Wortley was sent out to mediate.

The couple set off from London on 1 August 1716 to travel overland to Constantinople, en route taking in the sights in Rotterdam, Cologne, Nuremberg, Vienna, Budapest and Sofia before crossing into Turkey.[18] This six-month Grand Tour might have seemed foolhardy for a society lady, especially with a three-year-old in tow. For Lady Mary, it was a fabulously rich cavalcade of new cultures and experiences that helped to wipe clean the slate that had been defaced by smallpox – and might possibly jolt her husband back into her life.

The family did not travel light. Their retinue of 20 carriages conveyed numerous servants (five carriage-loads for Lady Mary), their own cooks, and a 50-year-old Aberdonian surgeon named Charles Maitland. On arrival in Constantinople, Maitland would look after the family in addition to the Embassy physician who had been there for nearly a decade – one Dr. Emanuel Timoni. Lady Mary's childhood friend, Sarah Chiswell, had been invited to join them but had declined – and thus lined herself up to receive one of the most famous letters in the history of medicine.

They were based initially in Adrianopolis, where Wortley was busy during the spring of 1717, trying to intercede between the

Turks and the Austrians. In May they moved on to Constantinople and set up house in Pera, the diplomatic quarter on the northern side of the inlet of the Bosporus known as the Golden Horn. Lady Mary immediately took to the place and the view across orange groves to the sea and the city beyond: "Perhaps, all together, the most beautiful prospect in the World."[19]

Wortley's three-year tour of duty was punctuated by frequent absences in his role as peacemaker. Lady Mary threw herself energetically into Turkish life and culture. She set aside a day a week to learn the language, wore Turkish clothes, prided herself in exploring where no European, Christian or woman had previously been allowed to set foot, and generally went native (see Plate 1). She fell in love with Turkish art, music, dance and architecture. By travelling incognito, concealed in full Turkish dress and only accompanied by an interpreter until she herself became fluent in Turkish, Lady Mary managed to infiltrate most of the higher strata of Turkish society.

She recorded her exploits and impressions in a series of letters sent home to various "Persons of Distinction, Men of Letters & c."[20] Lady Mary took pains to point out that they were "drawn from sources that have been inaccessible to other Travellers" and are quite different from the patronising ramblings of "common voyage-writers who are very fond of speaking of what they don't know". Her 'Turkish letters', part travelogue and part social commentary, have weathered well during the last 180 years and are still great fun to read today – an amalgam of the *National Geographic* and a *Rough Guide*, spiced with a dry and wicked humour. She describes people and their surroundings in near-photographic detail while the sketches of the places she visits glow with colour and atmosphere.

From her letters, we see Lady Mary comfortably holding her own in a bewildering series of unlikely situations.[21] She is the first Christian to pass – by special dispensation of the Grand Vizier – beyond the Black Eunuch who guarded his harem, to be entertained by his wives and their ladies. Robed up as a Turkish man, she is shown every corner of a mosque by a guide who has almost certainly seen through her disguise. She gallops on a white charger past open-mouthed men who had never seen anyone ride side-saddle; climbs into a huge cypress to inspect the house of the tree-dwelling schoolmaster of Kujük Cekmege; and outwits unscrupulous dealers who try to fleece her when she wants to buy some ancient Greek medals and a mummy. When the time comes to deliver her baby daughter, she does it *à la turque*, without fuss, and is up and about within days – a regime that would have scandalised the midwives and doctors of England.

Lady Mary tried one experiment to emulate the Turkish ladies' flawless complexion – "the loveliest bloom in the world" – using "the best sort" of the Sultans' fabled Balm of Mecca, which had been given to her as a gift.[22] She applied it to her face "with great joy ... expecting some wonderfull Effect to my advantage". Unfortunately, even the 'best sort' could evidently induce a severe allergic reaction. "The next morning the change indeed was wonderful; my face was swelled up to a very extraordinary size and all over as red as my Lady X's." Wortley was unsympathetic and "reproved my indiscretion without ceasing". Luckily, her face settled back to normal after a few days. No miracle cure, then; she concludes philosophically: "Let my complexion take its natural course, and decay in its own time."

A tale of old women

"I am going to tell you a thing that I am sure will make you wish yourself here."

Lady Mary Wortley Montagu, Letter to Sarah Chiswell, 1 April 1717

Lady Mary had been in Adrianople for only a few weeks when smallpox again lifted its head, but this time with a strangely defensive twist that immediately caught her attention. She had already noticed the peculiar lack of smallpox scars on the faces of her new Turkish acquaintances and now learned that this was put down to the practice of inoculation with smallpox (she borrowed the term 'ingraftment' from gardening, as the nearest translation of the Turkish word). She probably heard of this custom from her new Turkish friends. Later, in the diplomatic quarter in Constantinople, she would find that the practice was also familiar to other expatriates, including Dr. Timoni.

She first mentioned inoculation in her famous letter of 1 April 1717, to Sarah Chiswell.[23] She began by describing a near-miss with plague which had nearly killed one of her cooks during the last leg of the journey to Turkey. Then she went on:

> Apropos of distempers, I am going to tell you a thing that I am sure will make you wish yourself here. The smallpox, so fatal, and so general amongst us, is here entirely harmless by the invention of ingrafting, which is the term they give it. There is a set of old women who make it their business to perform the operation every autumn, in the month of September, when the great heat is abated. People send to one another to know if any of their family has a mind to have the smallpox; they make parties for this purpose ... commonly fifteen or sixteen together .... .

Lady Mary's choice of words was characteristically playful, but the stark contrast between the dread of smallpox in England and its casual dismissal in Turkey would have struck Sarah Chiswell instantly – especially as she had seen what it had done to her friend. The apparently absurd notion that people might choose to catch smallpox – and at parties! – rather than trying to shut themselves and their families away from it, must also have seized her attention.

> the old woman comes with a nut-shell full of the matter of the best sort of smallpox, and asks what veins you pleased to have opened. She immediately rips open that you offer to her with a large needle (which gives you no more pain than a common scratch), and puts into the vein as much venom as can lie upon the head of her needle, and after binds

> up the little wound with a hollow bit of shell; and in this manner open four or five veins.

The whole process was a casual seasonal ritual, so straightforward that it was entrusted to old women. 'The best sort of smallpox' might refer to milder cases, or like the '*best sort*' of the Balm of Mecca, might simply be a flicker of her dark humour. Technically, the incisions were not deep enough to open a vein, although this was the usual target of a sharp instrument wielded over an arm in England. Meanwhile, the party continued:

> the children or young patients play together all the rest of the day, and are in perfect health til the eighth. Then the fever begins to seize them, and they keep their bed two days, very seldom three. They have very rarely above 20 or 30 in their faces which never mark; and in eight day's time they are as well as before their illness. Where they are wounded, there remain running sores during the distemper which I don't doubt is a great relief to it.

After this trivial procedure, a successful inoculation caused minimal inconvenience, with a few pocks on the face, which, echoing Timoni's letter, "never mark". The dramatic contrast with the death-dealing ravages of 'ordinary' smallpox, and from the hundreds of pocks that had flayed the skin off Lady Mary's own face, leaps off the page. She acknowledged that the inoculation sites suppurated and discharged, but believed that this probably brought relief during the brief illness, and left only minimal scarring.

Not surprisingly, the practice had caught on in Turkey:

> every year thousands undergo this operation; and the French Ambassador says pleasantly, that they take the smallpox here by way of diversion, as they take the waters in other countries. There is no example of anyone that has died in it .... .

Lady Mary made no claims about how well engraftment actually protected against smallpox, but it must work – otherwise, why would thousands submit to it every year? If there were any risks, she did not mention them, and instead stated boldly that inoculation has never killed.

Her glowing appraisal of variolation – all good and no harm – led her into the declaration that must have made Sarah Chiswell sit up.

> you may believe I am very well satisfied of the safety of the experiment, since I intend to try it on my dear little son.

After dropping this bombshell, she continued in an evangelical spirit but well aware of how difficult it would be to spread the word of this breakthrough:

> I am patriot enough to take pains to bring this useful invention into fashion in England; and I should not fail to write some of our doctors very particularly about it, if I knew any one of them that I thought had virtue enough to destroy such a considerable branch of their revenue for the good of mankind. But that distempter is too beneficial to them not to expose to all their resentment, the hardy wight that should undertake to put an end to it ... perhaps, if I live to return I may, however, have courage to walk with them. Upon this Occasion admire the Heroism in the Heart of your Friend.

The letter might have left Sarah Chiswell wondering if this was an April Fool's story. It is widely quoted as the opening shot of a campaign for variolation, and Lady Mary certainly made her intentions clear. Later, when she moved to Constantinople, she would discover that Dr. Timoni's own daughter had already been inoculated, as had the children of Mr. Hefferman, the secretary to Wortley's immediate predecessor, Sir Robert Sutton. But now, a year before the experiment took place, her intention to try out inoculation on her son must have seemed a leap into the unknown, to her as well as Sarah Chiswell. Lady Mary neglected to mention an important hazard of inoculation, namely that it could spread smallpox among unprotected people. She may not have known about it then but certainly did later a year later, when she decided against inoculating her baby daughter, in consideration to the girl's nurse who had not had smallpox and so could have caught it herself.[24]

We can be sure that Lady Mary's letter had its intended effect on Sarah Chiswell, and that news of the daring experiment due to be carried out in Constantinople leaked out into English society during the summer of 1717. For Wortley, it may just have been an old wives' tale.

It may be just be coincidence that Wortley was out of town, at the Grand Vizier's camp in Sofia, when Lady Mary decided to press ahead with her experiment in the following March, 1718. Her timing was curious: she had told Sarah Chiswell that September was the traditional inoculation season, although Timoni's letter of four years earlier had mentioned spring as well as autumn.

On 18 March 1718, Charles Maitland summoned an elderly Greek nurse, one of Constantinople's most experienced 'old women' inoculators, to the Embassy.[25] She was instructed to inoculate the five-year old Edward Montagu, using blister fluid from a case of mild smallpox whom Maitland had selected. Maitland watched carefully as the old woman cut and inoculated one arm and then – dismayed at the 'torture' inflicted on the boy by her shaking hand and rusty needle – took over and repeated the procedure on Edward's other arm.

Maitland may or may not have approved of all this. His pointed use of italics in his own account suggests both some scepticism and a degree of coercion by the formidable "Ambassador's ingenious Lady", who was "so thoroughly convinced of the Safety of it, that *she* resolved to submit *her* only son to it … She first of all order'd me to find out a fit subject to take the Matter from … and then sent for an old Woman who had practis'd this way for a great many Years."[26]

Luckily, it all passed off uneventfully. On schedule, young Edward developed a fever, and a hundred or so red spots appeared – probably more than Maitland and Lady Mary had bargained for. The spots went through the classical stages of the smallpox rash, and this must have been a tense time for Lady Mary. Finally, the blisters shrivelled and dried up into scabs, which – to Lady Mary's huge relief – fell off without leaving scars, and the lad was back to normal.

Lady Mary reported the barest bones of this *fait accompli* to her still-absent husband some five days later, on Sunday 23 March in just one sentence that follows her description of a wrangle over some pearls with a "perfect mad" Dutch woman:[27]

> The Boy was engrafted last Tuesday and is at this time singing and playing and very impatient for his supper.

After a couple of *non sequitur* reminders that Wortley must claim his full allowance as Ambassador, she adds in closing:

> I cannot engraft the Girl; her Nurse has not had the small Pox.

It is curious that such a prolific letter-writer delayed telling her husband about a potentially dangerous encounter with foreign medicine. Perhaps she waited to be sure that the boy was doing well, although she wrote a couple of days before the fever and the rash were due to break out.

There is no record of whatever Edward Wortley thought of his wife pressing their son into service as a guinea-pig. Oddly, Lady Mary made no further reference to inoculation in her remaining letters from Constantinople, apart from the factual and dry sentence that opened her next letter to Wortley on 1 April 1718, exactly one year after first revealing her intentions to Sarah Chiswell:

> Your Son is as well as can be expected, and I hope past all manner of danger.[28]

Wortley's response – if any – is unreported.

The wanderer's return

"I am easy here, and as much as I love travelling, I tremble at the inconveniences attending so great a Journey with a numerous family and a little Infant hanging at the breast."

Lady Mary Wortley Montagu, Letter to the Countess of —, May 1718

Wortley had one notable success. He negotiated the original agreement which brought the Austrians and the Turks together and led to the Peace Treaty signed in July 1718. However, his own star took a nose-dive just as his achievements were reaching fruition. Back in London, there was a sudden change in political favour and the axe came down on Edward Wortley Montagu's term as His Majesty's Ambassador to the Ottoman Empire.

The Montagus returned to London later that year. The family began the pilgrimage home on 18 June, on board the warship *Preston*. Landing at Genoa, Wortley and Lady Mary completed the trip by land, leaving the two children, five-year-old Edward and babe-in-arms Mary, in the charge of their Armenian nurse for the remaining six months of the sea voyage.[29]

Back in town, Edward immediately dedicated himself to his rehabilitation while Lady Mary was rapidly swept back into the social whirl, with a wealth of exotic and outrageous experiences to recount. She became a regular and popular member of the Court circle and built a close friendship with Caroline of Anspach, the Princess of Wales.

The inoculation experiment was in hibernation. Lady Mary was laying the ground, priming influential contacts and waiting for a grand opportunity to roll out "this useful invention" to its best advantage.

Her chance came three years later in 1721, when smallpox broke out again in the metropolis and started to pick off various acquaintances on the fringes of high society.[30] Lady Mary contacted Charles Maitland, recently repatriated and retired, and asked him to inoculate her daughter, now just past her third birthday. Maitland reluctantly agreed, on the condition that this bold, first-for-England experiment would be witnessed by respected senior physicians. He undoubtedly wanted the inoculation and his role in it to be noted by those at the top of the profession; also, should anything go wrong, he was determined to spread the responsibility and cover his back against any accusations of dangerous foreign quackery.

In late April 1721, Maitland collected blister fluid from a young girl with discrete smallpox and, in the presence of Lady Mary and three medical worthies nominated by the Royal College of Physicians, inoculated young Mary Wortley in both arms.[31] Day by day, the little girl was studiously observed by the eminent physicians and "several Ladies and persons of Distinction", who noted a transient fever after ten days, followed by a mild rash with relatively few pocks. With the rash at its peak, they saw "Miss Wortley playing about the Room, chearful and well, with the pocks raised upon her." Within a few days, the rash crusted and the scabs fell off, and after three weeks, she was completely well and unscarred.

The medical eminences were suitably impressed by the mildness of the inoculated disease and by little Miss Wortley's rapid and full recovery. They must have also have been happy to assume that the procedure had been successful, even though the crucial evidence – that the little girl could not catch smallpox when exposed to it in the usual way – was entirely lacking. One of the attending physicians was Dr. James Keith, who had seen two of his sons die of smallpox. Keith was sufficiently impressed that he asked Maitland to inoculate his surviving four-year-old son.[32] This was done in early May – and the transformation of a bizarre heathen custom into a respectable medical innovation had begun.

Trials and tribulations

Word spread rapidly within high society, no doubt helped by Lady Mary's top-level connections and her personal networking skills. By the summer of 1721, she had recruited some big names to her cause – powerful enough to out-trump any members of the medical mafia who might try to strangle inoculation at birth. The most conspicuous jewel in Lady Mary's crown was the Princess of Wales, a kindred free spirit marooned in the stuffy routine of Court, and now a firm friend of Lady Mary.[33] The Princess carried huge influence, due as much to her personal qualities as to her title. Unusually bright for a royal, she was widely admired and was a key leader of English opinion and fashion. She also had an interest to declare: she had nearly lost her eldest daughter to smallpox earlier that year. So when the Princess of Wales suggested to a group of senior physicians that they might like to approach the King for royal approval to put inoculation to the test, the procedure took another significant step towards respectability and acceptance – even before the experiment had been conducted.

The clinical trial was duly carried out with six near-volunteers, three men and three women, on 9 August 1721.[34] Its main purpose was to confirm that inoculation was feasible and safe; its ability to protect against smallpox was tested later in only one subject.

Informed consent of a sort was obtained, in that the subjects could decline to participate. However, all six were condemned prisoners in Newgate prison, where conditions were so appalling that even those who had not been formally sentenced to hang were effectively consigned to Death Row. The offer of a royal pardon in exchange for having a few drops of pox fluid scratched into the skin therefore had its attractions – especially for one man who only pointed out after the pardon had been granted that he had already had smallpox.

Maitland again performed the honours, in front of an audience of the illustrious and inquisitive that included the Princess of Wales and Sir Hans Sloane. The personal interest of Sloane (who designed and supervised the test) represented the highest possible endorsement by the medical establishment. He was then one of the two physicians to the royal family as well as President of the Royal College of Physicians and one of the brightest stars in the Royal Society, where he would succeed Newton as President.

Figure 4.1 Sir Hans Sloane. Image of a commemorative medal reproduced by kind permission of the Natural History Museum, London.

All six subjects – except for the man who had previously had smallpox – reacted as both of Lady Mary's children had done: a transient fever, a mild rash, an uneventful recovery and no scars outside the inoculation sites. For five of the experimental subjects, that was the end of the story, and a remarkably happy one.

The sixth subject, 19-year-old Elizabeth Harrison, took part in a further study to test whether inoculation really protected against smallpox.[35] Sloane made up for the small sample size with a high-intensity exposure to smallpox. He paid for the girl to be sent to Hertford, where Maitland practised, to nurse a young boy with active smallpox. She was locked away with him for several weeks until after his rash had scabbed and faded. At night, she had to share his bed, which must have been a gruesome experience for both of them. The girl was followed up, remained free from smallpox, received her pardon and passed safely back into obscurity.

Sloane reported favourably on the experiment. Lady Mary would have been pleased with its positive outcome and its endorsement from the country's top physician-scientist. However, the Princess of Wales was not quite fully persuaded. Her reaction remained the crucial litmus test for English society, and was probably more important than any erudition from Sloane and his colleagues.

Lady Mary may well have been frustrated by her friend's hesitation, although the points at issue were reasonable ones that many would need to be reassured about. So far, fewer than ten subjects had been inoculated under English supervision, so there was still a chance that the good results were just a fluke. Also, only two children – both Lady Mary's – had been inoculated, and only one under the rigorous gaze of the English medical establishment. The essential next step was to confirm that inoculation was safe in children and babies, who were at the greatest risk of smallpox and were the age group in which (as in Turkey) the procedure should logically be performed.

The necessary experiments were done in late February 1722, with twelve subjects.[36] Ethical approval would not even have been thought of, but would anyway have been impossible to obtain in six cases, who were all orphans from the the Parish of St. James, Westminster. All the inoculations, performed by Maitland and at the Princess' expense, produced the same short-lived fever and mild rash as in the earlier cases and all the children made an untroubled recovery. The result of the experiment was announced with pomp from Whitehall on 10 March 1722:[37]

> Their Royal Highnesses the Prince and Princess of Wales, being desirous for Confirmation of the Safety and Ease of this Practice .... six persons more had the Small-Pox inoculated on them .... and the Curious may be further satisfied by the sight of these Persons at Mr. Forster's House in Marlborough Court.

This tipped the scales for the Princess of Wales and set the stage for the wider spread of inoculation through the English upper classes. Then, as now, society was susceptible to the celebrity culture, and every stage of the Princess' conversion to the peculiar cause of inoculation was noted and discussed with keen interest, and primed her followers and admirers to follow suit.

Once persuaded, the Princess soon proved that her conviction went beyond mere words. On 17 April 1722, Maitland was summoned to the Palace and, assisted by Claude Amyand, the Royal Surgeon, inoculated both Princesses Amelia and Caroline, aged eleven and nine.[38] This powerful expression of confidence matched another that would already have caught the eye of the thinking classes. Some weeks earlier, the grandchildren of Sir Hans Sloane had also been inoculated.[39] Both events were carefully publicized, and fortunately, all these high-profile guinea-pigs received their smallpox successfully and without incident.

From East to West

"I called the New World into existence, to redress the balance of the Old."
George Canning Speech, 12 December 1826

So far, the clinical trial of variolation had got off to a flying start in England, thanks to a combination of strong personalities, social engineering and good fortune. Inoculation had commandeered enthusiastic, top-level support from doctors and scientists. The English experience of over 20 cases had revealed nothing to contradict the seemingly miraculous claims from far-off Turkey that inoculation was safe and never failed: 100 per cent of the cases had successfully been given smallpox in a controlled fashion, 0 per cent had died or suffered any significant side-effects, and 0 per cent had severe residual scarring. Moreover, 100 per cent of the subgroup deliberately exposed to smallpox were convincingly protected against it (although admittedly this latter subgroup contained only one subject). Just one serious complication or a death among these first cases, which was well within the bounds of statistical risk, would have killed the putative breakthrough on the spot. Fortunately this had not happened.

At this stage, Lady Mary undoubtedly felt satisfied and possibly smug. She had succeeded in both the aims that she had set down in her letter to Sarah Chiswell four years earlier: to bring "this useful invention into fashion into England", and to "have courage to war" with the doctors who would oppose it. Also, inoculation had survived its first real tests of medical, scientific and public opinion. It must have appeared that inoculation was poised to fulfil its promise of knocking smallpox off its pedestal of fear and making it "entirely harmless", as it was in Constantinople.

Inevitably, though, as more subjects were inoculated and experience of the new technique grew, its risks and dangers became more apparent. In spring 1722, soon after Princess Caroline's daughters were inoculated, the first evidence began to emerge that the technique could be dangerous, and within weeks, bad news was suddenly coming in thick and fast. Soon after, the honeymoon was well and truly over and Lady Mary found herself struggling to keep her campaign on the road, and fighting to save her own reputation.

This is a bad moment to leave London as events are on the point of unravelling into chaos, but in deference to chronology, the story now needs to follow the main action and shift 3,000 miles west to the New England port of Boston.

5 Curiosa Americana

"I write the wonders of the Christian religion, flying from the deprivations of Europe ... I report the wonderful display of His infinite power, wisdom, goodness and faithfulness, wherewith his Divine Power hath irradiated an Indian Wilderness."

Reverend Cotton Mather *Magnalia Christi Americana*, 1702

Boston, 1721: the largest town in the American Colonies and a prosperous port with a population of about 11,000 souls.[1] The souls in question were ardently God-fearing and proud to live up to the Puritan principles of their Founding Fathers. Boston was the spiritual, administrative and social capital of New England and an established centre of learning. Even then, half a century before its elevation to full university status, the 60-year-old Harvard College in nearby Cambridge was building a formidable reputation throughout the Colonies.

For the settlers, 'New' implied improvement as well as novelty. The umbilical cord linking them to the original England was sometimes under tension, but as yet far short of breaking point. The eponymous Tea Party was still 52 years in the future, and the Declaration of Independence another three years beyond that. American thinkers and scientists were already creating their own community but most still looked towards the old country for guidance and inspiration, and the Royal Society of London remained the highest authority.

The New World pioneers are commonly perceived as hardy as well as resourceful and courageous. Four generations of survival in their adopted homeland had tended to weed out

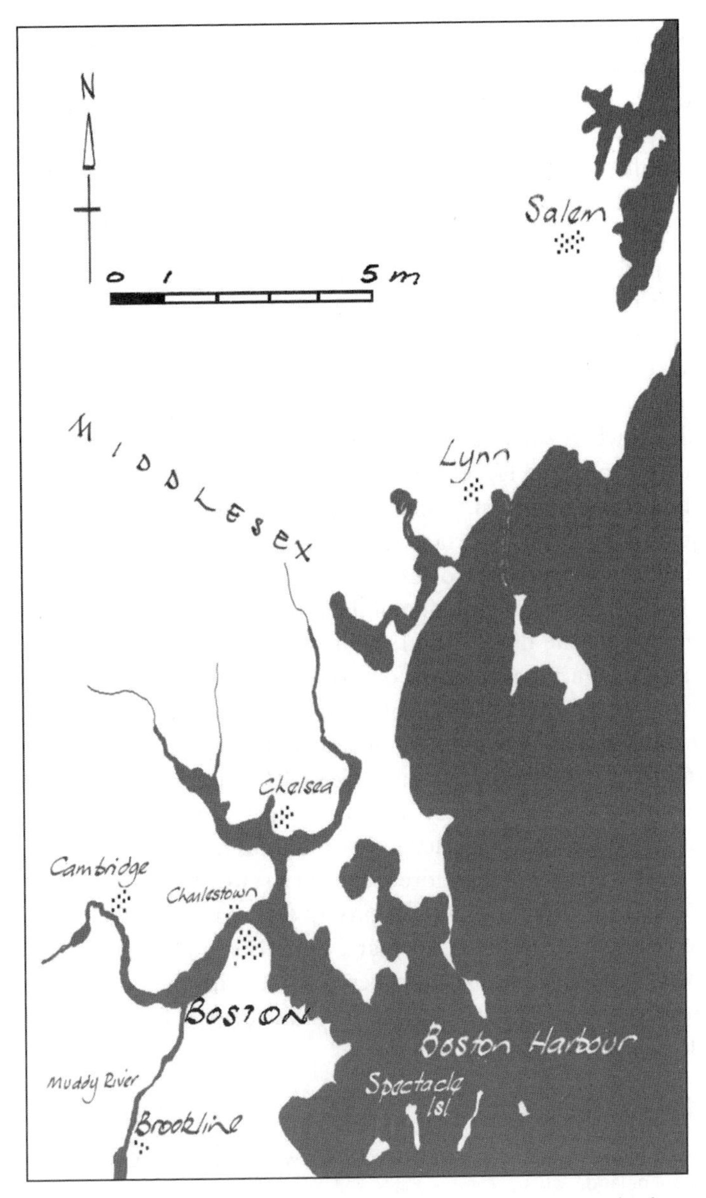

Figure 5.1 Map of Boston and the surrounding parts of New England.

the weak but many lives were blighted or cut short by disease, especially the top killer infections of diphtheria, scarlet fever and smallpox. As in London, smallpox was a nightmare that never went away. Boston enjoyed lucky runs of smallpox-free years but was always at risk when the susceptible population grew back to a critical level after the previous outbreak. Then, the town became a highly efficient incubator for smallpox. Outbreaks swept through the town with depressing regularity every dozen years or so: 1677, 1689–90 and 1702.[2]

Boston's strength as the busiest port in North America was also its weakness. Smallpox had a ready-made portal of entry through its poorly accountable sea trade, notably with the Caribbean Islands where smallpox was endemic. Rudimentary public health measures were in force to contain imported infections but all too often broke down. Incoming ships were isolated in the outlying quarantine area at Spectacle Island before being allowed to dock in the inner harbour. However, quarantine generally lasted only a few days – a small fraction of the Bible-inspired 40-day period implied by the name – and the health inspections were often careless. On the numerous occasions when infections slipped past these defences and broke out on shore, the response of the authorities was both pragmatic and resigned. The victims and their immediate contacts were sealed in the building where they were found. Outside, a warning flag carried the chilling, last-ditch appeal: "God have mercy on this house."[3]

Bostonians were therefore locked in a state of suspended fear, waiting for the blow of smallpox to fall again. In 1721, with no outbreak for almost 20 years, optimists might have hoped that the pattern had been broken. In fact, with the numbers of susceptible youngsters greater than ever before, Boston was a disaster waiting to happen.

Man of the Cloth

"I have indeed set myself to countermine the whole Plot of the Devil against New-England."

Reverend Cotton Mather *Wonders of the Invisible World*, 1693

The notion that smallpox could be given deliberately to protect healthy people against the disease had already reached Boston 15 years earlier – eight years before Timoni contacted the Royal Society and twelve years before Lady Mary Wortley Montagu sat down in Adrianople to write to Sarah Chiswell. This spark of an idea could easily have been snuffed out on the spot because it came from a source that New Englanders regarded with even greater misgivings than the English had for the Turks. It survived because it was implanted into the fertile but complicated mind of a Puritan preacher, the Reverend Cotton Mather.

Figure 5.2 The Reverend Cotton Mather.

Mather, aged 58 in 1721, was already well known in New England, both for his powerful ancestry and in his own right.[4] His name celebrated the fact that both his grandparents (John Cotton and Richard Mather) had been two of the most illustrious founding clerics of early Boston. Cotton Mather's father, the splendidly named Increase Mather, was a fire-and-brimstone preacher who had left smoking footprints across New England. Uncompromising in his sermons and prolific writings, Increase

Mather was also respected as a man of letters, and in 1692 was appointed the sixth President of Harvard College. In 1721, when events pushed his son into centre-stage, the 82-year-old Increase was still active, respected and feared.

As a third-generation product of this daunting dynasty, Cotton Mather had difficult acts to follow, made more challenging by a debilitating stutter that could have killed off his chances of a career in the pulpit. However, the fire in his forefathers' bellies burned just as fiercely in his own. He was also ferociously intelligent, and another puritanical trait – the unshakeable conviction that he could never be wrong – was even more strongly developed. A precocious and brilliant child, he was fluent in Greek and Latin by the age of eleven, when he was enrolled at Harvard for a Bachelor's degree in Classics and Divinity. Mather was also curious about nature in its widest sense, and informed himself particularly about science and medicine. He stayed on at Harvard for a Master's in Divinity, which he completed in 1679 at the age of 18. During his time at Harvard, he matured in other ways, fighting down his stutter so convincingly that his fellow students elected him their top orator.

At the age of 19, Cotton Mather took up post as assistant pastor to his father at the Second Church in North Boston, where he rapidly established himself as a fiery and sometimes incendiary preacher. His reputation soon came to rival that of his father, whose main base was at the First (South) Church across town. Undaunted by the length and temperature of Cotton Mather's sermons, his congregation were devoted to him. Years later, just before Christmas 1706, they would demonstrate that devotion by clubbing together to buy him a Negro slave. This gesture was matter-of-fact for the time but gains irony from the fact that Mather's flock included both native New Englanders and their slaves, whose souls he was equally determined to save. Mather's newest slave (he already had two) was to play a pivotal part in the story of inoculation in America.

Like Lady Mary Wortley Montagu, Mather wrote copiously. His output was prodigious to the point of incontinence: some 450 pamphlets and books on religion, science and medicine.[5] In contrast to Lady Mary, there is not an atom of humour, charm or humanity to be found in Mather's writing – the disordered, sometimes near-schizophrenic outpourings of a curious but naïve spirit locked inside a Puritan's narrow mind. Many of his works have peculiarly uninformative titles that give no hint of what lies within, almost as if there had been some confusion at the printer's about which cover belonged to which text. *Angel of Bethesda* turns out to be a medical treatise, while *Wonders of the Invisible World* is about visitations by the Devil. Unusually, *Curiosa Americana* is what it purports to be, namely a collection of puzzling observations from the New World, which he reported to the Royal Society. Mather's religious works – the majority – are an assortment of oddball tracts on topics ranging from the occult to the everyday battle with Satan for the souls of New Englanders. They present a binary view of the world: things were either Good (i.e. approved by Mather), or Evil (everything else).

An exception is *The Negro Christianised*, a persuasive appeal to "All you that have any Negroes in your houses" to save their slaves' souls.[6] His opening statement seems bigoted and prejudicial: he aims to transform the "most Bruitish of Creatures upon Earth ... the *Blackest* instances of *Blindness* and *Baseness* into admirable *Candidates* of *Eternal Blessedness*". A possibly charitable interpretation is that he did this to ingratiate himself with his intended readership and to smooth the way for the potentially inflammatory statements that follow, such as "Thy Negro is thy *Neighbour* ... my *Brother* too." Mather tailored his strategy for bringing the Bible to the slaves according to the learner's perceived intellectual capability – "the greater their Stupidity, the greater must be our Application"

– and he produced a simplified catechism for "Negroes of a limited capacity".

Mather's condescension was mild for the times and he had some sympathy with the slaves. This was in stark contrast to the contempt that he, like almost all New Englanders, held for the Native American Indians. Mather made clear his satisfaction with the 95 per cent mortality that smallpox had inflicted as it scythed through the original inhabitants of New England two generations earlier:[7]

> The Indians in these Parts had newly, even about a Year or Two Before, been visited with such a prodigious Pestilence, as carried away not one-tenth but Nine parts of Ten (yea 'tis Nineteen of Twenty) among them; so that the Woods were almost clear of these pernicious creatures to make Room for a better Growth.

Mather had a long-standing obsession with the Devil's interest in New England. To New Englanders, the Devil was not some biblical abstraction but a palpable presence, typically a short black man with cloven feet and about as tall as a walking stick.[8] In the early 1680s, when he was barely 20 years old, Mather investigated the 'possession' of the children of the Goodwin family in Boston, bringing the 13-year-old Martha into his own house for closer observation. His account of this episode – *Remarkable Providences* (1684) – did little for the disturbed children but helped him to crystallise his thesis that the Devil had his sights set on New England.[9]

Remarkable Providences also gave him a track record for diagnosing and combating Satan's works, and this levered him into a prime position to deal with outbreaks of bizarre behaviour, mostly in girls, in some outlying villages. Without his later interest in variolation, it is here that the Reverend Cotton Mather's reputation would have foundered on the sands of history – as one of the Puritan preachers at the centre of the Salem Witch Trials of 1692.

Man of Justice

"When the Accused had any motion of their Body, hands or mouth, the Accusers would cry out, as when she bit her lip, they would cry out of being bitten, if she grasped one hand with the other, they would cry out of being Pinched by her, and would produce marks ... if she stirred her feet, they would stamp and cry out of pain there."

Robert Calef *More Wonders of the Invisible World*, 1700

The Salem Trials remain a shocking reminder of the evil done in the name of religion and of how easily bigotry can reduce scriptural values to absurdity and brutality. After three centuries, the events at Salem still have the power to chill and horrify.[10]

Of the 29 people from Salem and neighbouring settlements who were convicted of witchcraft, 19 were hanged, one was crushed to death and five died in prison. Thanks to a strong steer from the preachers who acted as judges, it was easy for the accusers – mostly girls and young women – to get rid of almost anyone they wanted. All they had to do was fling themselves screaming to the ground and mimic the accused's postures and movements, which proved that they were in the grip of demonic possession beamed at them by the 'witch'. Many cases hinged on so-called 'spectral evidence', namely that the Devil could assume the form of an apparently innocent person.[11] This concept was applied erratically but always led to the finding of guilt. Supporting evidence, where needed, included the Devil's marks on the skin (this was a bad time to have moles or an extra nipple) and eye-witness accounts of a "Prodigious Prank or Feat" such as flying through the air on a stick or turning into a bird. The inquisitors also tortured confessions out of some of the 'witches', and encouraged neighbours and members of the accuseds' families to denounce them – which numerous spiteful people, with various personal or financial scores to settle, were happy to do.

Such were the inquisitors' powers of persuasion that only one of the accused, Giles Corey, finally refused to plead guilty.

Special treatment was reserved for him: the unhappy distinction of being the first person in New England to be "press'd", a euphemism for being crushed to death with boulders. This failed to extract a confession – even when, just before he died, one of the town's officials encouraged him to speak by using a cane to push his tongue, forced out by the pressure on his guts, back into his mouth.[12]

One of those hanged was the Reverend George Burroughs, another Harvard College alumnus who had graduated nine years ahead of Mather.[13] It was harder to convict the clever and articulate Burroughs, but they nailed him when a witness swore that he had lifted a seven-foot fowling gun with one finger – a feat that such a "very puny Man" could only accomplish with the Devil's own muscle power. Burroughs was taken out to be hanged on 15 August 1692. A large crowd had gathered on Gallows Hill, many of them shocked by his conviction and sentence. With the noose around his neck, Burroughs all but won them over by reciting, word-perfect, the Lord's Prayer – something which the inquisitors said was impossible for a witch. At that moment, an inquisitor rode up to witness the hanging. From horseback, he ranted at the crowd until the dissenters fell silent, when Burroughs was immediately strung up. Judicial contempt for Burroughs was completed by burying him with two other victims in a shallow grave, with his chin and one hand left protruding as an open invitation to the town's dogs.

So where in all this was the Reverend Cotton Mather? Surely the scholarly man of God, champion for the struggle of Good over Evil, and silver-tongued orator had the force of character to rein in such obvious travesties of justice, and steer proceedings back on to a fair track?

Not a bit of it. Mather was one of the chief inquisitors, right at the heart of their worst abuses. It was he who arrived on horseback to watch the execution of a fellow Harvard Divinity graduate and then turned an ambivalent crowd into a lynch mob. Mather was undoubtedly convinced that God and Boston's founding fathers were on his side, but in Salem he personified the very evil that he claimed to be stamping out and brought a passable version of Hell to New England.

Mather's centre-stage role at Salem brought him notoriety but certainly did not damage his standing in Boston and probably helped to further his career. In 1693, after the last executions, he published another book with an apparently randomly-generated title, *Wonders of the Invisible World*.[14] This showed that New England was no longer a land of Saints and had degenerated into a battleground with Satan. One of the few who dared to criticise Mather was Robert Calef, a cloth merchant. Under a parody of Mather's own title, *More Wonders of the Invisible World* (1700), Calef cleverly rubbished Cotton's arguments and highlighted Cotton's own conduct during the trials, with a damning account of George Burroughs' execution.[15] Calef hoped to stir up a storm of public uproar against Mather, but this came to nothing and the Bostonian version of calm was soon restored.

Man of science?

"It were to be wish'd the Writer had given an exact Figure of these Teeth and Bones ... He takes notice of vast Flights of Pigeons, coming and departing at certain Seasons; And as to this, he has a particular Fancy of their repairing to some undiscovered Satellite accompanying the Earth at a near Distance ... ."

Dr. John Woodward, Secretary of the Royal Society
Commentary on letters by the Reverend Cotton Mather,
Philosophical Transactions, 29: 63–64, 1714

Like his father before him, Cotton Mather strayed far into territories that many felt were beyond his capacity and authority. His incursions into medicine eventually angered the medical fraternity in Boston, notably its self-appointed leader. All but one of the doctors in New England at the time had learned their clinical skills (including surgery, midwifery and pharmacy) as apprentices to jobbing doctors, most of whom had also received no formal medical training. The exception was William Douglass, who had been awarded the coveted degree of Doctor of Medicine (MD) in his native Edinburgh.[16] On arrival in Boston a few years earlier, Douglass had initially been grateful for introductions to the town's movers and shakers. These included Cotton Mather, who professed to have a passion for medical and scientific matters. To begin with, the two got on well and evidently discussed current topics such as the papers in the Royal Society's *Philosophical Transactions*.

Mather had a personal interest in this journal, as he had been bombarding successive Secretaries of the Royal Society, first Dr. John Woodward and then Dr. Richard Waller, with a stream of letters describing New World oddities that had caught his attention. These *'Curiosa Americana'* appeared in the *Philosophical Transactions* as two bundles, abridged and pre-digested (possibly with dyspepsia) by the long-suffering Secretaries.[17] 'American curiosities' is a reasonable description for this haphazard miscellany. Most items were anecdotal – a "Prodigious Worm", a "Rain of Frogs", a "Monster (of two Children United)". Many stretched credibility as well as the imagination, such as the rattlesnake bite that instantly caused a steel axehead to crumble. Mather promised an inventory of herbal medicines plundered from the American Indians (which never materialised), in the meantime listing various plants claimed to cure a vast range of diseases, including one ("for which he does not give the name") that was effective against gangrene. As it seems fair to assume that Mather had no detectable sense of humour, he must have submitted all his offerings in good faith – in which case, some of his assertions are embarrassingly naïve. His was the proposal, mentioned earlier, that pigeons migrating from New England actually left the planet.[18] How the Royal Society allowed this drivel to be printed in the world's most illustrious scientific journal is a mystery. Whatever the reason, one corner of the page devoted to van Leeuwenhoeck's landmark drawings of the microscopic appearance of muscle fibres is taken over by Mather's sketch of "unaccountable Characters, not like any known Characters" carved into a rock.[19]

One gem stands out from the dross but even this is flawed. Mather described a cache of huge teeth (weighing up to five

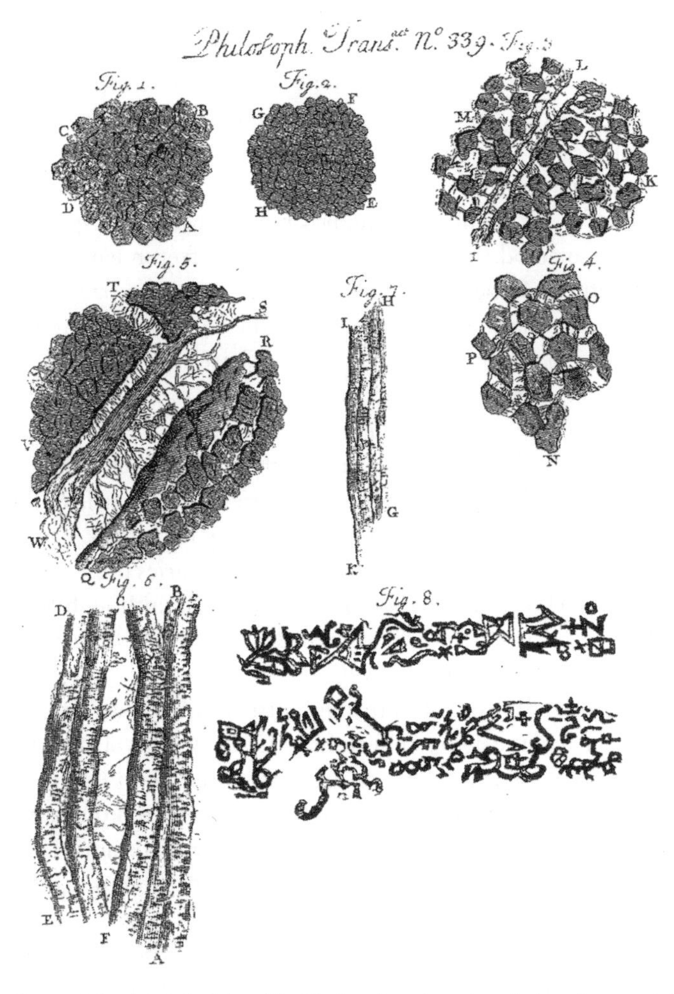

Figure 5.3 An indecipherable Native American inscription (lower right), from Cotton Mather's *Curiosa Americana*. Reproduced by kind permission of the Royal Society of London.

pounds) and gigantic bones, including a 17-foot long femur, from the banks of the Hudson River near New York.[20] These fossils played an important part in the early history of American palaeontology. The bones' original owner, initially called the Incognitum ('unknown'), was later identified as an extinct ancestor of the elephant and named 'mastodon' by the French zoologist Cuvier, after the breast-shaped cusps on the molars. Mather managed to embed himself in the history of this find, even though his account was sketchy, second-hand (the information came from the Governor of Massachusetts), and riddled with embarrassing errors This was literally a tall story: the mastodon's femur was up to eight feet in length, not 17. Whether Mather misreported this measurement or inflated it to impress those left behind in the Old World, these are not the actions of a conscientious scientist. Worse still, he stated the bones to be human and therefore evidence of a race of giants that were wiped out by the Great Flood.[21]

In all, Mather's correspondence with the Royal Society spanned eleven years (1712–23) and three Secretaries (Woodward, Waller and James Jurin), and during its less constipated spells, ran to one letter every day for some weeks. The Secretaries presumably dealt with Mather's submissions in good faith, although their statement that they were "entertained" by *Curiosa Americana* may hint at some reservations.[22]

In the end, and possibly through attrition, Mather's correspondence had the desired effect. In 1713, he was nominated for Fellowship of the Royal Society – the first native-born American to receive that distinction – although for some reason the ballot to confirm his election was not held then. It was not until 1723 that his name appeared on the list of those entitled to designate themselves FRS.[23]

These, then, are the decidedly dubious scientific credentials of the Reverend Cotton Mather. His insatiable curiosity led him far beyond his expertise and he lacked the scientist's crucial qualities of objectivity, self-criticism and insight. The events at

Salem also cast a long shadow and left him stigmatised as a loose cannon and a tunnel-visioned zealot who was prepared to hang old women because disturbed adolescent girls claimed to have seen them flying through the air.

Mather made an interesting contrast with the 30-year-old William Douglass, MD (Edinburgh). Initially, Douglass probably saw Mather as a gifted if odd polymath and an intellect worth cultivating. Soon, however, Mather's forays into medicine began to threaten Douglass' standing among the town's doctors and the wider community. The top-notch physician expected to dominate the medical hierarchy, looking down on all his colleagues who had simply scrounged their knowledge while on the job. Well below them lay the amateur Mather, who had no right to preach about medicine.

So when Mather started to do that, and people listened to him, it was inevitable that the two men would soon be locked in conflict.

A pretty Intelligent Fellow

"And it is to be desired, that the *Negroes* may not learn to say their *Catechism* only by rote, like Parrots; but that their Instructors may put unto them other *Questions* relating to the points of the Catechism, that by their *Answers* (at least of YES, or NO) it may be perceived that they *Know* what they *Say*."

Reverend Cotton Mather *The Negro Christianized*, 1706

The pre-Christmas gift of 1706, presented to their preacher by the grateful congregation of the North Boston Church, was a young man originally from North Africa, possibly Fezzan near Tripoli in present-day Libya.[24] We do not know his real name. Upon acquisition, the Reverend Mather christened him Onesimus, after a runaway slave whom Saint Paul had converted to Christianity. Onesimus and Mather got on well. As well as his obsessive mission to save African souls (as instructed in Psalms 6:31 – "Ethiopia shall soon stretch her hands out to

God"), Mather seems to have respected Onesimus as a man and a thinker. He taught him to read and write, paid him for his work and eventually allowed him to buy his freedom.

Smallpox was a serious concern for slave owners, who routinely asked newly-acquired slaves if they had ever had the disease. When Mather put this question to Onesimus, the response baffled him. As Mather later reported, Onesimus "answered, both, *Yes*, and, *No* ... then told me that he had undergone an Operation, which had given him something of Ye Small-Pox, & would forever preserve him from it".[25] Mather was not any more enlightened when Onesimus showed him his inoculation scar to prove that he told the truth.

The Harvard-trained Puritan could easily have dismissed Onesimus' claim as an unsubstantiated heathen custom from one of "the most Bruitish of Creatures upon Earth", but

TO BE SOLD, on board the Ship *Bance-Yſland*, on tueſday the 6th of *May* next, at *Aſhley-Ferry*; a choice cargo of about 250 fine healthy

NEGROES,

juſt arrived from the Windward & Rice Coaſt. —The utmoſt care has already been taken, and ſhall be continued, to keep them free from the leaſt danger of being infected with the SMALL-POX, no boat having been on board, and all other communication with people from *Charles-Town* prevented.

Auſtin, Laurens, & Appleby.

N. B. Full one Half of the above Negroes have had the SMALL-POX in their own Country.

Figure 5.4 Newspaper advertisement for slaves, from South Carolina, c. 1780.

something about the story hooked Mather's attention. He asked several other slaves if they too had been deliberately inoculated with smallpox. They had, and Mather concluded that the practice was common in parts of Africa. Yet nobody else in Boston appears to have picked up this snippet, or recognised its potential importance. Boston was in the lull between smallpox outbreaks, so Mather may have considered that there was time in hand to organise his thoughts. Apart from consulting the other slaves about inoculation, he does not seem to have pursued this further at the time.

The notion then lay dormant in Mather's mind for several years until the spring of 1714 brought number 339 of the *Philosophical Transactions*. Mather had a particular interest in this issue, as pages 62–70 contained the Secretaries' digest of his first 16 letters about the wonders of the New World.[26] Beginning on page 72, immediately after his own article, was Timoni's letter about inoculation in Constantinople. This account, with its sweeping claims of success in thousands of subjects, not only corroborated Onesimus' story but also set it instantly in a vastly broader perspective – and one in which Mather suddenly saw a place for Boston and the outbreak of smallpox that must soon come.

Mather immediately wrote to Woodward at the Royal Society. After a three-year respite, Woodward's heart may well have sunk at the prospect of another series of Matherian ramblings, but this letter was mercifully brief and to the point. Mather explained that he had already heard of the practice of inoculation from "my Negro-Man Onesimus, who is a pretty Intelligent Fellow".[27] He asked Woodward an obvious question: if this breakthrough really was all that it claimed to be, why was it not being used in England? Mather also stated his intention to put inoculation to the test, should smallpox ever return to Boston:

> For my part, if I should live to see ye *Small-Pox* again enter into our City, I would immediately procure a Consult of our Physicians, to Introduce a Practice, which may be of so very happy a Tendency.

Later that year, Mather's resolve was strengthened when he saw Pylarini's letter about inoculation in the next issue of the *Philosophical Transactions*. Mather discussed his ideas with Douglass, who did not take them seriously. Mather was disappointed but the two agreed to differ and remained friends for the time being.

Like Lady Mary, the Reverend Mather had to wait for his moment. Warned off by Douglass' sceptical reaction, he bided his time until Boston's doctors could be forced by the greater threat of smallpox into running a risky experiment that they would never have thought of trying.

By then, the town was already into borrowed time, but it was another seven years before the critical ingredients for the necessary disaster were brought together.

Return of the nightmare

"When the Destroying Angel has received his Commission from God, and his Sword is stretched out over a Land, it shall not return Empty."

Samuel Grainger, *The Imposition of Inoculation as a Duty Religiously Considered*, 1721

On 21 April 1721, His Majesty's warship *Seahorse*, leading a fleet of merchant vessels from the Dry Tortuga Islands (British Honduras) in the Caribbean, sailed into Boston's outer harbour.[28] According to regulations, the fleet remained in the holding area at Spectacle Island but the *Seahorse* sailed on, through a legal loophole that exempted the King's ships from quarantine, and docked at Long Wharf in the town's inner harbour. The no-nonsense yet graceful lines of the 20-gun *Seahorse* were much admired, but all was not well below decks. Two of the crew were laid low with non-specific symptoms but, as yet, no rash. If he knew about it, the Captain probably hoped that it would not turn out to be serious. On arrival, the two men were still well enough to leave the ship, and with the rest of the crew, set off into town to do whatever sailors do on returning to dry land.

Within a couple of days, both men were gravely ill with obvious smallpox. Wrong-footed, the town's authorities did their best to contain the damage. The two cases were sealed up in their newly-flagged sick-houses and an urgent search started to find their shipmates. The town held its breath, especially those who could remember the previous smallpox outbreak a generation earlier. Over the next two weeks, other crew members were rounded up, but it was already too late: nine of them were ill with smallpox. As predicted from the disease's incubation period, the first cases appeared among native Bostonians a couple of weeks later, and from there the sparks of infection began boring steadily into the brushwood of a whole generation, untouched by smallpox, who were ready and waiting to welcome the virus.

The outbreak followed the pattern of its predecessors, steadily taking grip over weeks and months and flaring up periodically – more like the smouldering of an underground peat fire than the incendiary blaze of a torched forest. By early June, five weeks after the *Seahorse* had discharged its cargo of virus, even those blinded by optimism or ignorance had to face the grim realisation that their defences had failed and that the nightmare of smallpox was once more stalking the streets of Boston.

For most Bostonians, only God could stand between them and the disease. Indeed, Cotton Mather himself wrote,[29]

> Because of the destroying angel standing over the Town, a day of prayer is needed that we may prepare to meet our God.

While Mather's congregation tried to trust in the power of prayer, others took more pragmatic action to keep out of the Destroying Angel's way. Some 1,000 people, about 10 per cent of the total population, left town. Those remaining included older people who had survived smallpox. Their own immunity did not give them much comfort as they watched smallpox ploughing through Boston, picking off family and friends. A daily tally of the death toll was published by Boston's two main newspapers, the established *Boston Globe* and the recently-founded *New England Courant*.[30] The newspapers rapidly polarised into sounding boards for the two sides of the arguments over inoculation, with the *Globe* in support and the *Courant*, under the direction of its outspoken editor, James Franklin, increasingly vehement in opposition.

A Consultation of Physicians

"A man guilty of such *Absurdities*, is no good voucher for an Experiment of Consequence."

Dr. William Douglass, Comments about Reverend Cotton Mather, 1721

In early June, with the death toll climbing, Cotton Mather decided that it was time for man, as well as God, to act. On 6 June, he followed up the commitment he had made to Woodward and sent a circular letter to all the dozen or so doctors practising in Boston, including Douglass (who later claimed, falsely, that he had been deliberately bypassed).[31] Mather explained about inoculation and summarised the experiences of Timoni and Pylorini. He also mentioned Onesimus' contribution, with a clumsy verbatim rendition – "People take Juice of *Small-Pox*; and cutty skin, and putt in a Drop; then by'nd by a little *sicky, sicky* ..." – that could well have undermined his case, especially as few of the doctors would share his view that this came from "a pretty Intelligent Fellow".

Mather suggested that they should all meet to discuss whether inoculation could be used to contain the outbreak. He wrote with unusual tact and deference:

> My request is that, you would meet for a *Consultation* upon this Occasion, and so deliberate upon it, that whoever first begins the Practice (*if you*

> *Approve that it should begun at all*) may have the countenance of his worthy Brethren to fortify him in it.

Well used to leading congregations, kangaroo courts and lynch mobs wherever he wanted to go, Mather was probably taken aback by the coldness and hostility of the doctors' response. All but one refused to have anything to do with Mather or his proposed experiment.

The most vociferous objector was Douglass, whose friendship with Mather was now in ruins. He was angered by what he saw as Mather's clumsy attempt to usurp his own authority and seize leadership of Boston's doctors. Douglass led the profession's attack on "this meddlesome priest" and his ignorant interference in medical matters. United against Mather, rather than the disease which was killing off their patients, the town's doctors fell into line behind Douglass.[32]

There was just one exception: a surgeon in his early forties from Muddy River (now Brookline) called Zabdiel Boylston.[33] Boylston was the antithesis of Douglass and indeed of Mather: quiet, thoughtful, methodical and always at pains to avoid confrontation. Mather must have been relieved to find someone medical prepared to back him, even though the mild-mannered Boylston would have struck him as puny beside the combined might of Douglass and all the other doctors of Boston. However, Boylston was held in high esteem by many townspeople. English-born, he had an MD from Oxford (which trumped Douglass' own degree) but his subsequent training as a surgeon while apprenticed in his father's practice in Boston had left him on the professional tier below Douglass.

The Mather–Boylston alliance proved remarkably effective and durable in the face of serious challenges ranging from character assassination to attempted murder. As an opening act of decisiveness – or provocation, in the eyes of their opponents – Boylston inoculated his own six-year-old son Samuel, together with his Negro slave Jack and Jack's two-year-old son, Jacky, on

26 June.[34] He adapted Timoni's method, dropping blister fluid from a mild case of smallpox into a short skin incision made with a sharpened toothpick and quills. All three subjects reacted as Timoni had indicated, with minor symptoms and a limited rash. Technically, at least, variolation in America had got off to a promising start.

However, this news was received with a clamour of outrage and hatred, with clerics as well as doctors weighing in against the unholy union of the meddlesome priest and the jumped-up surgeon. This was orchestrated by Douglass, who then drummed up a petition from the doctors of Boston to the town's 'Selectmen' (councillors), demanding that they outlaw inoculation because it was immoral and dangerous.[35] In particular, they claimed, inoculated subjects could spread smallpox while their rash was active. This was true – hence Lady Mary's decision against inoculating her baby daughter in Constantinople because the little girl's nurse had never had smallpox and so might catch it from her. Strangely, this complication did not appear to register with Boylston or Mather and, to their discredit, neither took it seriously after it was pointed out by Douglass.

On 21 July, the Selectmen heard evidence from a French physician, Dr. Lawrence Dalhonde, who swore that he had seen variolation practised in various European countries and had been appalled by its complications, notably the spread of smallpox to others.[36] Dalhonde was best known in Boston for his short stature and fractured English, and his account provided little hard evidence to support his claims. Nevertheless, it convinced the Selectmen, who to Douglass' delight, ordered Boylston to stop inoculating. By then, Boylston had inoculated ten subjects, with no complications. He did as he was told but put an explanatory notice in the *Gazette*, which made it clear that he intended to continue.[37] In early August, when the fuss had died down and the outbreak was flaring up again, he quietly resumed his inoculations.

Later that month, Mather had his son Thomas inoculated.[38] This was an act of pragmatism, not propaganda, and was kept secret. Thomas, a student at Harvard, had fled home in panic because his room-mate had died of smallpox. The thought of losing his son in the outbreak also touched a sensitive spot for Mather, who by then had buried eleven of his fifteen children; the decision to inoculate was probably as reasoned as it could have been in the circumstances.

By mid-September, Boylston had inoculated 35 people, with no mishaps. When the news broke, there was fury from Douglass and his associates. Then the gloves came off on both sides, and the inoculation debate descended into a bare-knuckle brawl with increasingly vitriolic abuse flung in from both sides and, before long, threatened and then actual violence.

Douglass comes across as the main villain of the piece: petty, snide, manipulative and ultimately a two-faced opportunist. Initially, even the Reverend Cotton Mather, still trailing his appalling baggage from Salem, appeared moderate and balanced by comparison; before long, though, he reverted to type.

Boylston was a softer target than Mather, and Douglass set out to undermine this upstart from a lower level of the profession. Under the name of "W. Philanthropos", Douglass published an article in the *Courant*, which belittled 'Doctor' Boylston by calling him "a certain Cutter for the stone".[39] This piece of condescension, tossed down by the clever Edinburgh physician to crush the plodding sawbones, may well have misfired. Boylston was renowned as an expert and pioneering surgeon who winkled out the biggest and spikiest stones from the bladder. He had also successfully removed a cancerous breast from a woman who, several years later, was still alive and well and happy to sing his praises.[40]

The task of demolishing Cotton Mather's reputation was largely left to his fellow preachers, although Douglass also fired off a broadside that was designed to blow Mather's credibility as a scientist out of the water. For those gullible enough to be taken in by Mather's pretensions to Fellowship of the Royal Society, Douglass pointed out that this was the man who believed that pigeons migrated to an invisible moon, and that his name didn't even appear on the Royal Society's own list of Fellows (as explained above, this was true).[41]

Both sides continued to trade insults in the newspapers, alongside the daily updates of the numbers of smallpox cases affected and the steadily mounting death toll. James Franklin at the *Courant* published a stream of anti-inoculation articles, mostly anonymous but often recognisably in Douglass' style. Franklin also printed one-off pieces of complementary propaganda, including pamphlets by Douglass and his friend John Williams.[42] Williams dismissed inoculation as "a Violence to the Law of Nature and the Pattern set by God" and "a Delusion of Satan, covered over with more cloaks than the Doctors". Also, inoculated smallpox was worse in every respect than "when it comes on in the good old way", even to the fouler smell of the inoculation sore compared with the natural rash. For good measure, Williams lunged at Mather by comparing inoculation to the horrors of Salem.

Boylston was the only one who conducted himself with relative dignity throughout, keeping his head down and carrying on with his inoculations as best he could. Boylston kept detailed notes of the outcomes, which later showed clearly that the mortality of inoculated smallpox was much lower than that of the naturally acquired infection.[43] Over the summer of 1721, he steadily built up his clinical experience with "the Operation": 17 cases by mid-August and 35 by late September, of whom only a few had developed a serious rash and none had died.

On 25 September, Douglass took his campaign overseas, to London and the Royal Society. He wrote a damning account of inoculation in Boston, addressed to "Dr. A— S—, MD & FRS" (Alexander Stuart, a physician at the Westminster Hospital in London) who read it to the Society on 16 November.[44]

Douglass stressed that the experiment was conducted in defiance of the town's doctors and councillors, who were all deeply concerned that the procedure was dangerous and would spread smallpox, rather than contain it. By then, Boylston had inoculated over 60 people but his safety record was no longer unblemished: several of his subjects had been seriously ill with a confluent rash and two had died. Elsewhere in Boston, however, the outbreak was still in full swing and had just claimed its thousandth victim.

If Douglass had hoped for top-level endorsement of his stance and a put-down that even Mather could not ignore, he was disappointed. The Royal Society recorded his communication in its internal minutes book but not in the high-visibility pages of the *Philosophical Transactions*. Perhaps sensing that they would not get a balanced and dispassionate account of the wrangle in Boston, the Society published only one of the next four reports that reached them from the town over the next 18 months, all of which were sent in by Mather and supporters of inoculation. This was Mather's paper entitled 'The Way of Proceeding in the Small Pox Inoculated in New-England', read to the Royal Society in April 1722 and later printed in the *Philosophical Transactions*.[45]

Mather also countered Douglass' transatlantic propaganda with a pamphlet which he published anonymously in London in 1722. 'An Account of the Method and Success of Inoculating the Small-Pox in New-England' was (for Mather) unusually clear and allowed him to pay fulsome tribute to his fellow evangelist for his courage and foresight:[46]

> One who had been a more successful Practitioner than most of them, and had, with singular dexterity in his Practice, perform'd things not attempted by any of them (namely Mr. Zabdiel Boylston) was prompted, by his enterprizing Genius, to begin the Operation.

And God spake unto them

"And is there not an appointed time for man upon the Face of the Earth? I am sorry to see any in New-England so ignorant ... None can stay His Hand, or say unto Him, what doest Thou."

John Williams *Several arguments, proving that inoculating the smallpox is not contained in the law of physick, natural or divine, and is therefore unlawful*, 1721

Religion was never far from any aspect of life (or death) in eighteenth-century Boston, and the Bible, God and Satan were rapidly dragged into the debacle over inoculation. Most of the heat came from the clerics ranged against Mather, who distinguished themselves by the energy with which they scoured the Scriptures for pronouncements that proved how much variolation disgusted the Almighty.

The simplest argument – that inoculation was ungodly because it was not mentioned specifically in the Bible – was rapidly dumped because it undermined more powerful evidence. On reflection, inoculation was clearly referred to in the boils which Satan raised on the hapless Job (Job 2:7–8). This proved that the Devil had been the first inoculator, and that all who followed his example were "wicked and blasphemous".[47]

Inoculation was also a direct affront to God's innate right to determine who was to die, and when and how; as John Williams put it, "A Violence to the Law of Nature and Pattern set by God."[48] To interfere with God's grand plan was not just blasphemy: it was dangerously foolish, as the Almighty would quite rightly take exception to those who flouted His will. Indeed, the more serious the pestilence, the greater the significance which God attached to it and therefore the more terrible the retribution. Meddling with something as mortal as smallpox was bound to call down catastrophic revenge from a Jealous God. This was entirely just, because visitations of smallpox were simply well-merited punishments which the tough but fair Almighty meted out to New Englanders for all their collective sins. Any attempt to resist Divine Judgement was futile and sacrilegious. In the words of Samuel Grainger, "It is impossible that any *Human Means*, or *preventive Physick* should defend us from, or over-rule a *Judicial National Sickness*."[49]

In addition, it was blasphemy to pay any heed to stories of inoculation, because these emanated from the lower, sub-Christian orders of humanity, the "heathen Africans and Infidel Mahometans". And finally, the perverted notion of using the products of a revolting disease like smallpox to try to protect against that same disease was an insult to logic. John Williams taunted the inoculators to apply the principle to treat the haemorrhagic diarrhoea of dysentery by giving "a Bottle of that fluxical bloody Excrement ... by a Clyster Pipe [enema] or any way else". He concluded that, like the pus of inoculation sites, "your Argument do Stink".[50]

Sticks and stones

"The Destroyer, being enraged at the Purpose of any Thing that may rescue the lives of our poor People, has taken a strange possession of the People ... They rave, they rail, they blaspheme; they talk not only like Ideots but also like *Franticks* ... I am also an Object of their Fury."

Reverend Cotton Mather, Diary entry, November 1721

As the winter of 1721 drew in, there was no sign that the war of words over variolation was cooling off, in harmony with the weather. With the opposing factions' horns solidly locked, the whole pathetic spectacle was spiralling down into the gutter.

The smallpox epidemic and the inoculation battle both peaked in October and November of that year. While accusations, slander and threats flew thick and fast, the Selectmen ordered restrictions on the tolling of the traditional funeral bell; with up to 20 deaths each day, it rang almost continually day and night, causing disturbance and great distress to victims and their families.[51]

Few of the contributions from either side were at all edifying. Francis Archbald, writing an anonymous *Letter from One in the Country to His Friend in the City*, attempted to mediate with a sensible suggestion.[52] If people genuinely believed that inoculation could protect them against smallpox, then that was their choice – but he added that they must "withdraw from the Community into such places where there can be no danger to their Neighbours". He appealed to the Selectmen to set up "Proper Pest Houses in solitary places" in which to isolate those inoculated until they were no longer infectious. Boylston and Mather did not respond, and the Selectmen did not pursue the notion.

There was increasing militancy in Douglass' attacks, possibly in anger over Boylston's persisting defiance of him and the Selectmen or the fact that the tide now appeared to be turning against him. A pocket of support for inoculation had emerged in neighbouring Cambridge and Roxbury. Two Cambridge preachers, the Reverends Colman and Cooper, were now firmly aligned with Mather.[53] They spoke of inoculation as "an astonishing mercy" and begged "let us use the light which God has given us and thank Him for it". Even worse, some local doctors had received instruction from Boylston and had started to perform variolations themselves.

Douglass condemned inoculation as a "felonious crime" and demanded that all physicians who performed it should be hanged.[54] The inoculators soon lowered themselves to his level. Increase Mather said stiffly, "Known Children of the Wicked One are generally against Inoculation", but his son went further and said what both undoubtedly felt – that Douglass deserved to die:

> The Church ought to deliver him over to Satan, for he deserves the highest Censure, Deserves to be Scourged out of the Country ... he should be pillori'd and afterwards ston'd by the People.[55]

As this came from Mather, who had actually made such things happen in Salem, Douglass might not have seen this as empty rhetoric.

In November, events took an ugly turn. Boylston and Mather were already used to being showered with abuse, threats and worse whenever they ventured outdoors. But then Boylston found himself being mobbed in the street by a crowd who had come to find him. Boylston ran home and went to earth in his house for two weeks, sneaking out in disguise to visit patients at midnight and on several occasions having to hide in fear of his life when a lynch mob arrived to drag him out.[56]

Meanwhile, Mather was facing horrors of his own. He was tougher than Boylston and his reputation, and respect for the Mather dynasty probably shielded him to some degree. Nonetheless, he must have been shaken to find himself facing death threats and a narrow escape from a violent end. In the small hours of 14 November, a grenade was thrown through his window.[57] It may have been intended to terrify rather than kill and anyway failed to explode because the fuse fell out – which meant that the note tied to it, ungrammatical but with unmistakable intent, could be read:

> Cotton Mather, you Dog. Damn you! I'll inoculate you with this a pox to you.

Mather exploited his brush with the "Unknown Villain" (who was never caught) in a sermon entitled, "This Night there stood by me the Angel of God, whose I am & whom I serve."[58]

These acts of violence temporarily shocked both sides into a more reasoned state, but verbal hostilies resumed in the New Year of 1722. Douglass, as a thinly-veiled 'Anonymous', wrote another vitriolic anti-inoculation pamphlet addressed to Dr. Alexander Stuart of London, in which he again sneered at Mather and his credentials and accused all priests who supported inoculation of "vile calumny".[59]

Douglass' attack provoked the publication of a pamphlet purporting to report *A friendly debate* between 'Academicus' and two 'eminent physicians' identified as Sawny and Mundungus.[60] This was a rare ray of light in the confrontation: sparkling satire with lethal intent, which would have delighted Lady Mary and Alexander Pope – and exactly what would be expected from the future Harvard Professor of Natural Philosophy (Isaac Greenwood) who took shelter behind the scholarly pen-name. 'Sawny' (slang for Scotsman), speaking the transliteration of a broad Scottish accent, is instantly recognisable as the Aberdonian Douglass, while the gibberish-spouting Mundungus ('tripe') is Dr. Alexander Stuart, the addressee of Douglass' own pamphlet. Academicus deftly cut both Douglass and Stuart down to size:

> Thou, Sawny, can't spell the word *Philosophy*, nor construe the word Hades, tho' thou hast sent so many people thither ... apart from Brother Mundungus, I know not another conceited Coxcomb [simpleton] in the World.

Academicus raised smiles, for example with his helpful English/Mundungian glossary, but also fired deadly serious accusations at Douglass: liar, hypocrite, rabble-rouser and unfit to be a doctor. Douglass, he claimed, had admitted in private that inoculation protected against smallpox, had written in a letter that he was "thankful to GOD for the late Deaths of several Inoculated" and had demanded the death penalty for his colleagues who performed inoculation. And although Douglass tried to wriggle away from these accusations, all were correct.

The dust settles

"When the Honour of the Religion, and the Safety of a People, in general are concerned ... Who can be silent?"

Samuel Grainger *The Imposition of Inoculation as a Duty Religiously Considered in a Letter to a Gentleman in the Country*, 1721

As 1721 approached its end, so too did the smallpox outbreak. The number of new cases fell away after the New Year. The Selectmen's announcement on 26 February that the disease had left Boston was premature, but it was all but over by Easter. People and business gradually returned to the town and the combatants in the inoculation battle reflected on what they had achieved.[61]

Both sides had behaved appallingly and both had helped to boost the death rate from smallpox. Those opposed to variolation had obstructed the introduction of a measure that after a few weeks was showing obvious promise, while Boylston refused to accept that he could spread smallpox by failing to tell his inoculated subjects that they were infectious and must isolate themselves until the rash had crusted and the last scab had fallen away. Boylston and Mather did not recognise the risk initially, perhaps because Timoni did not mention it, but they should have taken it seriously when it was drawn to their attention – even though it meant conceding a point to the loathsome Douglass. Victims of "good old" smallpox were mostly very sick and it was intuitive to them and their families to remain in isolation until after the rash had gone. By contrast, inoculated smallpox was often mild and seen as innocuous by the subject and others. Many immediately resumed normal social contacts while they still had an active rash, and some threw parties to celebrate how well they felt. Later, in 1730, Boylston acknowledged that the inoculated rash was "capable of Infecting and producing the Small Pox in the ordinary way, on others" but did nothing to mitigate that risk at the time.[62]

By summer of 1722, the results of the great Boston inoculation experiment were laid bare for all to see. The two obvious questions that underpin every clinical trial and that are potential show-stoppers for every new treatment are: Does it work? and Is it safe? The trial, conducted under battle conditions, was far from perfect but Boylston's careful records allowed the outcomes of natural and inoculated smallpox to be compared.[63] Of the 10,000 or so people who had remained in Boston, 5,980 (about 60 per cent) caught smallpox, of whom 844 died – a mortality rate of 15 per cent. Boylston and colleagues had inoculated a total of 281 subjects, with only six deaths. This mortality rate of 2 per cent was less than one-seventh of that with natural smallpox. None of the inoculated people went on to get a second attack.

These figures indicate that inoculation worked and that it was much safer than natural smallpox. However, they also exaggerate the benefits. Much tighter and longer follow-up would be needed to confirm complete protection against smallpox. Several of Boylston's patients developed a severe confluent rash and although they survived, were left badly scarred. Finally, we do not know how many collateral cases occurred through contact with inoculated subjects, or how many of these died.

Nevertheless, inoculation had successfully passed its first proper trial and, bolstered by the favourable experience in England, was ready to be rolled out on a wider scale. However, all the medical and religious objections that had been tried out in Boston were being actively nurtured on both sides of the Atlantic in readiness for the next stage of the battle to kill off inoculation.

In many ways, the Boston of 1721–22 could have been a glorious chapter in the history of mankind's struggle against the tyranny of smallpox. Instead, the image that endures is both pathetic and surreal. At centre-stage, doctors and preachers are at each others' throats, hell-bent on doing each other down. Neglected in the background – almost as though it were happening somewhere else – is the worst catastrophe ever to hit Boston.

And notable for its absence is the tolling of the funeral bell that had to be silenced because it signalled all too clearly the supremacy of smallpox.

6 From Passion to Fashion

"That the small-pox raised by inoculation, is much milder and far less mortal than the natural sort, is now sufficiently confirmed by experience."
Conrad Sprengell *Aphorism XXIV of Hippocrates*, 1737

Most new treatments follow a set sequence of events. Initial scepticism gives way to enthusiasm which increases (often too rapidly to be credible) until something goes wrong. Confidence then collapses. If it recovers, the pros and the cons then fight it out until the use of the treatment finds its level – which may or may not reflect the true balance between its risks and benefits.

Variolation followed this curve, although the pattern was hard to see because events in London and Boston were not synchronised. And on each side of the Atlantic its supporters were preoccupied with excitements and problems of their own.

Teething troubles

Back in London in the spring of 1722, variolation had barely started up the initial phase of enthusiasm when it ran into difficulties. In fact, the first problems had already occurred when the two Princesses were inoculated in April – but luckily off-scene and mostly out of mind.

Maitland's fingers had been burned the previous autumn. Emboldened by his recent successes and new-found veneration, he was asked to inoculate the daughter of a wealthy patient in his home town of Hertford.[1] For eight-year-old Mary Batt, it was a fuss about nothing: like her namesake, Mary Wortley Montagu, she was only mildly ill and had only 15 pustules when the rash broke out. As in the Montagu household, the guinea-pig and her trivial spots became the centre of attention. Unfortunately, however, some of the Batts' servants had not yet met smallpox. Six of them went down with it and, to confirm that this was 'good old' smallpox, one of them died.

This mini-outbreak might have pricked Maitland's conscience but did not deter him. Neither did he learn from the experience: he still took no precautions to prevent his inoculated subjects from passing on smallpox to non-immune contacts. How he fell into this trap is a puzzle, especially as the Ambassador's 'Ingenious Lady' who had pushed him into variolating her son had decided against doing the same to her daughter in case her unprotected nurse caught the disease. Perhaps he thought that the act of inoculation somehow robbed smallpox of its lethality – hence the mildness of the 'artificial' disease – but that fatal experiment in Hertford should have told him that smallpox breaking out on a variolated subject was just as dangerous as usual. It is only 30 miles from Hertford to London and bad news travels fast if unimpeded, but whatever intelligence filtered through to Court failed to derail the variolation bandwagon.

This was the first item in a run of bad luck that played out behind the conspicuous successes in the foreground. The next, in April 1722, was a triumph of bad timing as well as a family catastrophe. The Earl of Sunderland, excited by variolation (or perhaps the desire to be seen emulating the Princess of Wales), volunteered his son to be variolated. The Earl was ill and feverish when he rode into Piccadilly to watch Maitland inoculating the two-year-old boy. This turned out to be a bad day for both father and son. Both died 18 days later on 21 April; the father from the pleurisy which had been taking hold, and his son apparently of fulminating smallpox.[2]

News of the double tragedy reached Court immediately and at a particularly bad moment – just four days after Princesses

Caroline and Amelia had been variolated. What should have been a routine run-up to another successful result turned into an agony of waiting; suddenly, there were important lives and not just doctors' reputations hanging in the balance. Fortunately, both Princesses sailed through their smallpox and the credibility of variolation survived with them.

Naturally, the success with the Princesses outweighed any unfortunate problems in the Sunderland family. Maitland and Dr. Claude Amyand, the Royal Surgeon who had assisted him at the inoculation of the Princesses, began to receive requests for inoculation from the rich and influential. This led to the next setback, a higher-profile replay of the unlearned lesson in Hertford. In early May, Amyand inoculated the children of the Earl of Bathurst.[3] All went well until a 19-year-old "strong, hail young man" who was footman to the Bathurst family fell ill a couple of weeks later. Maitland evidently suspected smallpox and stepped in to inoculate the footman. This was another valuable learning point. It takes several days for inoculation to raise defences against smallpox and once the natural infection is well established it is already too late. The footman developed full-blown smallpox and died just over two weeks later, on 19 May – coincidentally, just as the smallpox outbreak in Boston was claiming its first victims.

Ripples reached Court but were smoothed out by the combined efforts of Maitland, Amyand and Sloane and did not coalesce into a significant threat. Maitland and Amyand continued to variolate and by the end of July had performed another dozen inoculations. As reported in the *Weekly Journal* of 21 July, these included the four children of Samuel Brady, physician to the Garrison at Portsmouth. This was the article that struck a familiar chord with Dr. Perrot Williams in Haverfordwest.

Now that variolation was becoming respectable, Richard Mead – Lady Mary's failed gladiator Mirmillo with his gold-headed cane – went back to the *Philosophical Transactions* and decided to try out nasal insufflation.[4] His experiment was poorly designed: the only interpretable outcome would have been the subject's death. Mead only tried it once, perhaps because he thought it was not really worth doing or might go wrong and kill off variolation altogether. Newgate again provided a willing volunteer, a young woman. Ignoring centuries of Chinese technical refinement, Mead simply stuffed some cotton soaked in smallpox blister fluid up her nostrils. He did not bother to expose her to smallpox to see whether it had worked. This half-baked experiment won a royal pardon for the guinea-pig but failed to persuade anyone that the nose had it.

Growing pains

By the summer of 1722, strong opinions were emerging both for and against variolation.

The medical profession seemed mostly unmoved but Dr. William Wagstaffe, a powerful London physician from St. Bartholomew's Hospital, took against it. His open letter to his colleague Dr. Freind in June 1722 lambasted the "danger and uncertainty of inoculating the small-pox" and especially the "prodigious mortality" caused by infecting non-immune contacts.[5] Citing evidence from Boston (the letters from William Douglass to A— S— and Dalhonde's testimony to the Selectmen), Wagstaffe also poured vitriol on Turks, old women and those who had brought this laughable custom "practised only by a few *Ignorant Women*, amongst an illiterate and unthinking people" to "one of the most learned and Polite Nations in the World".

Wagstaffe's attack was more effective than Douglass' and cast a longer shadow, even though he had much less experience of inoculation. His letter became a book, reprinted several times and translated into French when it was instrumental in stirring up anti-inoculation feeling on the Continent.[6] Indeed, he poisoned the waters so effectively that the French lagged several years behind the rest of Europe in adopting the practice.

Fortunately for the variolation evangelists in London, help was at hand. Thomas Nettleton, a physician from Halifax in Yorkshire, had seen Timoni's and Pylarini's papers and, like Cotton Mather, had decided to wait for the opportunity to try this out. Encouraged by the reports from London and facing a smallpox outbreak in December 1721, he set about persuading his fellow Yorkshiremen to undergo this peculiar foreign treatment. Working off Timoni's description and without having seen it performed, his do-it-yourself version was unconventional. Rather than the shallow scratch which Maitland and Amyand were using, Nettleton made deeper knife-cuts that bled freely and pushed a cotton pledget dipped in blister fluid into the incision. It might have been improvised but it worked, and by early April 1722 he had performed over 40 inoculations – many more than the London experience.

Nettleton's account, flagged up in the *Philosophical Transactions* as coming from "this learned and ingenious gentleman", is measured and impressive for its clarity and honesty.[7] Several of his cases had complications including a severe eruption and convulsions, and one girl had died. All are described clearly and dispassionately and he asks for the reader's "impartial judgment" to decide whether the fatality should be blamed on inoculation or whether, as seems likely from the timing, the girl was killed by smallpox which she had already caught from her dying brother. Like Timoni, however, he does not mention isolating his patients or the risk that they might infect others.

Nettleton bolstered the case for variolation in another crucial respect, with his desire to prove that inoculation was safer than natural smallpox. He explained this to James Jurin, the Royal Society's Secretary and later President, who was formidably qualified in both medicine and mathematics and had a particular interest in the mortality statistics for smallpox. Jurin rapidly set up a correspondence network with almost all the known variolators in Britain and gathered in annual returns of their results. This study, one of the first prospective surveys of any medical treatment, enabled him to compare the death rates from natural versus inoculated smallpox. His first results, published in 1724, were clear and convincing: during this period, natural smallpox killed two in eleven (19 per cent) of its victims, whereas only one in sixty (1.6 per cent) died of smallpox after variolation. Jurin's subsequent surveys showed even lower death rates (1 per cent) from inoculated smallpox, possibly because severe cases were no longer used as donors.[8]

As in Boston, God soon entered the debate. Prompted by his anti-inoculation brethren in Boston, the Reverend Edmund Massey, Rector of Colne Engayne in Essex, delivered an incandescent sermon from the pulpit of St. Andrew's Church in Holborn, London on 8 July 1722.[9] Set around the unfortunate Job and entitled *The Dangerous and Sinful Practice of Inoculation*, this railed against this "diabolical operation" which "promotes the increase of vice and immorality". Strong stuff, especially as Massey intended 'diabolical' to be taken literally. It was the Devil who had smitten Job with boils; this affliction was obviously smallpox and therefore the Devil had been the first to give the disease deliberately, that is, to inoculate. Alternative diagnoses were ignored, as was the later use of the same trick by Moses. Massey further argued that smallpox was sent by God to punish mankind for sin, to test his people's faith and to frighten them away from vice: "If men were more healthy, 'tis a great chance they would be less righteous." Therefore, only the Atheist, Scoffer, Heathen and Unbeliever could dare to interfere with the Almighty's wish to hit His targets. To summarise: "Let them *Inoculate* and be *Inoculated*, whose Hope is only in and for this Life!"

Massey's work, like Wagstaffe's, was skilfully exploited and did the rounds, first printed as a pamphlet and later recycled in Boston as the leading article in a volume of anti-inoculation propaganda.[10] It helped to harden opinion against variolation and acted as a nucleus for other religious objectors. Before long, variolation was bracketed together with suicide and duelling as cowardly, atheistic and immoral.[11] And of course its anti-Christian origins among the heathen Muslims also provided useful ammunition.

English newspapers and magazines also weighed in and, as in Boston, were polarised from the outset. Early shots in the exchange were fired in mid-August 1721 by *Applebee's Journal* which criticised inoculation and mocked the Newgate experiments as an easy way to allow condemned felons to escape justice. In the tradition of Franklin's *New England Courant*, *Applebee's* claimed in February 1722 that those in Boston "have had but bad luck with the project of Inoculation". It also pointed out that the Selectmen had slated clerics who supported inoculation and that an Act to outlaw the practice had been passed by the lower House of Representatives in Massachusetts (but failed to admit that this Act had been rejected by the higher Council).[12]

By contrast, the *Post-Boy* aligned itself with variolation and leapt into battle against *Applebee's* and the like-minded *London Journal*, threatening to expose any future misreporting of this "safe and universally useful Experiment" and to name and shame "the Persons therein concerned".[13]

The combination of arrogant doctors, loud-mouthed preachers and bent reporting provoked Lady Mary to sail back into battle. She had withdrawn from the front line, possibly sensing that her publicity campaign had achieved all it could for the moment. Now, she was angry and dashed off a fiery and libellous piece under the see-through pseudonym of 'A Turkey Merchant'. This had to be edited (possibly with asbestos gloves) to remove some of the more colourful insults against the medical and clerical bigots who opposed "this useful invention". The watered-down *Plain Account of the Inoculation of the Smallpox* that appeared in the *Flying-Post* of 11–13 September 1722 was mostly a plea for common sense.[14] This simple, effective and low-risk procedure had prevented thousands of deaths (albeit in far-off Turkey) and deserved to be given the benefit of the doubt and tested properly. Strangely, for one so fearless and articulate, this was Lady Mary's only (almost) public writing about variolation.

At this point, things on both sides of the Atlantic quietened down. In Boston, the outbreak had died away and now that variolation was no longer needed, interest in it subsided. When Boylston's analysis of the Boston experiment reached England, it essentially strengthened the prejudices of both camps. Maitland, Nettleton and Amyand got on with inoculating and others joined the cause. By the end of 1722, 15 inoculators were active in England and Wales and 182 people had been treated at a cost of two deaths. The total number of inoculations reached 481 a year later and over 1,300 by the end of 1729.[15] Uptake was therefore steady but by no means the explosion that might have been expected for the first ever truly effective weapon in the struggle against the most terrible of the Ministers of Death.

Over the next half-century, variolation steadily overcame opposition and established itself as a mainstream medical treatment across the world. Part of its success was built on the careful statistical studies of Jurin and others, which confirmed time after time that (at least for the consumer) it was much safer than natural smallpox, bolstered by the assumption that it protected against smallpox for life. But more important for its general acceptance was skilful marketing by doctors who saw its potential to make money. They seized control from the evangelical do-gooders who had brought inoculation into the Western world, and the main driving force for variolation changed from passion to fashion.

Trust me, I'm a doctor

"I know nobody that has hitherto repented the Operation tho' it has been very troublesome to some Fools who had rather be sick by the Doctors' Prescription ..."

Lady Mary Wortley Montagu, Letter to Lady Mar, July 1723

Change has never been comfortable for the medical profession and doctors soon realised that variolation brought both opportunities and threats. The opportunities were obvious. This was the first real chance to fight back against the Angel of Death and, for many doctors, the chance to make lots of money.

Its threats lay in its simplicity, at least as practised in Turkey: people simply turned up, bared their arm and went away a couple of minutes later. Turkish physicians didn't even bother with it; if an old woman with a rusty needle and a walnut-shell of venom could variolate, then anybody could. English doctors therefore faced a terrible dilemma. Managing smallpox gave them a good living but only because they dressed it up with bleeds, purges and vomits. The new wonder of variolation could kill off that income stream by promising something better – and with a tiny scratch that did not need medical training.

The doctors' solution was predictable and effective. They quickly turned variolation into an exclusively medical procedure that enabled them to keep their monopoly over the treatment of smallpox. The well-tried tricks of the profession – secrecy, mystery and ritual – helped to medicalise variolation and made it such a complicated business that the uninitiated would not dare to dabble in it. Before long, variolation was fully Anglicised, with bleeds, purges and vomits.

From the start, it had been assumed that variolation would be doctors' property. In his original letter to the doctors of Boston, Cotton Mather had carefully stated that inoculation would only be done by a skilled physician "who will wisely prepare the Body before he performs the Operation".[16] Zabdiel Boylston bled one of his early cases to 'purify' the boy's blood[17] and other doctors were soon tempted to fiddle. Henry Newman's update from New England to the Royal Society in March 1722 showed that symptoms after inoculation were treated intensively: nausea with "a *gentle Vomit*'" fever with "a little bleeding" and any delay in the appearance of the rash with caustics to raise blisters on the skin.[18]

Next came pre-emptive action. Thomas Nettleton's first letter (printed immediately after Newman's in the *Philosophical Transactions*) described rhubarb purges in children and bleeding in adults to prepare them for variolation. Nettleton insisted that inoculation should "never be undertaken without the Advice of Physicians to direct a proper Method of Preparation before the Incision is made, as well as a just Regimen afterwards".[19]

Further 'refinements' were added by physicians, surgeons and apothecaries, all jostling to stamp their brand on variolation and then milk it for all it was worth. Some may have believed they were doing good by bringing the primitive technique up to English standards, while others just kept a cynical eye on the patient's wallet. Many recipes became heavily-guarded secrets, such as the 'Suttonian' method described below. Some favoured deep incisions, which also discouraged amateurs. Most used clear or milky blister fluid from mild cases rather than the thick yellow pus that developed later. 'Arm-to-arm' variolation, using fluid from blisters in a variolated subject rather than a victim of natural smallpox, was first tried in 1721 by Maitland and was felt by some to be gentler and safer.[20]

The preparation of the patient became an elaborate ritual that would have baffled the old women of Constantinople. Bleeding, often to faintness, was supposed to purify the blood and prevent fever, and aggressive purges and vomits and patent cures were also popular.[21] These all added mystique, tightened the doctor's grip on the patient and inflated the bill. With these measures, the run-in period eventually stretched to three weeks or more, and some professional variolators such as the Suttons set up specific inoculation houses. Here, according to the doctors' taste, the client could be fed or starved, heated or frozen, exercised or locked away in the dark, and with a price tag that also covered food, wine and fresh bed-linen. The prime mover was income generation but the introduction of inoculation houses at least helped to isolate the patients – except for visitors, of course.

The rituals of variolation were best developed in England, where two names stood out in the mid-eighteenth century: Dimsdale and Sutton. Thomas Dimsdale was a fashionable London physician and skilled self-publicist who made his name with a book, *Present Method of Inoculation*, that saw several successful editions and a high-profile invitation to Russia, which figures later.

The Suttons were a triumph of family enterprise and clever marketing.[22] The patriarch, Robert, was a surgeon from Debenhall in Suffolk who became interested in variolation in 1757 after it went badly wrong for one of his sons. He experimented to make inoculated smallpox as mild as possible and by 1762 was advertising "A new Method of Inoculating for the Small-Pox" in the county's newspapers.[23] Shrewdly, he kept his method secret. This mystique was a brilliant advertising ploy as well as a necessary safeguard against commercial espionage – essential because, when the secret was finally revealed, it turned out to be so simple that anyone could have followed the recipe.

Robert Sutton, helped by three sons who eventually followed him into the business, worked hard to spread the word. Its particular selling points were the guaranteed mildness of the inoculated smallpox (patients were rarely in bed for more than two days), the excellent care offered and, of course, mystique. The business became phenomenally successful, expanding in monopoly-fashion from Robert's first inoculation house in Suffolk to a chain of houses and eventually an enterprise that stitched up several counties in eastern England and stretched to Ireland, Holland and France.[24] The Sutton sons, especially Daniel, were all experts in inoculation and salesmanship; Daniel had a church built to house a tame vicar who preached anti-Massey sermons.[25] They set up a network of inoculation houses, clinics and offered franchises to other inoculators for a share of the profits and on condition that the secret would not be revealed. Even the Suttons' ex-housekeeper cashed in, advertising her business near Winchester and her personal skills in the "perfect art and mystery" of Suttonian inoculation for two to three guineas (extra for tea, sugar and linen).[26]

Their caseload also became phenomenal as their fame spread. Robert Sutton's 2,500 inoculations in his first decade were eclipsed by the 8,000 performed by Daniel in 1766 (including 700 in one improbably heroic day). By 1770, the Sutton collective had treated over 300,000 satisfied customers. The cost was five guineas for the standard three-week package and one guinea for a home inoculation, with a reduced rate for the poor and whenever the parish paid. Daniel himself raised funds to inoculate over 400 poor inhabitants of Malden during an outbreak in 1763; with an annual income of 2,000 guineas (roughly the combined salaries of 40 surgeons), he could afford it.

The Suttons made extravagant claims for the safety of their method, admitting to only five deaths in their first 40,000 cases – compared with the overall mortality rate of 1–2 per cent.[27] This sounds incredible, but the Suttons' later offer of 50 guineas for anyone who could name a fatality was never taken up and the Royal College of Physicians accepted their estimate.[28] Their version of inoculation was certainly mild: the children at Thomas Coram's Foundling Hospital in London had an average of just 20 pustules each when variolated in 1767.[29]

It was Daniel who eventually lifted the lid on the family secret in his book, *The Inoculator*, published in 1796.[30] The mysterious 'art' in fact went back to basics: a shallow scratch, careful selection of only mildly-affected donors and no bleeding – just a gentle laxative and sometimes a secret powder, later shown by its side-effects to contain mercury and antimony. After this revelation, the legend gradually faded but the Suttons left their mark, including (probably) a reduction in smallpox in heavily-inoculated Maidstone that lasted until 1800[31] and a poem by Henry Jones that put their "nobler and unmatch'd discovery" above Columbus' discovery of America.[32]

Suffer the poor

Initially, variolation did not come cheap and the cost was far beyond the reach of the poor majority – several weeks' pay for a labourer, let alone three weeks of enforced unpaid idleness. As a result, variolation became a socially divisive force and, in areas where it was widely used, began to divert smallpox from the haves to the have-nots. The wealthy could afford to have themselves and their families variolated but seldom bothered to isolate themselves afterwards. Recently variolated people triggered countless outbreaks and these tended to target and kill the unprotected poor – the archetype being the unfortunate footman of the Batt family in Hertford.

Because of this risk, many English parishes banned the entry of anyone freshly variolated and even of variolators themselves.[33] Other parishes cautiously allowed the mass inoculation of their poor, such as the 300 treated in the Cotswold village of Wotton-under-Edge (near Jenner's home in Berkeley) in 1756. The cost, a job lot at two shillings a head, was covered by the Poor Rates raised by the parish.[34] This was not necessarily a purely humanitarian decision, as various parishes had complained about being bankrupted by the expense of burying their impoverished smallpox victims.[35]

Some towns were bullish in encouraging inoculation. The parish register for Mickleover in Derby lists the names of nine children who died of smallpox in July 1788.[36] Beside their names is a grim collective epitaph that added to their families' grief:

> These children died of the smallpox from the supreme and superstitious folly in not inoculating and then these absurd parents excusing their impiety under a pretence that their time had come.

However, inoculation was not a magic solution in this setting as the poor had to continue working and so spread smallpox among their peers.

More positive developments helped to raise the protective umbrella of variolation over the poor. In 1746, following a successful campaign to raise public funds, the Smallpox Hospital opened in London to provide free variolation and care for smallpox victims among the deserving poor; others soon followed in the provinces.[37] This was part of a wider movement to flatten the gradient of social inequality, led by the prominent Quaker physician John Coakley Lettsom (whose influence spilled over into the early years of vaccination). For him and others, inoculation was in the vanguard of their idealistic crusade to give all access to life-saving treatments, irrespective of wealth.

Coming of age

"Smallpox is a terrible lottery: with inoculation, the conditions of this lottery are changed and the number of black balls diminishes; all the centuries to come will be jealous of ours for this discovery."

Charles-Marie de la Condamine *Mémoire sur l'Inoculation*, 1754

For some years after the twinned outbreaks of 1721–22, Boston and London remained at the forefront of variolation and the cross-fertilisation of ideas – both for and against – continued between the two centres.

In 1725–26, Zabdiel Boylston visited London, carrying a letter of introduction from Cotton Mather FRS to John Jurin.[38] This trip went well. Boylston wrote his *Historical Account of the Small-Pox Inoculation in New England*, which the Princess of Wales graciously allowed him to dedicate to her and was published in 1726 to great acclaim (at least by the supporters of variolation).[39] He was also elected to Fellowship of the Royal Society, having been proposed by Johann Georg Steigerthal, Surgeon to the King. Sloane evidently took to Boylston: he made the necessary royal introductions and may have paid for the publication of the *Historical Account*.

When Boylston returned to Boston, much of the excitement of innovation had evaporated. Variolation became a dormant technology, reawakened periodically by the deadly kiss of another smallpox outbreak. The town was hit in 1730, 1752 and then (back into its twelve-year cycle) 1763 and 1775. Each time, the number of people inoculated increased and its life-saving effect was reconfirmed. Boylston's data from the 1721–22 outbreak had shown 281 people (just 4 per cent of the population) inoculated with a death rate of 2 per cent, compared with 15 per cent mortality for natural smallpox. In 1752, the number inoculated had shot up to 2,113, nearly 14 per cent of the population, with just 30 deaths (1.4 per cent).[40]

This was a good outcome for those treated but, as the opponents of variolation insisted on pointing out to the Selectmen, the mildly-infected inoculees who continued to walk the streets simply helped to spread smallpox among the non-immune. Therefore, they argued, its true death toll was much higher, although how many of the overall smallpox deaths it had caused could not be calculated. The Selectmen accepted the argument and banned inoculation between outbreaks until proper isolation facilities were introduced. Initially, treated patients were confined to home and later, those who could afford it were isolated in the inoculation hospital in Brookline. Boston also tightened up its quarantine arrangements, all but closing the loophole that had allowed the *Seahorse* to penetrate its defences in 1721. As a result, smallpox began to retreat from Boston, whereas Philadelphia (which had not mounted any of these defences) continued to be ravaged.[41]

Back in England, the climate steadily warmed in favour of variolation, as confirmed by two landmark seals of public and professional approval. In 1738, inoculation made its entry into the second edition of Chambers' *Cyclopaedia*, the fount of all knowledge for the literate classes.[42] Then in 1754, it received the sanction of the Royal College of Physicians. This was a couple of years after a further demonstration of its usefulness during a smallpox outbreak in London and a posthumous tribute to Sloane, the College's former President, who had died the previous year. The College seemed to put nationalism above science but still gave variolation instant respectability:[43]

> The College, having been informed that false reports concerning the success of inoculation in England have been published in foreign countries, think proper to declare ... that the arguments which at the commencement of this practice were urged against it have been refuted by experience, that it is now held by the English in greater esteem, and practised among them more extensively than ever it was before; and that the College thinks it to be highly salutary to the human race.

The same year, 1754, saw a spectacular transatlantic collaboration between a Founding Father of America and William Heberden, a leading light of English medicine and the first to distinguish properly between smallpox and chickenpox. Benjamin Franklin, formerly the teenaged dogsbody on the *New England Courant*, had grown into a scientist, writer, politician and leader who made the term 'polymath' look pedestrian. He had been converted to variolation by the death of his uninoculated four-year-old son in 1736. Franklin wrote a sober piece about the benefits of variolation, backed up by hard evidence from Boston, and Heberden tacked on his recipe for a "plain" method of inoculation.[44] Like the Suttonian 'art', this was the sensible basics, stripped of all padding except for the occasional gentle purge. Strangely, despite the benefit of 30 years' experience, neither of these great minds mentioned the risk of collateral infection.

Seeing the world

All this made England the international centre for variolation, attracting visitors eager to learn the 'new' methods and exporting its expertise through a network of roving ambassadors. The first was Maitland, invited to inoculated the royal offspring in the Court at Hanover in 1723.[45] His rapturous reception and generous reward set the standards for others, while royal endorsement sparked off enthusiasm for inoculation across Germany. In 1775, Dr. William Bailies of Bath followed a similar trail to Berlin at the invitation of Frederick the Great.[46]

The most spectacular export was Thomas Dimsdale, who had developed a modified technique of inserting into a cut a thread that had been pulled through a smallpox blister. Dimsdale's reputation percolated as far as Russia and in 1768, he answered a summons by Empress Catherine the Great to inoculate herself and her children.[47] The Empress' confidence in Dimsdale was nearly complete; luckily, there was no need for the chain of getaway horses that she had lined up to enable him to escape from Russia if he killed anyone important. Dimsdale returned to London with an Imperial Baronetcy, a 10,000-pound thank you, and a generous annual pension. Although the Empress led by example, her people did not follow as willingly as the citizens of Germany had done, and variolation was taken up patchily in Russia.

London also acted as a magnet for those wishing to introduce the benefits of inoculation into their own countries. In 1745, Théodore Tronchin, a suave man-about-town in his native Geneva, came, learned and returned, converting Amsterdam to variolation en route.[48] Tronchin was a persuasive force who later was crucial in breaking down French resistance to inoculation. Another well-travelled evangelist was the accomplished Dutch physician and scientist, Jan Ingenhousz, who was destined for Fellowship of the Royal Society (1771), the discovery of photosynthesis (1779) and a nasty clash with Edward Jenner (1798).[49] In 1746, Ingenhousz learned variolation in London and carried the skill to Vienna and the Habsburg court, where Empress Maria-Theresa asked him to inoculate her three children. This unleashed the predictable surge of enthusiasm across Austria, although too late to prevent young Wolfgang Amadeus Mozart from catching smallpox.

France, typically gripped by medical conservatism and anti-English feeling (except for Wagstaffe), was the last major European country to embrace variolation. This only followed determined campaigning by Voltaire (François-Marie Arrouet) and Charles-Marie de la Condamine. Voltaire, best known as a philosopher and writer, had become fascinated by variolation and produced beautifully embroidered but completely fanciful accounts of its use in Georgia to preserve girls' complexions so that they could be sold into harems. His essay *Sur l'insertion de la petite vérole* (1733) was more convincing and argued strongly for the technique to be brought to France.[50] This appeal was reinforced by de la Condamine in his *Mémoire sur l'inoculation* (1754).[51] Condamine had seen variolation for himself in Constantinople and was reminded of it by Portuguese missionaries while collecting cinchona seeds in the Amazon jungle. His formidable reputation as a mathematician gave credibility to his assertion that one-tenth of deaths in France were due to smallpox – "La petite vérole nous décime" – and that variolation would save 25,000 French lives each year.

Medical and public complacency was shaken by a smallpox outbreak in Paris in 1752.[52] This was trivial beside the monster of 1723 that had wiped out 40,000 Parisians, but it struck deep into the nation's heart because it nearly killed the Dauphin, heir to the French throne. After some hesitation, Théodore Tronchin was recruited from Geneva to inoculate the Dauphin in 1754. This went well, news spread, the brake of suspicion was released and variolation took off with spectacular speed. Tronchin's private practice mushroomed and inoculation rapidly became a social must-have. The whole nation was soon swept into a carnival of gratitude for this miraculous intervention. This was marked by various creative outpourings inspired by variolation – music, poetry and even a new hairstyle (*pouf à l'inoculation*) and smart spotted headgear (*bonnet à l'inoculation*) dedicated to the technique.[53]

Similar frivolity also broke out in England in the late eighteenth century, with parties for variolation creeping into the social calendar much as in the Constantinople of Lady Mary's day. Across the Atlantic, things were much grimmer: the relationship with Britain had turned rotten with the Boston Tea Party and then collapsed into the chaos of the War of Independence. In America, smallpox and inoculation both conspired to write history. Smallpox kept George Washington's unprotected army out of British-occupied Boston and later wiped out the American forces besieging Quebec, where "the Army melted away in a little time as though the Destroying Angel had been sent on purpose to demolish them".[54] Washington (another scarred victim of smallpox) subsequently decided to order large-scale variolation of the American forces, which abolished their fear of the disease and was probably decisive in enabling them to chase the British out.[55]

Further afield, the Chinese merchant Li Jen-Shan found himself in a severe smallpox outbreak in Nagasaki, Japan in 1744. He told the city's governor about the Chinese tradition of intranasal variolation and this was promptly tried out on 20 volunteers (prostitutes rather than prisoners). This led a Japanese physician, Ogata Shunsaku, to variolate children during an outbreak in Chikuzen Province. There were no deaths and they appeared to be protected. There was some limited uptake of this innovation; variolation had nearly returned to its roots.[56]

Respectability

By the end of the eighteenth century, variolation had matured through passion and fashion and into respectability. People believed in it: when smallpox broke out in Boston in 1792, virtually everyone who needed it flocked to be inoculated.[57] Even the Almighty approved, according to senior clerics who added their voices to that of the vicar peeping out of Daniel Sutton's pocket. In the 1750s, Isaac Maddox, Bishop of Worcester, delivered a well-crafted sermon that outgunned Massey's clumsy twisting of the Bible. For extra impact, this was delivered in St. Andrew's Church, Holborn (the site of Massey's original) and reprinted for distribution throughout England and Europe.[58]

Variolation was one of the greatest successes of medicine at that time. Almost uniquely, there was hard evidence that it worked and no other treatment of a lethal condition (such as amputation for gangrene, cutting stones out of the bladder) could claim a mortality rate of just one in fifty. It became the subject of serious study by leading physician-scientists such as

Figure 6.1 William Woodville. © The Royal Society of London.

William Woodville. Woodville was already famous internationally for his phenomenally successful *Medical Botany*, a work of outstanding beauty as well as the greatest authority on the herbs that underpinned much of medical treatment.[59] The first volume of his *Inoculation* was published in 1796 but, even though variolation still had decades of life ahead of it, the second volume never appeared.[60] The reason will become clear.

Variolation also inspired some to think big, on a scale that would have been laughed down a couple of decades earlier. John Haygarth, an energetic physician from Chester, was first struck by variolation because its increasing use locally was paralleled by a rise in deaths from smallpox, notably the "great epidemic" that hit Chester in 1774.[61] Haygarth made the connection that had eluded many supposedly greater minds and pinned the blame on the failure of inoculated subjects to isolate themselves. He recognised the huge value of variolation but also that it was a double-edged sword, and suggested that it would work best if inoculated people were carefully confined under 'Rules of Prevention' that kept non-immune contacts at a safe distance. After another outbreak in 1777, he set up the Small-Pox Society to contain smallpox in Chester by combining a city-wide inoculation programme with his Rules of Prevention. Four years later, the Society reported that Haygarth's strategy had led to a "great fall" in smallpox deaths.[62] Once again, however, human nature triumphed over evidence and clear thinking. Despite its obvious success, the project had to be abandoned because so many of the grateful townsfolk refused to have their children variolated, preferring them to catch smallpox in the natural way. Haygarth noted tersely that his plans could well succeed in other countries and continued to refine his ideas.

In 1793, he published a hugely ambitious plan – a nationwide pre-emptive strike against smallpox based on mass variolation and effective isolation, with further aggressive variolation to stamp out the first stirrings of any outbreak. His goal was breathtaking, but Haygarth believed that his strategy would work, even in Chester. The title was self-explanatory: *A Sketch of a Plan to Exterminate the Small-Pox from Great Britain.*[63]

This should have been a turning point in how people thought about smallpox, but there was no interest in taking it forward. The notion was simply too bold and too incredible for the age.

Also, everyone's attention, including Haygarth's, was about to be dragged on to something even bigger and better.

Figure 6.2 John Haygarth. © Royal College of Physicians of London.

7 The Greatest Improvement?

All good things come to an end. If time had somehow frozen in the final years of the eighteenth century, mankind would have been justifiably pleased with variolation. However, medicine moved on and readers with an eye on the clock of history will know that vaccination is about to appear on the scene. This is a good moment to see how hindsight might judge variolation.

How good was it?

"Variolation, the greatest improvement ever introduced in the treatment of smallpox, although beneficial to the individual inoculated, has been detrimental to mankind in general. It has kept up a constant source of noxious infection which more than overbalances the advantages of individual security."

John Baron *The Life of Edward Jenner*, volume 1, p. 261, 1827

John Baron's opinion was clear: variolation was a fatally selfish treatment that did much more collective harm than individual good. However, he did describe it as "the greatest improvement ever introduced in the treatment of smallpox". Coming from Jenner's biographer and awe-struck disciple, this is serious praise.

Any treatment must be challenged for proof that it works and is safe. Variolation coasted along on the assumptions that it protected against smallpox for life and was unlikely to kill. Eighteenth-century practice cannot be expected to meet today's standards of evidence-based medicine, but even so, neither of these key assumptions was properly tested.

Early writers (Timoni, Pylarini, Perrot Williams) made anecdotal claims of lifelong protection which were simply swallowed and regurgitated by others, including otherwise critical thinkers like Sloane, Jurin and Heberden. Confirmation would have needed lifelong follow-up of a large group of inoculated subjects, exposed periodically to smallpox – which clearly was not feasible at the time. Protection may well have been lifelong for some people but there were others who caught smallpox after being variolated. Several cases were reported in 1746 by Pierce Dod, a colleague of Wagstaffe's at St. Bartholomew's Hospital in London.[1] Even if this was just ammunition for Wagstaffe's anti-inoculation campaign, the observation should have been followed up. Instead, Dod was lampooned by supporters of variolation in a spoof letter that sarcastically exposed the "low absurdity and malice ... falsely ascribed to that learned physician, the real and genuine Pierce Dod".[2]

Even natural smallpox sometimes failed to protect against a second attack; Jenner managed to collect 1,000 such cases.[3] Some may have been due to genuine lapses of immune 'memory' while others may simply have been misdiagnosed: even experts could confuse smallpox and chickenpox. Moreover, variolation was an art that required some skill and attention to detail, and could fail at the hands of the careless. The Suttons made a point of checking for local redness and discharge to make sure that the inoculation had taken and repeated it after three days if there was any doubt,[4] but few other inoculators took such precautions. Some 'variolated' subjects who later caught smallpox had never been adequately treated in the first place.

Safety was the fatal (literally) flaw in variolation but was long ignored as a side-issue. For the inoculated, this was an outstandingly safe treatment by contemporary standards, with barely one chance in fifty overall of dying. Careful operators undoubtedly did better and the Suttons may even have justified their claim of just five deaths in 40,000 cases variolated.[5] This represented a mortality of just over 0.01 per cent, about one-hundredth of the general level (and some 120 times the death rate from vaccination against smallpox during the 1960s).

It was the risk to others that ultimately damned variolation. Lady Mary understood this when she decided against inoculating her baby daughter so as to protect the girl's nurse. Countless lives would have been saved if others had picked up this cue. Instead, many who should have known better turned a blind eye to all the collateral smallpox cases spread by variolated people. They included not just jobbing doctors like Boylston and Nettleton but some of the brightest brains of the age: Sloane, Jurin, Franklin and Heberden. They all seemed mesmerised by the simple brilliance of the procedure and the therapeutic triumph of inducing mild smallpox. Newman set the trend early with his typical patient in Boston in 1722, who "sits up every day and entertains his Friends, yea, ventures upon a *Glass of wine* with them" and then "gets abroad quickly".[6]

Attention was repeatedly drawn to the lethal paradox of the smallpox which was strikingly mild in the variolated subject but could kill those who came into contact with it. These voices included Douglass and Dalhonde in Boston, Wagstaffe in London and Haygarth in Chester. Their motives were not necessarily pure, but neither were the excuses invented to disregard them. It was impossible to calculate the number of collateral deaths due to inoculation, but that was no reason to ignore the problem. As was often confirmed in real life (and death) it only took one infectious inoculee to ignite an outbreak.

Some of the great men of science and medicine later recanted. In 1801, Heberden admitted that variolation "occasions many to fall a sacrifice to what has obtained the distinction of the natural disease".[7] For the "many", this was too late. Luckily, some thoughtful people outside the profession had been taking precautions of their own for some considerable time. Black Bushe House was built in the early 1750s on a remote corner of Sir James Peachey's estate near Chichester to shelter members of his family and workforce who had been inoculated. They were kept there in comfort but strict isolation until the last scabs of their mild yet potentially deadly smallpox had fallen away.[8]

How did it work?

At the time, variolation was simply accepted as fact. There were vague notions that the introduced smallpox 'atoms' attracted blood particles, thickening the blood (hence the need to 'purify' it first by bleeding) and then encouraging the hidden contagion of smallpox to bubble out through the skin.[9]

For a more accurate explanation, we need to revisit briefly the battleground of a natural smallpox infection. What we saw previously was essentially the virus's side of the conflict. In fact, the body began to fight back almost immediately by recruiting an army of immune cells that were programmed to kill the virus and any cells unlucky enough to harbour them. In natural smallpox, the race between the viral invasion and the frantic struggle to create the immune army was a close-run thing that could tip either way. With variolation, the method of introducing the virus immediately put it at a disadvantage that gave the immune defences a head start. Ideally, the virus never got a proper grip and only caused a mild illness. Afterwards, the body's immune system added smallpox to its memory bank. If the virus ever reappeared, immune cells programmed to kill it were instantly recruited and then multiplied into a single-mission task force that wiped it out before it had any chance to spread.

Some more detail for those who are technically inclined.[10] The immune response which ultimately determined whether a first encounter with smallpox would kill or let live showed few signs to the naked eye. The milkiness and yellowing of the blister fluid was due to millions of immune cells – mostly lymphocytes – flooding in, summoned by powerful chemical signals (cytokines) released by their brethren which were already locked in combat with the virus.

All these immune cells were created for just one task: to seek out the variola virus wherever it hid and to destroy it, if necessary dying in the attempt. The number of cells in the immune army was vast – the millions in each blister were multiplied up in the millions of invisible skirmishes dotted throughout the whole body as well as in the skin. Incredibly, none of these cells had yet been created a few days earlier when, as far as the immune system was concerned, smallpox did not exist. Yet during that period, immune cells were recruited, trained to identify the virus and then stimulated by more cytokines to divide and produce countless copies of themselves, all bred specifically to recognise and attack the virus.

Every resistance movement needs a failsafe way of identifying the enemy. For the variola virus, the telltale markers that attracted the fire of the immune system were specific antigens, proteins in its outer coat which are peculiar to the virus and not found anywhere in the human body.[11] These foreign proteins were snipped up into bite-sized fragments (peptides) by specific immune cells, descriptively called antigen-presenting cells. These cells then handed the peptides on to lymphocytes, effectively branding them for life with the memory of an antigen which marked out the enemy. If a programmed lymphocyte subsequently met this antigen, it immediately swung into action, attacking the virus itself and more importantly releasing a spurt of various cytokines to bring in reinforcements. Some cytokines attracted other lymphocytes and antigen-presenting cells passing in the bloodstream or tissue fluid, while others induced the immune cells to replicate. The result was a rapidly-escalating cascade of immune activity, producing a new army of immune cells that specifically targeted the variola virus.

Killing the viruses posed problems because they were mostly hidden inside cells which, being part of the body, are naturally ignored by the immune system. Fortunately, cells infected with the virus managed to post fragments of the viral antigens on their surface, like hanging a distress flag in the window of a house under siege. These were recognised by lymphocytes, especially so-called 'killer cells'.[12] These live up to their name: they creep alongside the infected cell, drill a hole in its membrane and inject powerful signal molecules that freeze all the cell's normal activities and force it to commit suicide. This is a dignified death (called apoptosis, the Greek for 'leaf-fall') which barely disturbs the neighbours.[13] The nucleus shrivels up and the cell collapses in on itself – when its remains are promptly gobbled up by scavenging cells (phagocytes) also called in by cytokines. The phagocytes' digestive juices are also powerful enough to kill the viruses in various stages of replication.

And so the battle continued, repeated in every site of the viral invasion. It was the strength, speed and efficiency of the immune response which determined above all whether the patient survived an attack of smallpox.

So how did variolation work? The variola virus got off to a flying start when it entered the body through the usual route of the alveoli and lymphatics, and was already being scattered through the body by the time the immune response got off the ground. By contrast, the virus began with a handicap when it was implanted into the relatively unfriendly environment of the skin. Patrolling immune cells, automatically drawn to sites of tissue damage, rapidly picked up the foreign protein and used it to trigger an immune response. Thus the immune system was quickly alerted, with immune cells dispersing from the inoculation site throughout the body and prepared to do battle by the time the virus settled in the tissues.

The early activation of the immune response explained the mild clinical course and the low mortality rate of inoculated smallpox. Crucially, the virus itself was unchanged. The only reason it struggled was the relative hostility of the environment in the variolated host. If the smallpox virus then encountered a non-immune contact, it was strictly business as usual.

Fame, fortune and irritation

"If a man should throw a Bomb into a Town ... ought he not to die? So if a man should wilfully bring Infection from a person sick of a deadly and contagious Disease, into a place of Health; is not the mischief as great?"

Thomas Archibald *A letter from One in the Country to his Friend in the City*, Boston, 1722

Variolation was more than a controversial medical treatment and a useful phase in mankind's battle against smallpox. Wherever it was taken, it stirred up powerful feelings – from suspicion and hatred to hope and adulation – that infected the supposedly rational minds of doctors and scientists as well as the wider public.

Many fields of medicine had their dedicated followers of fashion, but the cult of celebrity that sprang up in many countries around the roving ambassadors of variolation was exceptional. Part of the euphoria was undoubtedly fuelled by relief and gratitude, with the realisation that the grip of the Angel of the Death really could be broken.

Perversely, this same realisation also explained why variolation touched such a raw nerve with some clerics. Here was something that could dissipate the threat of a disease so terrible that it could only have been sent by God to punish mankind. As long as smallpox was free to walk the world, it was a powerful enforcer of compliance with the Bible's directives and therefore valuable in maintaining the priest's stranglehold over his flock. Variolation was a direct invasion by medicine of the church's territory – or, as Baron put it, "an attempt to counteract the visitations of an all-wise Providence".[14] Later, variolation won over clerics such as Bishop Maddox, and some of these converts even took on the task of inoculating their congregations themselves.

Money and secrecy were powerful drivers for variolation, at least in Europe. These were to figure much less in the story of vaccination, mainly because Jenner deliberately brought his discovery into the open and made his knowledge freely available to all. Doctors still charged for their services but had less opportunity to medicalise vaccination and milk it to their own advantage.

For the first time, variolation gave people the chance to rebalance the odds in the lottery controlled by the Angel of Death. It also forced parents into making decisions that, either way, could kill or maim their children. Many made choices that they regretted. Mozart's father and Benjamin Franklin both hated themselves for deciding against inoculating their children; Mozart and his sister Nannerl survived smallpox but Franklin's son did not.[15] Others were also crippled by knowing that their own decision not to inoculate had killed a much-loved child. Queen Maria of Portugal lost her eldest son José in this way in September 1788. She was already a tortured soul, saddled with status that she did not want and revolted by the everyday cruelties of her country, but her guilt and anguish over José's death helped to tip her into the madness which gripped her for the last 20 years of her life.[16]

Variolation broke boundaries by stimulating people to think bigger than before, notably in Haygarth's grandiose plan to banish smallpox from Britain. The need to isolate inoculated subjects, always the millstone around the neck of variolation, would have made his plans unworkable. Nevertheless, the breadth of his canvas and the boldness of his brushstrokes make Haygarth's vision the legitimate forerunner of the global eradication campaign which eventually exterminated smallpox two centuries later.

Exeunt

Variolation swept up a colourful cast of characters and comfortably outlived all of them. In Boston, Zabdiel Boylston remained in practice in Brookline until the age of 70; he was still breaking in colts at the age of 84 and died in 1790, aged 87.[17]

William Douglass was quietly converted to inoculation when it became popular and lucrative, but never retracted his criticisms of Boylston or Cotton Mather.[18] Mather, another hardbitten non-retractor, lived to the age of 62 and practised what he preached about taking life's hard knocks as God's will: he buried two of his three wives and 13 of his 15 children.[19] Onesimus, who lived up to his name (which means 'useful'), bought his way out of slavery and enjoyed freedom of a sort, sweeping the streets of Boston.[20] Lawrence Dalhonde, whose garbled testimony to the Selectmen caused such a stir, lived on among Boston's children in the nursery rhyme which commemorates him as a "very small French doctor who spoke very bad English" (and whose horse was also of short stature):[21]

> Doctor Delone desires one prayer,
> Before he falling off his little black mare,
> He bruises de meat – no broken de Bone
> Please pray Sir, for poor little Doctor Delone.

Back in England, Sir Hans Sloane succeeded Newton as President of the Royal Society and died in 1753 – three years before the publication of his account of the Newgate experiment (which he somehow failed to write-up until 1736).[22] Daniel Sutton, a near contemporary of Edward Jenner, lived through the first quarter-century of vaccination without seeing much impact on his practice.

Lady Mary Wortley Montagu continued on course as a Comet of the Enlightenment. Smallpox still stalked her, killing a nephew in 1723 (her sister had refused to have the boy inoculated at the same time as little Mary) and then, in 1726, her old friend and correspondent from Adrianople, Sarah Chiswell.[23] The Comet's orbit wobbled badly in her later years.[24] Highlights included a catastrophic falling-out with Pope (whose portrait she had painted on the inside of her commode), a long affair with the bisexual Lord Hervey (who preferred another chap), a 20-year sojourn in France and Italy in pursuit of Love (which eluded her) and recurrent heartache and tooth-grinding over her prodigal son (to whom she left one guinea in her will). She died of breast cancer in 1761, aged 73. Her daughter Mary grew from the "chearful" spotty little girl into a woman who never answered her mother's letters but saved and published them (suitably edited) after her death; she also married Lord Bute, who later became Prime Minister.

To give the last word to Lady Mary, in a letter to her sister:[25]

> To say truth, people are grown so extravagantly ugly that we old Beauties are forced to come out on show days to keep the Court in Countenance.

Not quite the end of the story

Logically, the birth of vaccination should have sounded the death knell for variolation. Vaccination outperformed variolation and did not have the lethal drawback of spreading the disease it was designed to stamp out. But vaccination had problems of its own and almost immediately raised a storm of opposition. Many of the objections to Jenner's invention simply plugged into pre-existing arguments against variolation, with the added ingredient of bestiality because of the origin of cowpox. Human nature ensured that variolation clung on for much longer than it deserved.

Many tried to kill it off. In 1813, Lord Boringdon presented *An Act for Preventing the Spread of the Small Pox* to the English Parliament, which Jenner helped to draft.[26] It failed but led into a concerted campaign that dragged on after Jenner's death, some 30 years after he brought vaccination to the world's notice. Baron's attack on variolation, quoted at the start of this chapter, was intended to be its epitaph but the technique was kept alive in England by a "set of unprincipled, unfeeling and ignorant persons, reckless of the miseries which they spread" who "extort from the prejudiced parents a pittance sufficient to excite their cupidity".[27] It was not until 1840 that variolation was banned and made an imprisonable offence under the revised Vaccination Act.[28]

Elsewhere, variolation mostly died out when the benefits of vaccination were allowed to declare themselves during the first two decades of the nineteenth century. But old traditions die hard. 'Buying the smallpox' was still practised in Sudan in the late nineteenth century.[29] Incredibly, variolation survived even longer elsewhere. The bottle of smallpox scabs for variolation (see Plate 11) does not look antique and is not; it was collected in India in 1976, during the closing years of the World Health Organisation's global programme to eradicate smallpox. Here, and in Afghanistan, China and Ethiopia, traditional inoculators were still operating throughout the 1970s. Like Maitland, they did not bother to isolate their patients and repeatedly sparked off smallpox outbreaks which threatened to undermine the smallpox eradication campaign (see Chapter 15).

A terrible state of the disease

This chapter ends with a reminder of what variolation was like for most people – an intensive, drawn-out and deeply unpleasant ritual. The year was 1757 and the place Wotton-under-Edge, perched on the Cotswold escarpment and overlooking the Vale of Berkeley and the broad sweep of the River Severn. Coincidentally, this is barely 60 miles as the crow flies from Pembrokeshire, where Dr. Perrot Williams was in retirement in Haverfordwest and employment prospects remained buoyant for the leeches of Marloes.

The patient was an eight-year-old boy, orphaned at the age of five and brought up by his elder brother, who had to be variolated at his boarding school in Wotton.[30] The headmaster had an arrangement with a local surgeon and, together with some classmates, the lad was slotted into a well-tried routine. He was locked away with the others and put on a semi-starvation diet, with frequent doses of purgatives to induce 'purifying' diarrhoea. To further thin and cleanse his blood, he was bled every few days. This preparation lasted four weeks. At the start of it, he was fit and active; at the end of it, he was emaciated and so weak that he could barely stand.

The variolation itself was trivial by comparison and the prodromal symptoms of fever, headache and vomiting may well have added little to his misery. When the rash broke out, it was mild by a doctor's standards. The eight-year-old lad probably took a different view of the fiery pain that came with the eruption and the stinking pus that oozed out when blisters burst under the weight of the piled-up blankets. The rash resolved on schedule and, seven weeks after the start of the process, he and the others stumbled back into the daylight.

Everyone reacts differently to illness. Many simply shrugged off the unpleasantness of variolation but for this boy, it was a recurring nightmare that kept him awake at night throughout his childhood and that induced flashbacks and panic attacks which haunted him until well into his adult life.

There was a positive spin-off, however. The boy's horror of variolation primed him to find a better way to protect people against smallpox, and later kept pushing him to see his ideas through. His name was Edward Jenner.

8 Everyday Tales of Country Folk

Everyone knows how vaccination was discovered. Not in the intellectual powerhouses of the Royal Society or the Royal College of Physicians but in the quiet backwaters of Gloucestershire, renowned chiefly for its gentle green landscape and the quality of its cheeses. Not by one of the high-fliers of Oxford or Cambridge but by a country doctor. Not in a 'Eureka' moment of inspiration, but prompted by a farm labourer's account of a local folk tradition.

The sequence of events leading up to the first vaccination is also well-known. The doctor was told by a milkmaid that those who caught cowpox, an unpleasant blistering disease affecting cows' udders, were then protected against smallpox for life. The doctor decided to follow this up and soon found many people with an earlier attack of cowpox who had later been sidestepped by smallpox. Then the doctor wondered if he could spread the benefits of cowpox beyond the milking parlour. He collected fluid from cowpox blisters and scratched it into the skin of healthy volunteers – and the rest is history.

But not quite. The doctor in question was not Edward Jenner, even though he is generally regarded as the discoverer of vaccination, and the informant could even have been a middle-aged farmer rather than a milkmaid. Moreover, other people – including a farmer and a travelling tutor – were ahead of Jenner in deliberately inoculating healthy subjects with cowpox so as to protect them against smallpox.

This part of the story covers a broader canvas than might be expected, ranging from Jenner's patch in Gloucestershire to

Dorset and then to other cow-rich localities in France, Germany, the Andes and India. In keeping with the rest of the story of smallpox, the action includes petty jealousy, nationalism and fabrication as well as the pursuit of scientific enquiry.

Close to home

Edward Jenner's home county of Gloucestershire extends from the Forest of Dean in the north to the edge of the Cotswolds in the south and east and is bisected by the broad meanders of the River Severn. Jenner's birthplace, the town of Berkeley, lies 16 miles south of Gloucester and just off the main road to Bristol which, after another seven miles, passes close to the old market town of Thornbury (see Fig. 9.1, page 176).

The milkmaids and farmers of Gloucestershire had been talking about cowpox and its charm-like ability to protect against smallpox for decades before Jenner's work became known. John Fewster, a surgeon-apothecary in Thornbury, was sufficiently convinced by what he heard to make a trip to the capital to present a paper entitled *Cowpox and its ability to prevent smallpox* to the London Medical Society.[1] This was in 1765, almost 30 years before Jenner submitted his first paper on cowpox and vaccination to the Royal Society. Fewster never wrote up his lecture but its explicit title makes it clear that he understood the significance of the link between an attack of cowpox and the subsequent absence of smallpox.

Fewster himself never made any claim to have followed this observation through to the intentional inoculation with cowpox in order to prevent smallpox. This seems unthinkable now but variolation was then in its prime and Fewster, a dedicated disciple of the Suttons, was making a good living from their "perfect art". With two other local doctors, he had bought a large house at Buckover, on the Gloucester–Bristol road close to Thornbury, and turned it into a Sutton-style variolation centre. Fewster may well have regarded the cowpox effect as an intriguing medical snippet with no therapeutic or financial potential. This would be in line with comments he made later, in 1798, shortly after Jenner published his *Inquiry*, that described his experiments with cowpox.[2] Asked how the new procedure compared with variolation, Fewster wrote that cowpox "is a much more severe disease in general than the inoculated smallpox ... inoculation for the smallpox seems so well understood, that there is very little need of a substitute". He added, "It is curious, however, and may lead to improvements."

Fewster's dismissal of Jenner's work is especially odd, because a local man, John Player, later claimed that Fewster himself had done everything that Jenner did, and decades earlier.[3] Player was a gentleman farmer from Tockington, a village between Thornbury and Bristol, and revealed all in a letter addressed to John Coakley Lettsom in London. As early as 1763, Fewster had apparently been struck by the failure of one of his patients at Buckover to develop any sign of smallpox after being variolated – even after the third attempt. The patient, a man called Creed from north of Bristol, explained that he had suffered a serious attack of cowpox. Fewster wondered if there might be a connection and discussed this possibility that same evening over dinner with some local medical friends. Those tucking into the haunch of venison included a surgeon from nearby Chipping Sodbury, Daniel Ludlow, and his young apprentice, one Edward Jenner.

Whatever interest was generated evidently fizzled out. Player wrote that the observation "lay dormant for several years" – actually for 33 years, until mid-March 1796 when Player himself pointed out to Fewster that inoculation with cowpox could replace variolation in protecting against smallpox. According to Player, Fewster replied that he had thought of this and discussed it with Jenner (who at the time was still two months away from his first vaccination and almost two years before any public recognition), but had decided not to pursue the idea. However, Player reported, Fewster then set off to experiment with cowpox inoculation, followed up by variolation to check if the subjects were safe from smallpox – exactly the same strategy that Jenner would employ over the coming months. Fewster's first subjects were apparently three children from Thornbury, who all showed little reaction to variolation. These promising results so impressed "a gentlewoman at Tockington of extensive philanthropy" that she paid Fewster to inoculate "a great number" of poor children in the neighbourhood. Several of them were test-variolated by Fewster's son and again showed no sign of developing smallpox.

Fewster's experiments would have preceded or coincided with Jenner's own in the spring and summer of 1796. The significance of Fewster's results would have been immediately obvious and Fewster, at the heart of the local doctors' community (which also included Jenner), was perfectly placed to communicate his findings and their excitement. Strangely, he remained silent. Player's letter, written some six years later in 1802, contains the only known reference to Fewster's work with cowpox inoculation, and as far as we can tell Fewster returned to the safe and lucrative business of variolation as though it had never been interrupted.

Fewster was just one of many Gloucestershire doctors to hear the rumours about cowpox. Joseph Adams, a doctor in Gloucester who later became a physician at the Smallpox Hospital in London, wrote in his book about "morbid poisons", ulcers and cancer that cowpox was well-known to the farmers around Gloucester.[4] The disease presented as a spreading ulcer on cows' teats and could infect man, causing ulceration in the hand, a badly swollen arm and a fever, all of which soon settled. He also highlighted the claim that people who had cowpox never got smallpox – but with the rider that this was "as far as facts have hitherto been ascertained". Even if he believed it, the observation did not fascinate him enough to make him study it any further. This was in 1795; Jenner had already been collecting cowpox cases resistant to smallpox but had not yet published anything.

After Jenner's first paper – the *Inquiry* – was published in 1798, the neglected wisdom of Gloucestershire milkmaids was suddenly remembered and broadcast by many who previously had remained silent, possibly because they had regarded it as unverified or inconsequential. A Dr. Rolphe pointed out dismissively that this was common knowledge to all doctors in the county,[5] while even Théodore Tronchin, by now back in Geneva, claimed to have heard the intelligence from the milking parlours of Gloucestershire.[6] Presumably they all dismissed it as a peasant fantasy that was not worth looking into. As Tronchin reputedly said, "How can they be so superstitious?"

The first (known)

Just over 50 miles due south of Berkeley lies Dorset, another of England's great dairy counties and the home of an unassuming man who had a much stronger claim to have pre-empted Jenner. Benjamin Jesty (see Plate 2) was a gentleman farmer from the village of Yetminster in the heart of the county.[7] Jesty knew of the supposed protective effect of cowpox and had a personal interest as he had suffered an attack himself. He had not had the chance to see whether he was immune to smallpox but two of his maids – Ann Nutley and Mary Reade – had tested the hypothesis. Both had had cowpox and both had subsequently nursed relatives with smallpox without becoming infected. Jesty was prompted to experiment with the protective effect of cowpox in 1774, when smallpox broke out nearby.[8] Jesty was then in his late thirties, with two sons aged two and three and a baby daughter; his wife Elizabeth was expecting their fourth child. All were at risk of catching smallpox, but why he decided against having them variolated is not known.

None of his cows had cowpox at the time, but one of his neighbours, Mr. Elford, had a case in his herd at Chetnole. Leaving the baby girl behind because he thought she was too young, Jesty marched his wife and sons a couple of miles across the fields to the infected cow. Using a darning needle, Jesty burst a blister on the cow's udder and scratched the fluid into the skin of his wife's and sons' arms. The procedure went well for the little boys but not for Elizabeth Jesty, whose arm became badly infected, presumably from the intestinal bacteria that contaminate the unwashed underparts of a cow. She developed a high fever and the doctor who had to be summoned feared that she might lose the arm. Fortunately, she recovered, but her husband's reputation did not. Word spread rapidly, not of Jesty's innovative and courageous experiment, but of his cruel abuse of his wife and children, poisoning them with bestial filth from a sore on a cow's udder. Jesty was pelted with abuse and sometimes stones whenever he ventured out to market. Not surprisingly, he dropped any idea of doing the same to his other children – even though he could later prove that it had protected his sons and wife against smallpox.

There follows a 27-year gap in the story, during which Jesty built up his farming business and moved in 1787 to a larger estate at Worth Matravers, near Swanage in the Isle of Purbeck. Meanwhile, Edward Jenner, 50 miles to the north, performed his first vaccinations and published his *Inquiry* in 1798.

In 1801, a new vicar was appointed at Swanage.[9] He was the Reverend Andrew Bell, founder of the Free School movement for orphans and an enthusiast for the newly discovered miracle of vaccination, which he was determined to introduce throughout the Isle of Purbeck. During his travels, he heard about Jesty's experiment, was intrigued and met Jesty to find out more. What he was told convinced him that Jesty, not Jenner, had discovered vaccination. In August 1803, Bell wrote to the newly-established Royal Jennerian Society in London to point out Jesty's claim to fame – and triggered a malicious wrangle (one of many in the early days of vaccination) that entangled the good-natured Jesty and set him up against Jenner.[10] The outcome is described later in this chapter.

What Jesty did fell essentially one step short of Jenner's achievements. He checked whether his vaccination really did protect against smallpox. Both his sons were variolated in 1798 and failed to develop the infection, and Jesty watched them and their mother sail unharmed through various smallpox outbreaks.[11] Jesty would not have known that Jenner used variolation to test his first guinea-pig, James Phipps, in 1796. It is not clear why Jesty variolated his sons; he may have worked out that this was a neat way to test the boys' immunity, or may just have decided to give them extra protection.

Unlike Jenner, Jesty did not attempt to publish or broadcast his results. Having no medical or scientific background, this would not have occurred to him and his neighbours' hostile reaction would have discouraged any wider publicity. It was only with Bell's interest that his experiments were given any prominence at all.

Parallel evolution

Numerous other reports surfaced about the protective effect of cowpox and, in some cases of the deliberate inoculation of cowpox fluid. Not surprisingly, most came from areas with a high density of cows such as the dairy counties of England and Holstein in Germany. Some antedated Jenner's *Inquiry* and therefore have some claim to the coveted title of the first ever vaccination. A few were published at the time and in enough detail to give them credibility, but many others were dug up long after the event – either in the spirit of wishful thinking or as a deliberate attempt to undermine Jenner's claim to the discovery.

The first stage in the process leading to vaccination was the recognition that having had an attack of cowpox protected the individual from smallpox. Cowpox was unpleasant – a painful ulcer, fever, swollen arm – and therefore memorable, but even the initial step of linking this to subsequent freedom from smallpox was a major feat of clinical deduction. In 1668, a Dr. Goëtz is said to have reported the "astonishing" immunity against smallpox enjoyed by the herdsmen of Holstein. The original article in the *Allgemeine Unterhalterungen* is elusive but has been cited repeatedly since 1896.[12] More robust is the account in the same journal in 1769 by Jobst Böse, a government official in Göttingen, who stated that "many reputable people" had confirmed that cowpox protected against smallpox.[13] Similar anecdotal reports came from Oxfordshire and Ireland in the years leading up to Jesty's experiment in 1774.[14] Further afield, William Bruce described a long-established tradition of the nomadic Eliaat tribe in Persia that an affliction of the udders of cows (or sheep) could infect man; as this protected against smallpox it was regarded as a bonus of the job. Bruce's account was published in 1819.[15]

The next step was to confirm that cowpox really did prevent smallpox, either by waiting to see that natural smallpox passed the subject by or that variolation failed to produce a characteristic eruption. Jesty did both. In 1781, seven years after Jesty's unpublicised experiment, it was reported that Nicholas Bragg of Axminster and a Dr. Archer at the Smallpox Hospital in London, working independently, had each shown that people who had had cowpox did not react to variolation.[16] However, neither seemed to appreciate the significance of the observation.

There are very few substantiated reports of the next step, the deliberate introduction of cowpox material into a healthy subject to produce immunity. Like variolation, this was bold and counter-intuitive – substituting the innate revulsion of animal filth for the loathing and fear of smallpox itself. In England, there were vague reports that a Mrs. Randall of Whitechurch in Dorset had vaccinated herself and three children and even that a Dr. Nash had vaccinated over 50 people.[17] These were in 1781, seven years after Jesty's experiment. A question mark must remain hanging over these accounts; as explained later, they were pulled together by George Pearson who was hell-bent on demolishing Jenner's claim to have invented vaccination.

On a smaller scale but more convincingly documented were the efforts of Peter Plett, a travelling Dutch tutor whose wanderings took him first to Schönweise in Holland, where the local milkmaids told him of the miraculous protective effect of cowpox, and then to Hasselburg in the heart of Holstein's cattle country, where he decided to try it out.[18] For some reason, he was allowed to vaccinate his employer's two daughters, one of whom emulated Mrs. Jesty with a badly infected arm. This was in 1791, five years before Jenner's first vaccination. Both girls recovered and provided proof of concept by being among the few survivors of a smallpox outbreak in 1794. Also from Holstein came retrospective reports by Dr. Hellwag of Eutin that farm girls in the last quarter of the eighteenth century had deliberately sought out infected cows and then handled the udders, sometimes cutting their own fingers to ensure a good take.[19]

The final step of challenging the vaccinated subject with smallpox to prove that they were immune was undertaken by very few – Jesty, Plett, and of course Edward Jenner. Jenner was the only one who broadcast his work and brought it into the public arena.

All due credit

The invention eventually claimed by Jenner was both big news and hot property. Predictably, vultures gathered to tear it apart and pick through its bones for evidence that others had done this before. The vultures' motives ranged from nationalism and jealousy to a personal dislike for Jenner.

History was ruthlessly trawled for episodes that could have been early attempts at vaccination, which were then beefed up with embroidery or invention as required. Some instances appeared credible; others were heroically contrived. From Ancient India came news of the weakness of the Hindu god Krishna for milkmaids and their unscarred faces, backed up by an astonishingly detailed account of how to prick a blister on a cow's udder and scratch the juice into the skin of the arm.[20] This was attributed to Dhanwantari, a renowned physician from before AD 400 whom the Hindus regarded with Hippocratic awe. Strangely, though, the historic source document (the *Sacteya Grantham*) did not seem to contain this passage; the reason is explained later.

In Europe, a fascinating letter dating from 1705 is still held by the city library in Rouen. This was from Madame du Maine (a friend of the anatomist Du Verney) to an acquaintance, mentioning that "vaccine" had run out but would be available again the following Saturday.[21] Sadly, the term 'vaccine' was not coined until 95 years later and the archaic French was not quite archaic enough. Verdict: a nineteenth-century hoax, presumably intended to stamp French ownership on Jenner's invention.

Benjamin Jesty found himself used as a propaganda weapon in an anti-Jenner hate campaign masterminded by the jealous and vindictive George Pearson, who dedicated himself to undermining Jenner and his claim to have discovered vaccination. Pearson heard indirectly of Jesty's experiment and used this as an argument to try to stop Jenner from being given a large honorarium by the grateful government.[22] This ploy was frustrated because Pearson could not find out key information, including Jesty's name, in time. It was only after Jenner had received his money that Pearson identified the 'first vaccinator' and then, in 1803, had sight of the letter which the Reverend Andrew Bell had written to the Royal Jennerian Society in support of Jesty's recognition. Pearson had set up the so-called Original Vaccine Pock Institute, primarily to do Jenner down, and Jesty was perfect ammunition. Pearson first invited Jesty to visit his own Institution in London in 1803. This was 29 years after Jesty's experiment; Jesty, then aged 66, pleaded an attack of gout and declined. Two years later, Pearson tried again and offered an all-expenses-paid trip to London and the painting of a ceremonial portrait to celebrate Jesty's status as the original discoverer of vaccination.[23] Further riches may have been promised. At the time, Jenner was negotiating another bigger honorarium from the House of Commons and Jesty may have been tempted by the suggestion that some of this was rightfully his. Accompanied by his son Robert, he travelled on horseback up to the capital.

Thanks to Pearson's malice, we know what Benjamin Jesty looked like. His portrait shows a portly, ruddy-faced gentleman, well preserved for his 68 years, wearing an expression that could be either long-suffering or bemused.[24] He had packed a clean shirt for the trip; otherwise, his clothes were smart for rural Dorset but not London. The artist was the celebrated and famously expensive Michael Sharp, whose wife was a pianist. Both exercised their talents during the three-day sitting, as Jesty only stopped fidgeting when Mrs. Sharp played to him.

Other highlights of the trip included cross-examination by a dozen senior (and anti-Jenner) doctors and the public variolation of his son Robert to confirm the lack of reaction already noted in 1789. At the end of the visit, Pearson presented Jesty with a pair of gold-ornamented lancets and a verbose testimonial scroll stating that he, as the originator of vaccination, deserved the country's gratitude. In case Jenner had missed these events, Pearson and his colleagues summarised the key points in an article which rapidly appeared in the *Edinburgh Medical and Surgical Journal.*[25] Jesty's portrait was exhibited ostentatiously in Somerset House and then proudly installed in the Original Pock Vaccine Institute.

Probably still bemused, Jesty and his son returned to deepest Dorset. Jenner later admitted that he had been irritated by "the portrait of a farmer from the Isle of Purbeck ... and the farmer's claim to reward as the discoverer"[26] – which would undoubtedly have delighted Pearson. Interestingly, Jesty is not mentioned by name anywhere in Baron's otherwise exhaustive biography of Jenner.

Ignorance and bliss

Jenner was therefore one of many who were prompted to look for a protective effect of cowpox; one of the several who actively sought evidence that this was true; one of the few who inoculated healthy subjects with cowpox; and the only one who followed the experiment through to publication of the results. As such, he is not the 'father' of vaccination, but as the person who delivered it safely into the world he can reasonably claim to be its midwife.

But was he really the independent genius working in intellectual isolation that some – notably his adulatory biographer John Baron – claim him to be?

Jenner undoubtedly knew about Fewster's work. They were friends and sparring partners in the local doctors' dining club, the Convivio-Medical Society which met regularly at the Ship Inn at Alveston, just outside Thornbury.[27] There they ate, drank and talked medicine. We know from Baron's biography of Jenner, which kept the spotlight carefully trained on him as the discoverer of vaccination, that the protective action of cowpox was often discussed at the Society, to the point where other members threatened (presumably in jest) to throw Jenner out if he continued to talk about it.[28]

What about John Player's detailed description of Fewster's pioneering work? His 1802 letter to Lettsom was discovered in a bound volume of Lettsom's correspondence, now archived at the Royal Society of Medicine in London.[29] Strangely, Lettsom never mentioned it. He was intensely loyal to Jenner but was also renowned for his fairness and would have taken seriously a communication from a fellow Quaker such as Player. Lettsom may have had misgivings, and the letter certainly has peculiarities. Player apparently began it in January 1802, adding further sections in July and then finally October; there is no postmark. Headed *Narative* [sic] *of the discovery of the cowpock as preservative from and to prevent infection from the smallpox*, Player describes it as coming from the "Cranium Portfolio" of "Ruricola Glocestris" (a countryman of Gloucester). It is a rambling document, opening with a page on the role of cows in sustaining the crew of Noah's Ark and highlighting the benefits of their four nipples over the two of other species. Player misquoted a crucial date: Fewster could not have made the connection between cowpox and protection against smallpox at Buckover in 1763, because he did not set up practice in Thornbury until six years later. Player also claimed that he himself had first thought up the key ideas in Jenner's *Inquiry*, published three years earlier: not only that cowpox inoculation could replace variolation, but also that cowpox originated from a disease of horses' feet known locally as "Grese" [grease], transmitted to cows' nipples by stable boys who forgot to wash their hands before milking. Player had even foreseen the world-changing vision which Jenner had laid down a year earlier, namely that cowpox inoculation would "eventually extirpate the smallpox".

Player's closing paragraphs could only reinforce any reservations. Allegedly anxious not to deprecate Jenner's discovery, Player sketched out his imaginative plans for a commemorative statue: Jenner standing on the back of a long-horned cow, lancet poised in his right hand, with a bas relief on the pedestal showing Fewster and "the right hand of a man issuing from behind a curtain", both helping Jenner to rise from obscurity beneath the walls of Berkeley Castle. The tone of the letter suggests that the disembodied arm should rightly belong to John Player, Esq., of Tockington.

Significant omissions provide the strongest evidence against Player's assertions. Lettsom never alluded to them and, above all, neither did Fewster. He and Jenner remained friends until late in their lives. After Jenner's death, the son of the then-ailing Fewster told Baron that his father made no claim on the discovery of vaccination.[30] This sounds like a graceful acknowledgement of the winner, by someone who might also have run but did not finish the race.

What about Jesty, only 50 miles away to the south? Various oblique links between Jenner and Jesty have been suggested, but without any documentary evidence.[31] Moreover, neither man referred to any meeting or correspondence between them. Even if Jenner had wanted to cover up the fact that he knew "the farmer from the Isle of Purbeck", Jesty would certainly have mentioned it during his five days in London at Pearson's expense and Pearson would have leaped at any opportunity to publicise this and so weaken Jenner's claim of primacy.

Fewster and Jesty were born within a year of each other, a dozen years before Jenner. John Fewster died aged 86 in 1824, the year after Jenner, and lies beside his wife under the aisle of St. Mary's Church in Thornbury, seven miles away from Edward and Catherine Jenner in Berkeley. Elizabeth Jesty also died in 1824 and joined her husband, who had suffered a fatal stroke in 1816 at the age of 79, in the churchyard at Worth Matravers. When she composed his epitaph, she had evidently forgiven him for nearly killing her by infecting her arm. We cannot fault what she wrote:[32]

> An upright honest Man: particularly noted for having been the first Person [known] that introduced the Cowpox by Inoculation, and who from his great strength of mind made the Experiment from the [Cow] on his Wife and two Sons in the Year 1774.

9 The Disinterested Divulger of a Salutary Blessing

Like the Angel of Death, Edward Jenner was a creature of paradoxes. He was also a fine illustration of the principle that nobody is perfect. A conscientious doctor but a careless scientist. An intelligent observer but easily distracted, hopeless at planning and an accomplished procrastinator. Warm with friends but easily wounded and petulant towards his critics.

Few of his contemporaries saw him in the round and he had a knack for polarising others' views of him. Across the world, his devotees worshipped him as the visionary who brought salvation to mankind; to his critics, he was a second-rater, a glory-grabber or the Devil incarnate.

According to Milton, childhood shows the man, and so it was with Jenner.

Country boy makes good

Edward Jenner (see Plate 3) was the product of solid Gloucestershire stock that had already shown flashes of ambition and brilliance.[1] His forefathers lived around Slimbridge and Standish, a few miles to the north of Berkeley, and the occupation of baker and the name Stephen recurred throughout the Jenner family tree. A cousin of Edward's father first broke with tradition by going to Oxford to study Divinity and later became President of Magdalen College. Edward's father Stephen followed his footsteps in around 1720 but settled at a lower and more functional level. In 1729, Stephen returned to his roots as

Vicar of Berkeley with peripatetic duties as Rector at the nearby villages of Rockhampton and Stone.

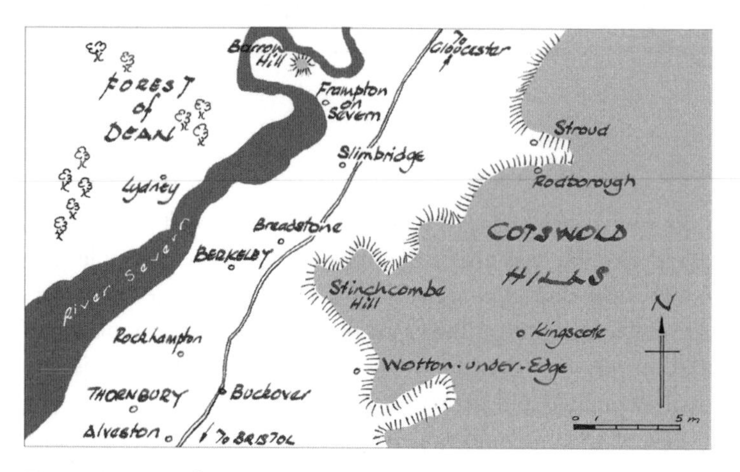

Figure 9.1 Map of Berkeley and the surrounding parts of Gloucestershire.

Thanks to the living raised by his parish, life was comfortable for the Reverend Stephen Jenner, his wife Sarah and the family which rapidly expanded to fill the vicarage at Berkeley. Sarah produced six children who lived beyond the age of five and three who did not. The survivors began with Mary and Stephen, born in 1730 and 1732, and ended with Edward who was born on 17 May 1749.

Edward was bright and inquisitive and developed an early obsession for collecting things: stones, fossils and the nests of birds and dormice. Childhood was easy and happy for him until October 1754, when the complications of a final pregnancy carried off his mother. By early December his father, who had been ill for some months, also died. This left responsibility for the youngest child with Stephen, then 22 years old and following the family tradition of Divinity at Oxford. Stephen eventually succeeded his father as Vicar of Berkeley and Rector of Rockhampton, but not until 1760.[2]

At the age of eight, Edward was packed off to the Reverend Clissold's boarding school at Wotton-under-Edge, overlooking the Vale of Berkeley from the Cotswold escarpment a few miles to the east.[3] He stayed there for less than a year; perhaps the delayed trauma of losing both parents and the six-week torture of variolation were to blame. In May 1758, he moved to another school in Cirencester, 25 miles further to the east, where he thrived.[4] Fossils and nests continued to pile up in his room and he met the like-minded Caleb Parry from Bath, who also went into medicine and remained a close friend for life.

At some point, someone decided that the boy should become a doctor. At the age of 13, Edward began working as a paying apprentice to Daniel Ludlow, a surgeon friend of Stephen who lived in Chipping Sodbury in the south of the county. Edward stayed with Ludlow for six years, an unusually long apprenticeship, and also served time with Ludlow's colleague, George Hardwicke.[5] His basic training as a surgeon covered diagnosis and simple operations and would have led Edward into the territory of both the apothecary, the forerunner of the pharmacist, and the physician who specialised in diagnosis and treatment using herbs, chemical compounds and procedures such as bleeding.

While apprenticed, Jenner would also have learned the dreaded art of variolation. According to tradition, a conversation about this led in around 1768 to one of the most famous (even if not unique) brief encounters in the history of medicine. Unlike Benjamin Jesty's informants six years later, we do not know the young woman's name but she had the same job and unmarked complexion. She explained, of course, that she could not catch smallpox because she had already had cowpox.[6] As pointed out in Chapter 8, the young Jenner could also have picked up this intelligence over a venison dinner with John Fewster and other medical colleagues.

Medicine still held many mysteries for Jenner, who was then only 19. But as with Cotton Mather, the milkmaid's remark stuck in his mind. Mather needed 15 years to translate Onesimus' revelation into decisive action; Jenner was to take nearly twice as long.

Ludlow and Hardwicke taught Jenner what was needed for a comfortable but undemanding practice as a rural surgeon but they evidently saw greater potential in their apprentice. They suggested that Edward should have further training at St. George's Hospital in London, where Ludlow's son had recently enjoyed inspirational teaching. Stephen Jenner, by now milking the income from three or four parishes, agreed to cover the costs. This would make Edward highly competitive, possibly even enough to make a living in the doctor-saturated capital. He would remain a surgeon and therefore on the tier below physicians, even though their hands-on (and hands-in) training was more limited and they usually had even less to offer.

The clever but raw lad from the depths of cattle country moved to London in late 1770. St. George's Hospital instantly put Chipping Sodbury in the shade, with its cutting-edge surgery, obstetrics and *materia medica* (drug treatment). Jenner learned anatomy at the famous school in Great Windmill Street, set up by the brilliant but iconoclastic Hunter brothers, William and John.[7] Here, comprehensive dissections covered each system of the body in 'fresh' subjects, i.e. recently dead and often exhumed. This was new anatomy that swept aside the classical teaching of Vesalius with its posed pictures of half-dissected corpses helpfully pointing out features of interest on themselves. Instead, there were meticulous dissections and an "inestimable treasure of preparations", graphic to the point of shocking, which showed the innermost workings of the body in health and disease. The anatomy museum contained hundreds of specimens bottled in alcohol: arteries injected with wax or mercury to outline the vessels ramifying through the organs, huge stones in the kidney and bladder, severed heads ravaged by syphilis and entire 'monstrous' babies. Altogether, an Aladdin's cave of wonders, curiosities and hard facts to fascinate and equip the surgeon for a life in practice.

There were also surgical procedures to learn, beginning with lancing boils and laying open abscesses and the emergency operations of tracheotomy and severing the temporal artery. Tracheotomy, cutting through the front of the neck and into the windpipe, was done to bypass a blockage higher up the airway such as the constricting membranes of diphtheria that could suffocate a child in minutes. The temporal artery (just in front of the ear) was cut and allowed to bleed in cases of stroke, in the hope of relieving pressure on the brain. As we shall see from Jenner's own story, tracheotomy saved life but cutting the temporal artery did not.

More heroic procedures included amputating limbs, trepanning (drilling holes in the skull, and often used in less desperate circumstances than might be imagined) and cutting stones out of the bladder. This slash-and-grab operation began with a decisive cut through the perineum between the genitals and the anus. The area was conveniently exposed by laying the patient on their back with their knees in the air and pulled wide apart by two assistants (they and others at the head end also helped to prevent the patient from getting off the table). Only an inch or two of soft tissue had to be sliced through to reach the bladder and hopefully the stone. A small, smooth calculus could be winkled out with a probe but larger or spiky ones had to be broken up first with a special crusher pushed through the cut. Speed was paramount as pain was part of the package: the perineum is generously supplied with nerves (as the interested reader can confirm in the bath by gentle prodding) and the only anaesthetics were gin and opium. A slick expert could pull out an easy stone in under a minute; an incompetent amateur could still be struggling after an hour.[8]

By the end of his two years at St. George's, Jenner was competent to do these operations and later performed them in his routine practice. There were no attempts to test his or other students' knowledge. Simply attending the lectures was enough to walk off with a signed certificate, the passport to set up as a surgeon. After graduating, Jenner remained 'Mister'. The title 'Doctor' was reserved for the graduates of Oxford, Cambridge or the Scottish medical schools of Edinburgh, Glasgow, Aberdeen and St. Andrew's, who were awarded the degree of Doctor of Medicine (MD, from its Latin initials). Jenner only became 'Doctor' when the University of St. Andrew's, encouraged by a letter of commendation from two of Jenner's friends and a large sum of money from him, granted him an MD in 1792.[9]

In late 1772, the 23-year-old Jenner returned to Berkeley to set up in practice. He was a fortunate young man, with a thorough grounding in surgery under his belt and a reasonable family income to buffer him against hard times.

And best of all, his time in London had brought him under the spell of one of the most charismatic thinkers and doers of the age.

Friend in high places

Of course nobody really has a brain 'the size of a planet', but occasional people make us understand why the expression was coined. John Hunter, the 'father of surgery', was one of those revelatory beings (see Plate 4). The packaging for the brain was not that prepossessing: short, squat, rough-mannered and often clumsy with the written and spoken word. Born in rural Lanarkshire in 1728, John grew up in the shadow of his more obviously brilliant and world-wise brother William. The start of John's career was also unpromising, dragged down from Scotland to London in 1748 at the age of 20 to help as a technician in William's new Anatomy School in Great Windmill Street. Initially, John fitted in badly, being awkward and more comfortable in the company of ruffians. The latter quality helped in his role as the procurer of fresh dissection material, which required a good working relationship with the city's graverobbers.[10]

This odd environment was exactly what John Hunter needed to flourish. He proved to be an expert craftsman, with a gift for dissection and preparing specimens for William's museum. He soon became the chief anatomy demonstrator and went on to formal surgical training at St. George's. The military campaign against the French at Belle-Île gave him the chance to perfect his operative skills and begin carving his own way through surgery, with a new 'conservative' approach to treating bullet wounds that he soon proved to be better than current wisdom. Belle-Île was also home to lizards that demonstrated how nature copes with battle injuries, by growing a new tail (sometimes two) when the original was torn off. This phenomenon seized Hunter's imagination as much as the puzzle of how to save human life and limb and sparked a passion for natural history that grew in parallel with that for surgery.[11]

On returning to London, Hunter's career as surgeon, naturalist and maverick took off.[12] He experimented with the transplantation of teeth, initially into a rooster's comb (to prove that the tooth survived) and then into the gums of the wealthy. He introduced a new operation to deal with the arterial swellings (aneurysms) that grew behind the knee and were treated by amputating the leg. Hunter simply tied off the artery above the aneurysm, allowing other arteries to open up and supply the lower leg. He applied the same methodical approach to natural history, describing structures and theorising about function in spectacular dissections, such as a 20-foot killer whale beached in the Thames, the charge-generating organs of the electric eel and various exotic species from circuses and zoos. All this brought him to the attention of the Royal Society and in 1765 he was elected a Fellow for his work on body temperature in animals. This shows how steeply John Hunter's trajectory rose: William, despite his ten-year head start, shrewdness and more conventional brilliance, was not made a Fellow until the following year.[13]

Not all of John Hunter's experiments worked. His conviction that life could be safely suspended by freezing was only overturned after the hypothermic demise of countless fish and mammals. Similarly, a cleric condemned to hang for fraud went to the gallows in the knowledge that Mr. Hunter would use his remarkable resuscitation skills to bring him back to life; despite improbable rumours to the contrary, the cleric's optimism was very rapidly shown to be misplaced.[14]

Jenner met Hunter as soon as he reached London. Hunter ran the course at St. George's and had also decided to offer accommodation for one or two students at his town house in Jermyn Street. Jenner was Hunter's first lodger and despite the 20-year age gap between them, they got on well from the start.[15]

Jenner's time in London, 1770–72, was a particularly busy period for Hunter.[16] As well as running his surgery course, he was vigourously courting Anne Home. John and Anne made an odd couple. Her poem, 'My mother bids me bind my hair', was set to music by Haydn and she was beautiful, graceful and cultured. Hunter was none of those things. He was also an obsessive workaholic with an 18-hour working day and four hours' sleep. When he ruptured his Achilles tendon by jumping up and down on his toes (possibly practising Scottish dancing), the accident occurred at 4 a.m. Hunter may also have had a dark secret. There is evidence that, six years earlier, he had deliberately inoculated himself with pus from a patient with gonorrhoea to see if the disease could be transmitted by that route. It was, and so was the syphilis which the patient had also picked up.[17] Hopefully, time and mercury treatment would have subdued the syphilis by July 1771, when he married Anne.

Hunter and Jenner were also an odd couple. Hunter was in perpetual motion, restless and bubbling with ideas which tumbled out too fast for anyone – even himself – to act on. Under his influence, Jenner's mind was stretched and his horizons broadened and, at least while within Hunter's reach, he was galvanised with new energy and excitement. Both worked to maintain their relationship after Jenner returned to Berkeley, mostly through letters. Only Hunter's side of the correspondence survives.[18] Jenner kept these letters safe until the end of his own life but his to Hunter disappeared, probably when Hunter's jealous brother-in-law Everard Home burned all his personal papers after his death.

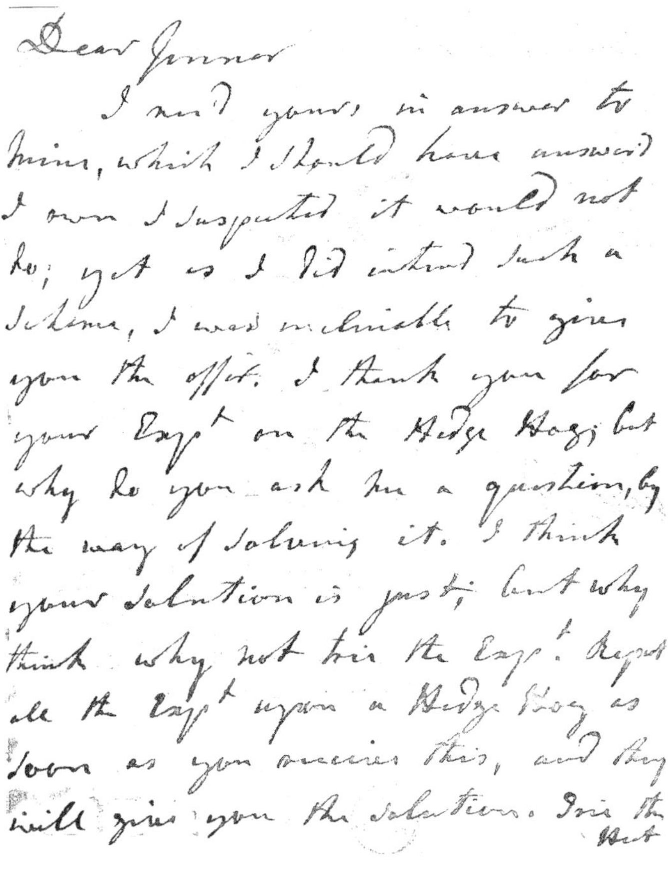

Dear Jenner
I rec'd yours in answer to
mine, which I should have answer'd
I own I suspected it would not
do; yet as I did intend such a
scheme, I was inclinable to give
you the offer. I thank you for
your Exp^t on the Hedge Hog; but
why do you ask me a question, by
the way of solving it. I think
your solution is just; but why
think, why not trie the Exp^t. Repeat
all the Exp^t upon a Hedge Hog as
soon as you receive this, and they
will give you the solution. Trie the
Heat

Figure 9.2 Facsimile of a letter from John Hunter to Edward Jenner, 1778. Reproduced by kind permission of the Royal College of Surgeons of London.

Hunter's letters to Jenner show a relationship that appears relaxed, respectful and affectionate, although Hunter's emotions can be as difficult to decipher as his writing. Most of the letters are an incontinent jumble of ideas and demands that spill out with little regard for grammar or spelling. Hunter bombarded Jenner with requests for experimental material that was hard to come by in London. As well as a general plea for "anything curious ... either in the flesh or fish way", Hunter wanted specifics. The easy ones included fossils, breeding toads and birds' nests ("I want a nest with the Egg in it also a nest with a young cuckow and also an old Cuckow"). More challenging were dozens of hedgehogs to replace those which Hunter went through in his continuing experiments on body temperature ("I am hedgehogless", he complained in March 1778), a porpoise and a large game bird, the great bustard. Somehow, Jenner found them and sent them to London, alive or carefully preserved: an eleven-foot bottle-nosed porpoise from the Severn Estuary and the great bustard, already heading for extinction, from one of its last enclaves on Salisbury Plain. As an encore, Hunter asked Jenner to tell him what fresh porpoise milk tasted like (answer: like cow's milk with added cream). As Hunter wrote, with or without insight: "I hear you saying there is no end to your wants."[19]

In return, Hunter sent Jenner the occasional gift – oil paintings, candlesticks – either free or at a reduced price. Jenner was not acknowledged in any of the papers that Hunter wrote (even though Hunter was angered when others omitted mention of his own contributions) but it does not seem that Jenner felt exploited. He was swept along by the fun and excitement of the science and the attention of this great man and friend – who by now had been awarded the entirely appropriate title of Surgeon-Extraordinary to King George III.

Hunter also tossed out ideas for experiments and snippets of philosophy. He sent Jenner a prototype mercury-glass thermometer with instructions for calibrating it, a plea not to break it with his "clumsy fingers" and suggestions about measuring the internal temperature of hibernating hedgehogs, possibly to check his own data.[20] This was a bad time for hedgehogs in Berkeley. Hunter wanted the thermometer inserted first in the rectum and then, through an incision in the belly, into the abdomen, chest and pelvis. Hunter's most celebrated advice about scientific discipline, delivered with a hint of exasperation in the summer of 1775, was essentially "Get on with it":

> I thank you for your Expt on the Hedge Hog but why do you ask me a question by the way of solving it. I think your solution is just; but why think, why not trie the Expt.[21]

Hunter undoubtedly felt that Jenner was wasted in the pastoral rut of Berkeley and in mid-1775 he tried to entice Jenner back to London and into partnership with him. For many surgeons in England, this would have been an opportunity from heaven. Jenner declined.[22]

There were undoubtedly other attractions in Berkeley, although Jenner initially kept Hunter in the dark. Hunter may have felt hurt when he wrote in June 1778:

> I was told the other day that you was married, and to a young lady with a considerable portion. I hope it is true, for I do not know any body more deserving of one, let me know if it is so or not.[23]

Unfortunately, it was not so and Jenner did not reply for some time. When he finally did, Hunter's response was both sympathetic and pragmatic:

> I own I was at a loss to account for your silence and am sorry about the cause. I can conceive of how you must feel, for you have two passions to cope with viz that of being disappointed in love and that of being refused, but both will wear out, perhaps the first soonest.[24]

Then he was straight back to science and advising Jenner to find solace in the warm entrails of hedgehogs. The identity of Jenner's well-off young lady was never discovered. She may even have been Catherine Kingscote, whom Jenner eventually married in March 1788, perhaps temporarily deterred by cold feet or family resistance.

Affection surfaces repeatedly throughout Hunter's letters to Jenner, as encapsulated in his remark: "I do not know anyone I would sooner write to than you." After the birth of the Jenners' first son Edward in March 1789, Hunter wrote apparently in haste but with obvious delight:

> I wish you joy, it never rains but it pours. Rather than the brat should not be a christian. I will stand godfather. for I should be unhappy if the poor little thing should go to the Devel. because I would not stand Godfather. I hope Mrs Jenner is as well. and that you begin to look grave now you are a father.[25]

Their correspondence lasted 20 years and ended with Hunter's death in 1793. Their personalities were part complementary, part antagonistic. Both were hugely intelligent and inquisitive about medicine and science, although Jenner's dithering and natural tendency to coast must have been maddening to the mercurial Hunter. If he could cram all his family life and diversions into the two hours left each day after work and sleep, why couldn't Jenner? But Jenner was probably his closest male friend as well as his star pupil and Hunter treated him with exceptional tolerance and warmth – compare the rudeness and criticism which Hunter directed at his own brother-in-law Everard Home, and which Home seemed to avenge after Hunter's death.

The Hunter–Jenner relationship was symbiotic but asymmetrical. Hunter would have blazed his own trail without Jenner, while Jenner could well have slipped into comfortable obscurity had he not met the brilliant and irascible "dear man". Jenner recognised his dependence and later became frightened by signs of weakness in his idol. When he realised that Hunter was having potentially dangerous attacks of angina, he was afraid to mention it to Hunter himself but instead tried to seek advice from another senior doctor, William Heberden.[26]

In the meantime, though, Jenner could refuse nothing that Hunter wanted – with the crucial exception of leaving Berkeley.[27] In return, Hunter fed him intellectual nourishment and fixed things quietly behind the scenes to establish Jenner's reputation as a serious scientist.

Nature calls

This was the age of the polymath, when gentlemen were expected to indulge their curiosity. Jenner and Hunter both had broad interests that took them from medicine into comparative anatomy, zoology and geology. Hunter had a Midas touch that could turned almost anything into the scientific gold of publications but Jenner, despite his mentor's encouragement, found it hard to plan experiments and see them through.

Jenner dabbled in many topics apart from vaccination: distemper in dogs, a transsexual chicken, earthworms, the migration of birds, the use of human blood as fertiliser, coronary artery disease and the cysts known then as hydatids.[28] From all this, he only produced a dozen publications, one anonymously and the last posthumously, and his scientific career was littered with half-finished and revisited studies. He also lacked Hunter's instinct for making the most of his discoveries. He found that 'tartar emetic', a poorly soluble antimony salt used to induce vomiting, could be easily dissolved in "Madeira ... or a good White wine" to a standard concentration that made accurate dosing easy. He published the outline method anonymously in a pamphlet in 1783. Hunter told him to burn all the copies so that he could patent his preparation. Jenner could not see the point and a decade later, to Hunter's exasperation, published the method again in full and under his name.[29]

In spring 1796, according to the entries in his notebook, Jenner was more preoccupied by hydatids than by cowpox.[30] The term 'hydatid' is still used today to denote the large cysts of the sheep tapeworm that grow in the lungs and liver of sheep and man, but to Jenner they were any cystic (fluid-filled) lesion. He first noticed them in the lung of a hare killed by a greyhound, then found lesions that he believed to be similar almost everywhere he looked: the lungs of sheep, cows and humans, 'measly' pork in the butcher's, even cysts that appeared on the faces of his wife and son. He explained everything in a letter to his brother Henry (who succeeded Stephen as Rector of Rockhampton). Hydatids were insects "whose character I will not attempt to describe to you", which invaded and grew in the tissues. They caused the "horrid malady" of consumption (tuberculosis in the lungs); in an earlier note, Jenner had also blamed hydatids for "all tumours truly cancerous".

In fact, Jenner's hydatids were due to several different infections, none of them insects: tapeworms in sheep and pig, tuberculosis in human and cow lungs and an infection called molluscum contagiosum (coincidentally caused by a poxvirus) on the faces of his family. To be fair, Jenner worked only with a hand lens (and occasionally his sense of taste) but he shows himself as an uncritical 'lumper', forcing observations together to make them fit into one unifying hypothesis. Jenner also took his conclusions too far, a fault that later dogged his studies of cowpox. He speculated that his discovery of the true cause of pulmonary consumption would lead to a "new and better mode of treating the complaint". There is particular poignancy in these words, as tuberculosis would later kill some of those closest to him.

Also disorganised were his experiments on the agricultural applications of human serum. These studies ran between 1780 and 1782 and each mopped up several pints of blood (perhaps four or five therapeutic bleeds) to irrigate assorted flowers and bushes. The design was poor and yielded the conclusion that ordinary manure was a better fertiliser. Perhaps wisely, Jenner never wrote these studies up as a formal paper but in July 1787 sent a detailed letter to Joseph Banks, the no-nonsense President of the Royal Society. Jenner may have been trying to curry favour, as he was being encouraged by Hunter to apply for Fellowship of the Society. Banks is unlikely to have been impressed by Jenner's rambling experiments or his apology for the five-year delay in reporting his findings: "a person engaged in business can't conduct these matters as he would wish; his pursuits are too often interrupted".[31]

Luckily, there was much better to come. In the same letter, Jenner remarked: "I hope to have the honour of presenting you with another paper in the autumn", and added that he would pursue the subject during the summer. For once he was as good as his word and the resulting paper did the trick. This was an observation "so much out of the ordinary course of nature" that it would probably be disbelieved – but it had nothing to do with smallpox.

It was the cuckoo that won Jenner his Fellowship of the Royal Society. Since Aristotle, the bird's odd behaviour had excited scientific interest. It arrived in spring, laid its eggs in other birds' nests, somehow made the rightful occupants disappear, duped the parents into raising the usurper as one of their own, then flew away again. It was suggested that the female cuckoo had a peculiarly shaped stomach which prevented her from incubating her own eggs, but the rest was a mystery. Prompted by Hunter to look into the cuckoo's habits, Jenner eventually did a thorough job. Here, he was in his element – no complicated experimental design, just dissections and fieldwork spread over the springs of 1787 and 1788 that he conducted himself with the help of his nephew Henry.

All his findings broke new ground. Careful dissection showed nothing unusual about the cuckoo's stomach that could prevent it from raising its young in the usual way. Instead, Jenner speculated that the cuckoo's short residence in England (mid-April to early July, half that of the swallow) would limit its chances of breeding successfully, hence its adaptation to laying many eggs and letting others raise its dispersed brood. His most spectacular finding solved the mystery of how the cuckoo chick ended up alone in the nest. By staking out nests containing very young cuckoos, Jenner and Henry were able to witness an amazing performance. Caught in the act, the tiny chick was seen to heave all its competitors – eggs and nestlings – out of the nest. It only did this between three and twelve days of age, thanks to a special depression in its upper back that enabled it to lever the enemy up and over the edge of the nest. More dissections showed that this purpose-built hollow filled in naturally after twelve days, just as the chick's killer instinct faded away.

Jenner's *Observations on the Natural History of the Cuckoo*[32] began and ended with thanks to John Hunter and expressed the hope that these observations would "throw some light on a subject that long laid in obscurity". Jenner sent the paper to Hunter just after Christmas 1787 and it was read to the society on 13 March the following year. It took another year for the Royal Society's processes to grind along to completion. On 25 February 1789, obviously satisfied for both of them, the old mentor wrote to his ex-pupil with the news that Mr. Edward Jenner could append the letters FRS to his name.[33]

Convivial and Medical

"For the united purposes of conviviality and improvement of medical science."

Aims of the Gloucestershire Medical Society, 1788

One might have expected that the return of 23-year-old Jenner to his native dairy county in 1772 would reawaken his interest in cowpox. Not so: it was not until 1796 that he performed his first vaccination. The intervening 24 years saw him settle into a comfortable country practice, enlivened by Hunter's influence and punctuated by major life events: marriage at the age of 39 and the birth of his first son a year later.

Jenner's version of work-life balance was much gentler than Hunter's. It was epitomised by the Medico-Convivial Society[34] (formally known as the Gloucestershire Medical Society) which convened at the Fleece Inn in Rodborough on the edge of the Cotswold escarpment, and the less focused Convivio-Medical Society that met at the Ship Inn outside Thornbury.[35] Jenner gave equal weight to both Societies' joint aims of being a doctor and enjoying life.

When relaxed, as he was most of the time, he was excellent company – generous, warm and witty. He played the flute and violin and had "a tolerable voice", certainly talented enough to have joined in Anne Hunter's musical soirées in Jermyn Street.[36] He also wrote poetry and songs. Most are fussy by today's tastes but were well regarded in the day. They were intended for the enjoyment of his friends and his 'catch' (singing) club and many are preserved in notebooks, hand-written by himself or his nephew William Davies.[37] They include 'The Robin' (two versions, the second an amusing parody of the first); odes to young women ('Betty Bunker' and 'Hannah Ball') and a local beer ('Ladbroke's Entire'); and an old countryman's shocked reaction to having his perianal abscess lanced by Jenner ('John's Story', transcribed in a broad Gloucestershire accent). There are also two verses on the death of a Dr. Wait, famous for his medicinal gingerbread nuts to treat intestinal worms (whose Latin names open the first stanza):

> Ascarides, Teres, Lumbrici and all,
> Ye chyle-sucking insects that tremblingly crawl,
> No longer be frightened – you're quite safe in our Guts,
> For Wait has done making his Gingerbread Nuts.

And, possibly after more Ladbroke's Entire had flowed:

> Alas! Poor Doctor Wait is killed,
> And carried to a pit-house –
> Who, lucky man, has often filled
> With Worms full many a ——.

The countryside around Berkeley provided other distractions and Jenner often rode off for some hours before seeing his first patient of the day. His favourite destinations included Barrow Hill, several miles to the north and with a magnificent view across a broad loop of the Severn to the Forest of Dean, and Hock Cliffs where he would spend hours digging out fossilised shells, teeth and bones.

His social and medical circles overlapped. During his bachelor years, his closest friends were Caleb Parry, his schoolmate from Cirencester and a fellow member of the Medico-Convivial Society, and Edward Gardner, a successful minor poet who by day was a wine merchant at Frampton-on-Severn, halfway to Barrow Hill. Gardner described the 25-year-old Jenner, confident and snappily dressed, striding out on the Green at Frampton. Below average height, stout but active, he wore a blue coat with yellow buttons, polished riding boots with silver spurs, and had his hair tied up in the fashionable 'club' under a broad-brimmed hat.[38]

Both Parry and Gardner helped Jenner with an exploit in 1784 which terrified some peasants, made the local papers and impressed Jenner's wife to be. Catherine Kingscote was the daughter of a well-off family at a village of the same name some ten miles north-east of Berkeley, whom Jenner had been pursuing for some time. The exploit concerned a hydrogen-filled balloon, recently invented in France. Parry had already launched one from the Royal Crescent in Bath in January 1784, one of the first such flights in Britain. Following his instructions, Jenner made a balloon from oiled silk, test-filled it with hydrogen (by throwing iron filings into a pot of sulphuric acid) in the Great Hall at Berkeley Castle, and then launched it from the Castle on 2 September. Carrying an uplifting poem by Gartner, the balloon drifted north-east and ditched close enough to the Kingscote's family home for it to be carried there for a relaunch, with much excitement and more poetry.[39]

Thereafter, Jenner closed in on domesticity and Catherine Kingscote. In 1785, he bought Chantry Cottage, a handsome house next door to Berkeley Castle. He married Catherine on 6 March 1788; he was then 39 and she 27. Catherine comes across as quiet and tending to the sickly (even before she caught tuberculosis) but she provided strong support for Jenner. He was devoted to her, even though he referred to her as "Mrs. Jenner" or "Mrs. J." Only rarely, such as following the birth of their first son, did she become "My dear Catherine" or "Dear Kate".[40]

Their first son, Edward, was born on 24 January 1789. He had a happy childhood but was slow to learn and had to be tutored at home – a necessity that ultimately cost him his life. The second child, Catherine, was born in 1794 and their second son Robert in 1797. Robert's place in history was assured by his experimental encounter with cowpox just after his first birthday.

There were intrusions on the Jenners' life in Berkeley. Catherine's health was poor, with several attacks of illness that probably included typhoid fever and left her depressed and debilitated. In 1791, Hunter wrote to Jenner, "I have been informed that Mrs. J has been extremely ill."[41] Extended trips to the spa town of Cheltenham helped her to some degree. They took a town house there and Edward built up a lucrative practice to which he returned periodically.[42]

A greater trial was lined up for Jenner two years later when two omens of disaster came together: Jenner's fear that Hunter's attacks of faintness were due to angina and Hunter's own realisation that these were brought on by anger and that his life was "in the hands of any fool who cares to upset me".[43] On 16 October 1793, during a furious argument provoked by surgeon colleagues at St. George's, John Hunter collapsed and died. He was 65 years old.

Jenner was devastated by the loss of his mentor, shield and above all his "dear man" and fell into a bout of deep depression. Fine feelings were less evident in London. Hunter's fellow surgeons at St. George's voted on whether to send condolences to his widow Anne; the majority was against.[44] Hunter's wish to be autopsied for the instruction of his students was respected but his request for his heart and healed Achilles tendon to be preserved for his museum was not. It was Everard Home who dissected Hunter and who threw away these specimens with the rest. He also wrote to his brother-in-law's favourite pupil to let him know that his diagnosis of narrowed coronary arteries had been confirmed at post-mortem.[45]

So much for Jenner, *bon viveur* and eventually family man. This quarter-century as a doctor also passed at his own pace. Thanks to his father's astute dealing in land, the family had a generous income which meant that Edward would never have to flog himself to death to live as he wished. He was undoubtedly a good doctor, respected and liked by his patients. According to the map, his catchment area in the Vale of Berkeley covered an impressive 400 square miles, but the population was thinly scattered and actually shrank in Berkeley during his lifetime.[46] He did his rounds on horseback and often stayed the night with more remote patients; we can imagine that he would have been an entertaining guest.

His diary for 1794 shows a tiny workload beside Hunter's 18-hour day and John Coakley Lettsom's completely unbroken decades of hard labour as a physician in London. Jenner nominally worked seven days a week and visited most patients at their homes, but there were many days when he had just one or no consultations at all.[47] In one four-week period that summer, he visited just eight patients and made up only two prescriptions. His diary gives few details but his consultations included trepanning, tapping a hydrocoele (fluid accumulation around the testis) and applications of "leeches and Co."

Jenner typically charged a guinea for a prescription and one or two guineas for a visit. His recorded weekly income ranged from nil to £37 and averaged £9 – nowhere near enough to run a substantial home and feed expensive tastes. That May, a Mr. Whithorn commanded much attention, with 36 home visits and even a trip to Bath with him. We do not know whether Whithorn was a complicated case, a friend or simply a demanding patient.

Jenner's diary does not show someone obsessed by detail or troubled to keep proper records, presumably because his family income meant that he did not have to. Even the £5 housekeeping payments to "Mrs J" are irregular. Jenner evidently had no need for the "Considerations on the value and importance of time to men of business" printed in the front of the diary:[48]

> Nothing should ever be left half done ... Never trust till to-morrow what may be done to-day ... Nothing should prevent a person's daily account of cash received from being attended to, before the close of the day's business.

Jenner was more focused in his medical/convivial societies. He injected some professionalism – officers, timetable, minutes – into the Gloucestershire Medical Society's proceedings.[49] This Society was active but had just five core members – Jenner, Caleb Parry (now working in Bath), Daniel Ludlow (the son of Jenner's first mentor), John Hickes from Bristol and Thomas Paytherus from across the Severn in Ross-on-Wye. They discussed cases and made some striking original observations, notably the first description of hardening of the coronary arteries in angina. Jenner was the first to note this and graphically reported the crunch of his knife cutting into the calcified artery wall during a post-mortem, which made him think that plaster had fallen from the ceiling into the dissection.[50] Parry was also interested and took the initiative in 1799 by writing a paper on the symptoms and causes of angina. However, he made it clear that Jenner had first discovered the link with "ossification, or some similar disease, of the coronary arteries".[51]

This was another trick that Jenner missed, but for good reason. By then he was in the thick of the excitement stirred up by his first paper on vaccination – and that harmonious balance between conviviality and medicine had been shattered for ever.

A pearl laid upon a rose leaf

And finally to vaccination. We have taken a long time to reach it but so did Jenner. He was 19 years old when the milkmaid's remark made him stop and think, 47 when he did his first vaccination and 49 when he published his finding that cowpox inoculation protected against smallpox. There was some activity between these milestones. While lodging with Hunter in London, Jenner told him the milkmaid's story and later showed him his drawing of a cowpox lesion. Presumably this was not confidential as Hunter later told Thomas Beddoes, a physician from Bristol, who wrote about the possible protective properties of cowpox in a book, but without mentioning Jenner.[52] Back in Berkeley and possibly prodded by John Fewster's observations at Buckover, Jenner began discussing it again, to the point of boring his friends in the Convivio-Medical Society (see page 172).

Sometime in the early 1790s, his interest was rekindled and he began looking for people who had apparently been protected against smallpox after an attack of cowpox. This was not an easy task: the folklore was widespread but cowpox was not, even in cow-rich Gloucestershire. To test the hypothesis properly, a large number of subjects with previous cowpox would have to be exposed repeatedly to smallpox under conditions where an unprotected person would be guaranteed to catch it, and then followed up for life. This was clearly impossible but Jenner made some attempt. He found a couple of people who had met smallpox but not caught it, 25 and 50 years after having cowpox. He then thought of testing other subjects using variolation, arguing that protection would be shown if they failed to develop the characteristic smallpox eruption and fever. This approach produced several more cases.[53]

The evidence was promising but still circumstantial. For direct proof, unprotected subjects had to be given cowpox and later tested by variolation to confirm that they were immune to smallpox. On 14 May 1796, Edward Jenner followed in the footsteps of Benjamin Jesty and John Fewster and inoculated a healthy subject with cowpox.[54] Sarah Nelmes (see Plate 5), the daughter of a wealthy landowner at nearby Breadstone, had developed the firm ulcers of cowpox on her right hand after milking a roan-and-white cow called Blossom. Jenner collected some fluid from the lesions and, as if variolating, scratched it into the arm of James Phipps, the eight-year-old son of the Jenners' gardener at the Chantry.

Over the next several days, Jenner followed the progress of what would become the classical response to vaccination, the "pearl laid on a rose leaf". A pale nodule grew out of the inoculation site, surrounded by a spreading red halo of inflammation, then broke down after ten days or so into a shallow ulcer that scabbed over a few days later. James had a transient fever but otherwise remained well. The cowpox inoculation appeared to have taken but the crucial confirmatory test still had to be done. Jenner left it several weeks, then variolated the boy on 1 July. He reported the result in a buoyant letter to Edward Gardner on 19 July:[55]

> As I promised to let you know how I proceeded in my Inquiry into that singular disease the cowpox ... I have at last accomplished what I have been so long waiting for, the passing of the vaccine Virus from one human being to another by the ordinary mode of Inoculation ... I was astonished at the close resemblance of the Pustules in some of their stages to variolated Pustules. But now listen to the most delightful part of my story. The Boy has since been inoculated for the small pox which as I ventured to predict produced no effect. I shall now pursue my Experiments with redoubled ardour.

This letter is interesting for several reasons. First, Jenner only needed a few days to know that the smallpox eruption was not going to break out, yet waited 18 days to tell his close friend this exciting news. Second, despite his eye for detail, he focused on some superficial similarities between the lesions of inoculated cowpox and variolated smallpox rather than their clear differences, a point that would come back to haunt him. Finally, he wrote that he would continue his experiments "with redoubled ardour".

At this point, any scientist would have shared Jenner's excitement but most would set out to repeat the experiment to convince themselves and their peers that the result was not just a fluke. Jenner did not do this and instead tried to rush into print. He wrote up his circumstantial cases and the single direct test on James Phipps and sent the paper to Joseph Banks, asking for it to be considered for immediate publication in the *Philosophical Transactions*.[56] The paper was more exciting than the non-fertilising properties of human blood, but Everard Home, whose opinion Banks sought, argued strongly against publication. Banks wrote back to Jenner in May 1797, rejecting his paper and passing on the referee's comments: the notion was interesting but the conclusion premature and another 20 or 30 experiments like the one on James Phipps were needed to confirm the finding. Indeed, publishing such incomplete data could damage Jenner's good reputation as a serious scientist.[57]

On scientific grounds, the decision was entirely reasonable but Jenner was disappointed and probably angry. He may also have been suspicious (especially given the involvement of the vindictive Home) that publication of his paper had been stalled so that others could steal his idea.

"Redoubled ardour" finally set in. Throughout that autumn and the spring of 1798, Jenner chased up more cases of natural cowpox and variolated them. He also inoculated another dozen subjects, mostly children and including his own one-year-old son Robert.[58] Jenner was in a hurry and cut corners. He tested only some of his cowpox-inoculated subjects with variolation and threw in some cases inoculated instead with fluid from a blistering eruption on horses' heels known locally as 'grease'. Some lesions of 'grease' superficially resembled cowpox and this had persuaded him that this was the natural precursor of cowpox and was spread from horses to cows by human hands.

Jenner's expanded publication ran to over 70 pages and included his own coloured drawings of Sarah Nelmes' infected hand and the typical lesions of inoculated cowpox on the arm of a girl, Hannah Excell. Jenner dedicated the paper to Caleb Parry and took it to London for publication. He did not risk the Royal Society, perhaps because he was far short of the 20–30 extra cases which they had stipulated. Instead, he paid for the work to be published privately.[59] It was printed by Sampson Low of Soho, cost seven shillings and sixpence and bore the cumbersome title, *An Inquiry into the causes and effects of the Variolae Vaccinae: a disease discovered in some of the Western Counties of England, particularly Gloucestershire, and known by the name of the Cow Pox*. Almost as soon as it appeared for sale on 17 September 1798, the *Inquiry* whipped up great excitement with its promise of something safer and better than variolation. It also provoked puzzlement, suspicion and jealousy.

It should have been a straightforward paper with a simple but powerful message: inoculation with cowpox solved all the problems of variolation. Facial scarring never occurred because no pustules broke out beyond the inoculation site, nobody had ever died of cowpox and it did not spread to other people (at least as evidenced by the two children who had shared James Phipps' bed after he was inoculated). All Jenner had to do was describe cowpox (which was virtually unknown to the medical profession) clearly enough for others to identify and use it, then present evidence that the natural infection protected against smallpox and that the same immunity was achieved by inoculated cowpox. Instead, the *Inquiry* was a difficult read and obscure in places. Key information was missing and some of Jenner's ideas and claims were obviously overblown for the evidence that he presented.

He began confusingly with his theory (for which there was no evidence) that cowpox emanated from grease, transferred from horses' heels by the unwashed hands of stable boys to cows' nipples, where it somehow transformed itself into the familiar lesions. His own cases showed that people who had caught grease were not always protected against smallpox, but he still went on to inoculate a boy with blister fluid from a man infected with grease. Unfortunately, the boy could not later be variolated because he had "felt the effects of a contagious fever", so Jenner could not even determine whether inoculated grease was protective.[60]

The main thrust of the *Inquiry*, that inoculated cowpox protected against smallpox "for ever", was also undermined by Jenner's carelessness and his tendency to take conclusions too far. Natural cowpox was supposed to confer lifelong protection and Jenner reported five cases who who failed to react to variolation, between 25 and 53 years after catching cowpox. However, the protection following inoculated cowpox had only been verified after a few weeks, and this in only four of at least a dozen inoculated subjects (strangely, Jenner gave no details or even the number of "several children and adults" whom he had treated). Experienced variolators would also have spotted another curious inconsistency. Jenner described the lesions of inoculated cowpox as "indistinguishable" from the pustules of variolated smallpox, which have ragged edges. In fact, they looked quite different, as shown by his own careful drawing of the smooth, circular vaccination lesions on Hannah Excell's arm (see Plate 8).[61]

Finally, he sowed further confusion with his belief that some cases of apparent "cowpox" did not protect against smallpox. He termed this "spurious" cowpox but made no attempt to explain how it could be distinguished from "true" cowpox. The existence of multiple types of cowpox presumably explained why he used the plural 'variol*ae* vaccinae' (literally 'smallpox*es* of the cow') in the title of the *Inquiry* and his follow-up articles.

In its way, the *Inquiry* was a gem, but flawed and easy to fault. The gaps that Jenner left or tried to paper over caused particular concern. Why had he failed to variolate so many of his inoculated subjects, and why did he not even state how many he had experimented on? It later transpired that the boy inoculated with grease was unavailable for follow-up because he had died of his fever. Why had Jenner covered this up? Had the inoculation caused the fever that killed him?

Nevertheless, the *Inquiry* sold well and made Jenner's name. He was slow to react to the inevitable criticisms. The second edition, expanded with further cases to 182 pages, was published in 1799 with the King replacing Caleb Parry in the dedication.[62] It still featured the unsubstantiated claim that inoculation protected against smallpox for life, together with more thoughts about "spurious" and "true" cowpox, but still confused and tangled up with grease. His *Further observations on the variolae vaccinae or cow pox*,[63] published earlier in 1799, together with *Further Observations and A Continuation of Facts and Observations Relative to Variolae Vaccinae* (1800) were bound in with the third edition of the *Inquiry* published in 1801.[64] Even this solid body of work, from the now well-known and admired Dr. Edward Jenner MD FRS was an easy target for the increasing number of doctors, scientists and others who wanted to knock him down together with his invention.

A pearl before swine

The criticisms that blew up over the *Inquiry* were Jenner's first real test of having to defend his science against his peers. He was yanked out of his comfort zone and cut off from the protection of the late "dear man" and Surgeon-Extraordinary whose skilful string-pulling had smoothed his way in the past. Jenner now found himself up against scientists who were cleverer, harder working and more meticulous than himself, as well as numerous deeply unpleasant people with their own reasons for doing him down. The quiet life in Berkeley had given him few survival skills and he was easy prey in the ruthless jungle of London medicine. Fortunately, he also attracted some powerful supporters; without them, he and his claim to fame could have disappeared without trace.

An early critic, later converted to the cause, was William Woodville.[65] Another country boy (from Cumberland), Woodville trained as a physician in Edinburgh and Europe and was famous for his *Medical Botany*, a stunning work of art as well as a standard medical text.[66] In 1798, he was physician to the Smallpox and Inoculation Hospital in St. Pancras. He had recently published volume 1 of another monumental book, *The History of the Inoculation of the Small-Pox in Great Britain*[67] and was working on its successor when the *Inquiry* appeared. Woodville's first instinct was to see off this threat but he also wondered whether Jenner had stumbled on something important. He dropped *Inoculation* and began to investigate cowpox for himself.

Jenner would have come out badly in a head-to-head contest with Woodville. Cowpox was even rarer in the largely cattle-free capital but within three months Woodville had found a focus of infection at a dairy near Gray's Inn Road and six months later published a full paper on his first 310 cases (a year later, his total was 2,500).[68] His early vaccinations produced odd results, with over half the cases developing a generalised rash that looked suspiciously like smallpox – and undoubtedly was, presumably introduced by contamination from the Smallpox Hospital. Woodville corresponded with Jenner, who now wished that he had not described the cowpox lesions as looking like those of variolation and insisted that Woodville's eruption was smallpox. Woodville then took better precautions and his vaccinations behaved as predicted.[69] With the evidence before him, Woodville became an energetic ambassador for vaccination, although never a close friend of Jenner. Woodville travelled widely in Europe to spread the word – another contrast with Jenner, who never left England and rarely strayed outside the corridor that links Berkeley, Cheltenham and London.

Other early sceptics included Thomas Beddoes and John Haygarth. Beddoes thought that Jenner's notion that cowpox originated from grease was "quite indemonstrable", while his own observations were "not favourable" to Jenner's claim that cowpox gave complete immunity to smallpox.[70] Haygarth, whose grand plan to exterminate smallpox hinged on nationwide, enforced variolation with isolation, wrote directly to Jenner:

> Your account of the cowpox is indeed very marvellous; being so strange a history and so contradictory to all past observations on this subject, very clear and full evidence will be required to render it credible.[71]

However, enough "credible evidence" accumulated during the next several months to turn both men into supporters of vaccination and of Jenner.

The same could not be said of Jan Ingenhousz, the upwardly mobile Dutch variolator and a fellow-FRS, who saw the *Inquiry* in October 1798 while staying with the Marquis of Lansdowne in rural Wiltshire.[72] Ingenhousz was probably motivated by the fear that cowpox inoculation would sweep aside variolation and so demolish his livelihood. He immediately made enquiries locally and soon found an exception to Jenner's Law: a "respectable Farmer" who claimed to have had cowpox but had reacted so convincingly to variolation that he had fatally infected his own father with smallpox. Ingenhousz wrote Jenner a "friendly" letter to avoid "the disagreeable necessity of entering into a public controversy" but his message was clear: Jenner must retract the *Inquiry*.[73] Jenner tried to explain that the farmer had "spurious" cowpox which did not protect but Ingenhousz was unconvinced. They would probably have remained at loggerheads had Ingenhousz lived longer; he died in September 1799.

Jenner's main battles were with leading figures in London's medical establishment, notably the slippery George Pearson and what Baron called "the renowned triumvirate" of Benjamin Moseley, "Dr. R. Squirrel" and William Rowley.[74]

Pearson, a physician at St. George's, was the worst. Like Woodville, he soon realised the potential of vaccination, but only as a money-making career for himself. He found a separate source of cowpox (at Marylebone Fields) and set about cornering the market and squeezing Jenner into obscurity. Pearson was energetic and rapidly promoted himself as 'the' cowpox inoculator in London, presenting his work at Sir Joseph Banks' famous medical soirées and writing up his experience.[75] In March 1799, just six months after the *Inquiry* appeared, he wrote to 200 doctors across England, describing cowpox inoculation as if it were his own invention and enclosing a length of thread impregnated with cowpox lymph so that they could all try it for themselves.[76] Shortly after, he sent the same materials to doctors in Europe – Paris, Geneva, Hanover and Vienna – and New England. He made no mention at this time of Jenner.

Figure 9.3 George Pearson. © The Royal Society.

Plate 1 Lady Mary Wortley Montagu and her son Edward, dressed in Turkish clothes (detail). Reproduced by kind permission of the National Portrait Gallery, London.

Plate 2 Benjamin Jesty. Reproduced by kind permission of the Wellcome Library, London.

Plate 3 Edward Jenner. Reproduced by kind permission of the Edward Jenner Museum, Berkeley, Gloucestershire.

Plate 4 John Hunter. Reproduced by kind permission of the Royal College of Surgeons of London.

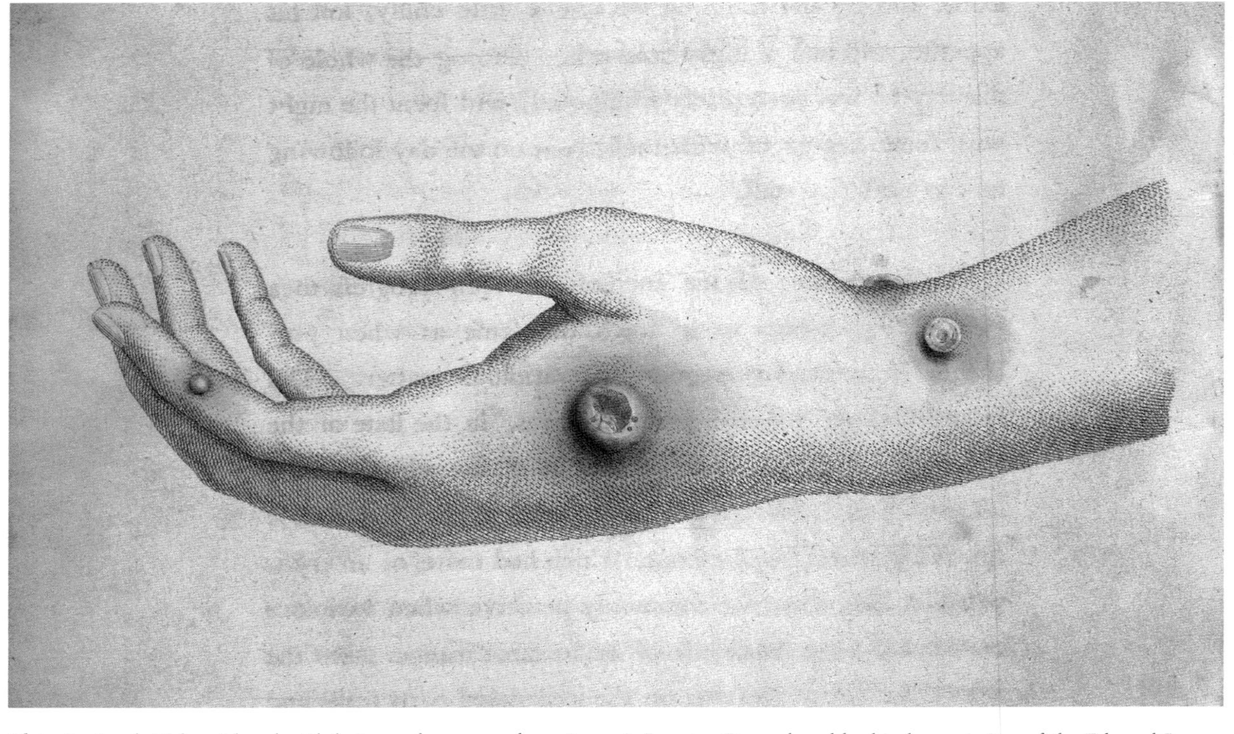

Plate 5 Sarah Nelmes' hand with lesions of cowpox, from Jenner's *Inquiry*. Reproduced by kind permission of the Edward Jenner Museum, Berkeley, Gloucestershire.

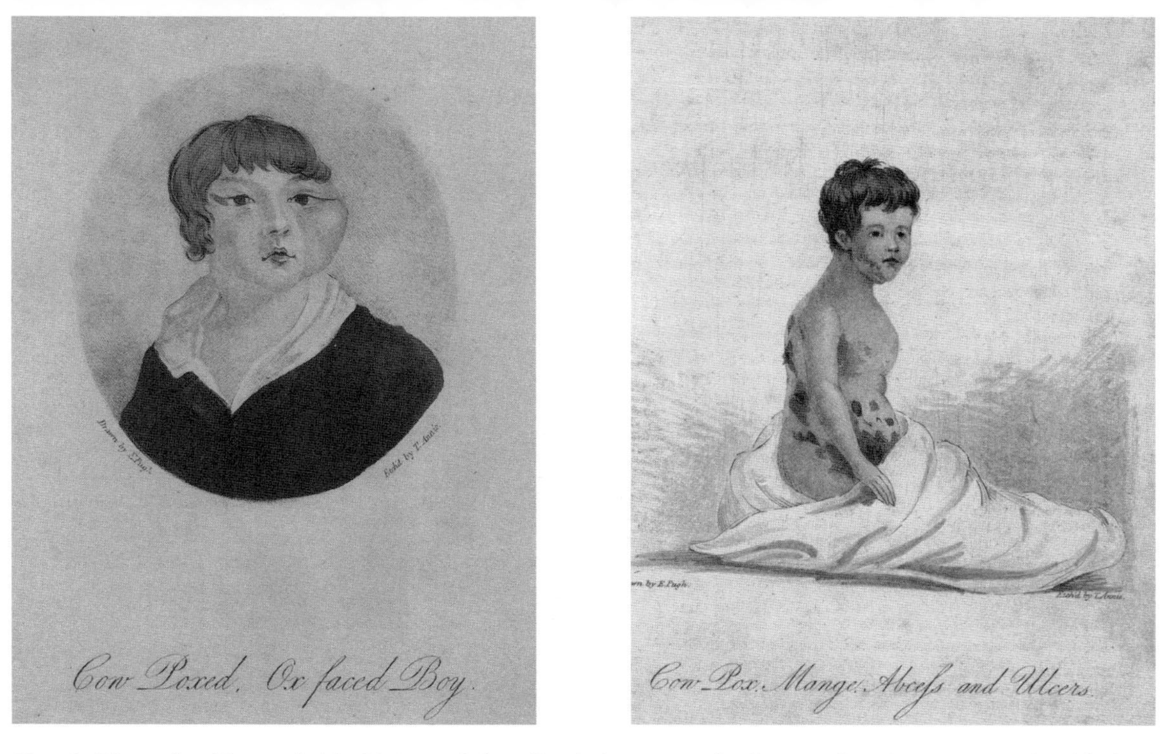

Plate 6 The ox-faced boy and girl with 'mange', from Rowley's paper on the dangers of vaccination. Reproduced by kind permission of the Wellcome Library, London.

Plate 7 Gillray's cartoon lampooning the imagined dangers of vaccination. Reproduced by kind permission of the Edward Jenner Museum, Berkeley, Gloucestershire.

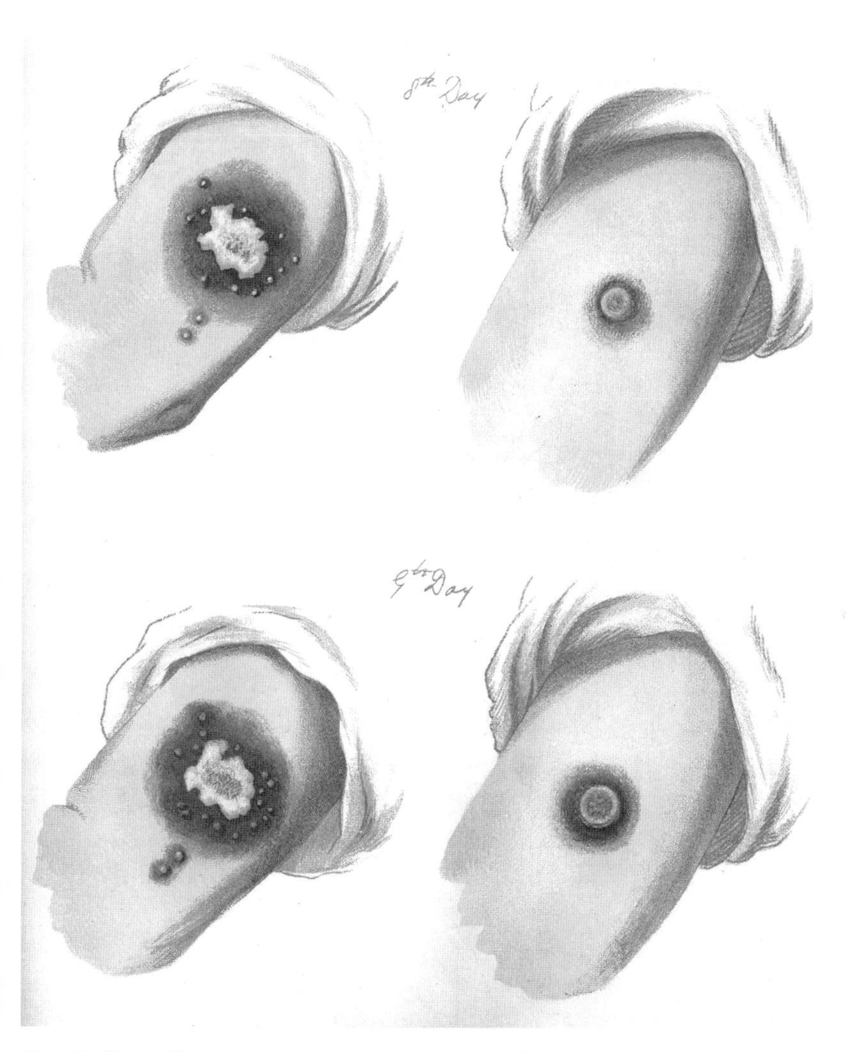

Plate 8 Kirtland's plates comparing the lesions of smallpox and cowpox, as published in the *British Medical Journal*'s issue celebrating the centenary of vaccination, 1898. Reproduced by kind permission of the Medical School Library, University of Bristol.

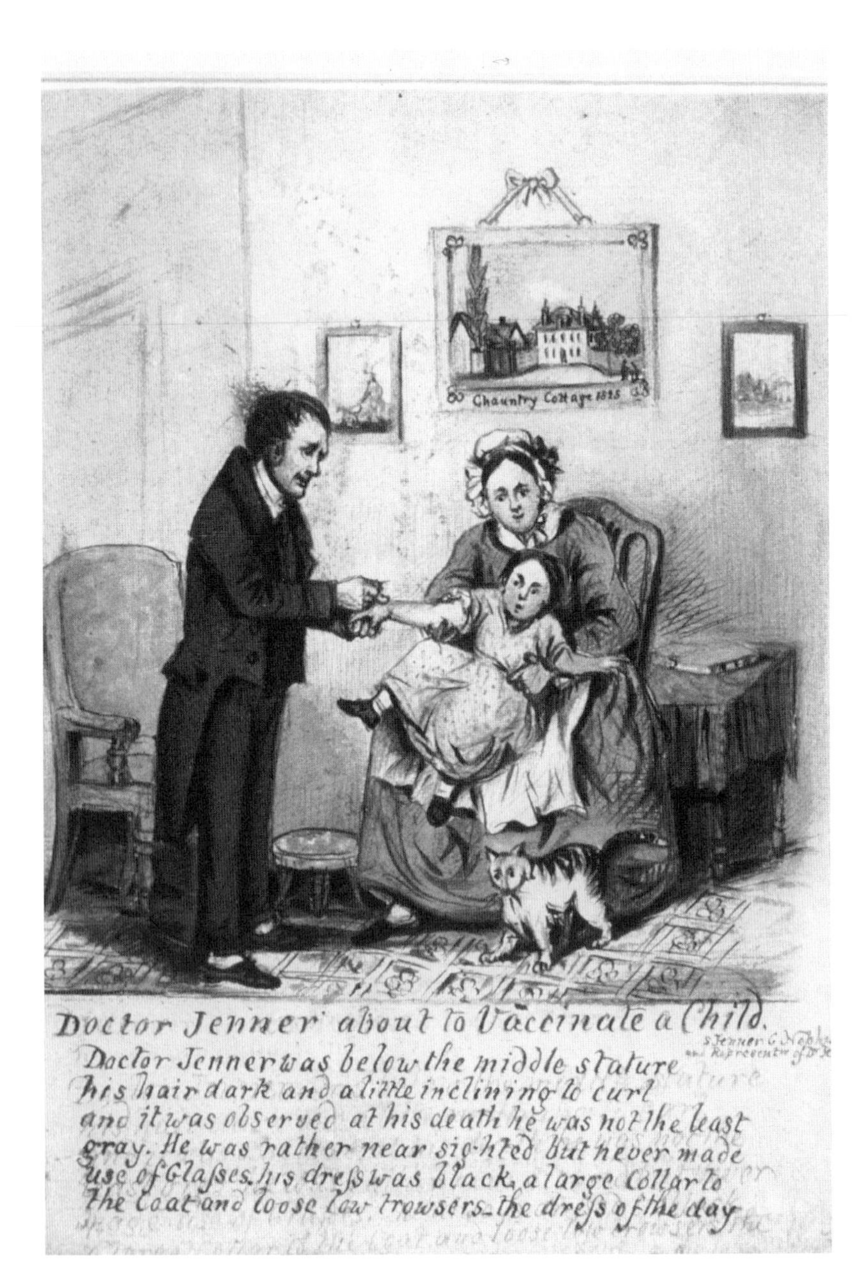

Plate 9 Drawing of Jenner vaccinating a child, by his nephew Stephen Jenner. Reproduced by kind permission of the Edward Jenner Museum, Berkeley, Gloucestershire.

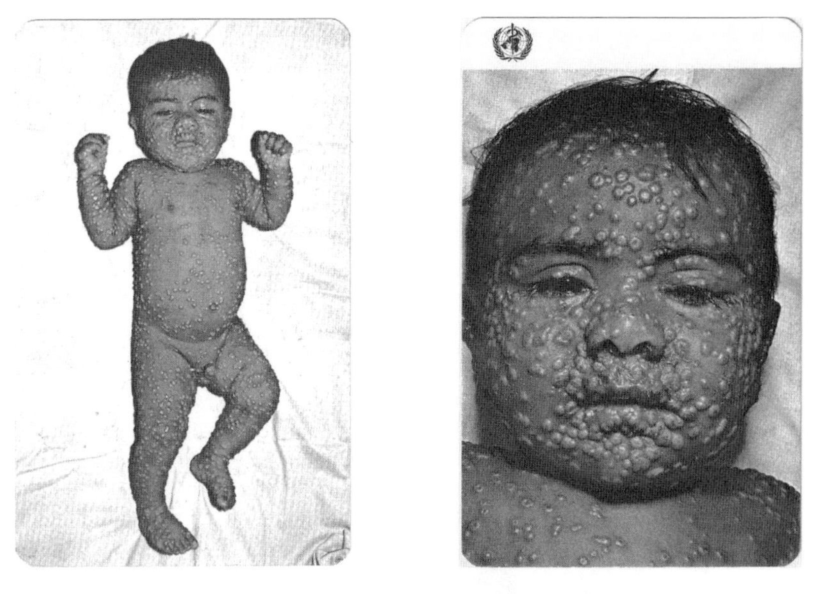

Plate 10 The WHO smallpox recognition card, used in the global eradication campaign. Reproduced by kind permission of the Edward Jenner Museum, Berkeley, Gloucestershire.

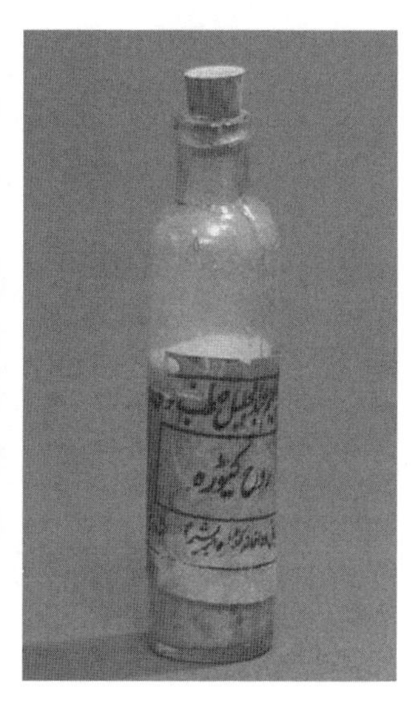

Plate 11 Bottle of variolation material from India, 1976. Reproduced by kind permission of the Centers for Disease Control and Prevention, Atlanta, Georgia.

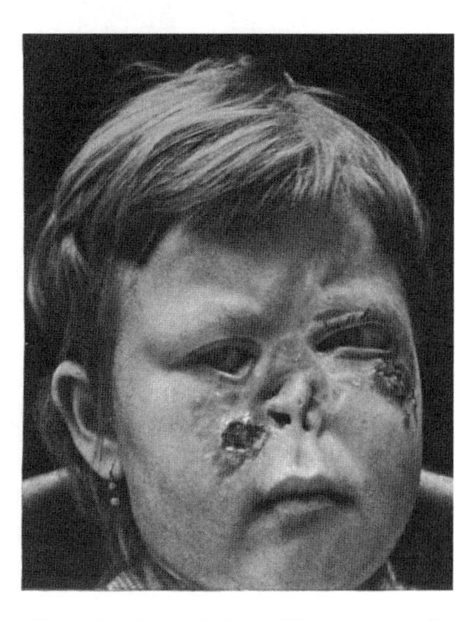

Plate 12 Congenital syphilis, showing the typical facial deformities in a girl. Reproduced by kind permission of the Wellcome Library, London.

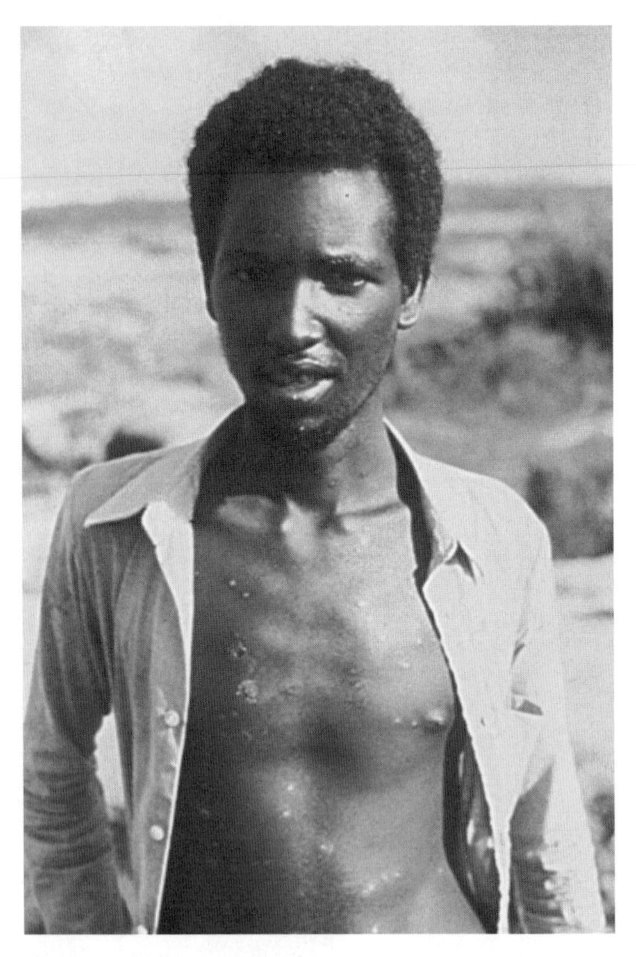

Plate 13 Ali Maow Maalin, the last victim of *Variola minor*, 1978. Reproduced by kind permission of The Centers for Disease Control and Prevention, Atlanta, Georgia.

Pearson's other great propaganda coup was to set up the Vaccine-Pock Institution, superficially modelled on the philanthropic Smallpox and Inoculation Hospitals but with one particular beneficiary in mind: George Pearson, self-appointed Director of the Institution.[77] He did this behind Jenner's back and completed the insult by offering him a role, not on the Board but as an honorary "Consulting or Corresponding Physician"; Pearson generously offered to waive the one-guinea subscription because Jenner would not be referring patients there. Jenner, angry and becoming paranoid (with good cause) that people were out to steal his recognition, replied with frosty restraint: "For the present, I must beg leave to decline the honour intended me." Pearson's campaign to undermine Jenner continued for many years and surfaces again later.

The Moseley–Squirrel–Rowley triumvirate, who "would have been quite laughable had they not been altogether founded in falsehood",[78] attacked Jenner by inventing dangers of cowpox. 'Dr. Squirrel' (whose real name was John Gale Jones) claimed that cowpox caused scrofula, a chronic foul discharge from infected glands in the neck that is actually due to tuberculosis.[79] Moseley and Rowley went further, by describing the horrific consequences of introducing the 'bestial humour' of cowpox into humans. Moseley invented a clever alternative to *Variolae vaccinae*, Jenner's Latin name for cowpox: *Lues bovilla* or 'syphilis of cattle', which piled on all the loathsome baggage of the 'great pox'.[80]

Rowley made more of this theme, with illustrated 'case histories' showing the unthinkable things that cowpox did to previously normal children. A boy from Peckham had started running around on all fours, bellowing like a cow. Another lad grew bumps on his forehead and his face (according to the artist's impression) took on a bovine appearance. A girl (Plate 6) had developed a nasty skin condition, diagnosed as cattle mange.[81]

Incredibly, many people believed this nonsense. The reports, coming from respectable doctors, reawakened primitive beliefs that animals could pass on their traits to humans. Paré, the sixteenth-century pioneer in surgery, had described fantastic creatures spawned by unions between man and other species, such as the goat-boy hybrid produced by a nanny goat that succumbed to the "*désir brutal*" of an errant priest.[82] Medicine had moved on but the dread of bestialism persisted. Ludicrous though these were, the reports by Moseley and Rowley helped to fuel public suspicion and fear of vaccination. This was manifested by the villagers of Hadleigh who stoned and chased away people who had been vaccinated.[83]

The stupidity of this notion was pointed out by Jenner and his supporters but the most effective foil was satire. In 1802, Gillray – one of the masters of the political cartoon – parodied the "wonderful effects" of cowpox, showing cows' parts sprouting out of all imaginable places on the recently vaccinated (Plate 7). Gillray was not mocking vaccination but rather those gullible enough to believe Moseley and Rowley. As well as its over-the-top humour, the cartoon caricatures the leading lights of vaccination – not Jenner, but Pearson (the stony-faced vaccinator) and Woodville (the boy assistant clutching a bucket of lymph, identified by 'St. Pancras' on his armband). In the same vein, William Ward lampooned Rowley in verse:[84]

> There nibbling at Thistles, stand Jim, Joe and Mary,
> On their foreheads (O horrible!) crumpled horns bud,
> There Tom with a tail, and poor William all hairy,
> Reclined in a corner, are chewing the cud.

Fortunately, Jenner had supporters. Henry Cline, a fellow-lodger at the Hunters' town-house and now a leading London surgeon, tried out some dried lymph that Jenner had collected from the vaccination blisters on Hannah Excell's arm (illustrated in the *Inquiry*). His subject was a boy with a painful hip. Cline wondered if the pain would improve with the counter-irritation of vaccination. It did not but when Cline then variolated the boy out of curiosity, there was no reaction. This experiment took place before the *Inquiry* was published and Cline helped to smooth the article's initial reception.[85]

A powerful ally was John Coakley Lettsom, the terrifyingly effective Quaker physician who fired off salvos of solid medical reasoning at Jenner's critics, sometimes under the anagram pseudonym of J. C. Mottles.[86] Lettsom also introduced Jenner into the medical establishment in London. He even arranged for him to be added to the official etching of the leading lights of the influential Medical Society of London. This meant cutting a hole in the engraving plate and patching in Jenner's image; in the doctored print, a rather ghostly Jenner floats beside the President's shoulder.[87]

Not all of Jenner's supporters helped his cause. John Ring, a surgeon from St. Thomas' Hospital and a dedicated vaccinator, began by collecting over 100 signatures from "nearly all the professional lights of London" who certified that Jenner's

Figure 9.4 The officers of the Medical Society of London, by Medley, with Jenner's head and shoulders added retrospectively. Reproduced by kind permission of the Medical Society of London.

invention was "a lifelong safeguard" against smallpox. He then turned to blasting Jenner's critics with heavy sarcasm. Unfortunately, some believed that Ring's vitriol came from Jenner and, to the great surprise and hurt of his devoted disciple, Ring eventually found himself disowned by Jenner.[88]

Two quieter landmarks were laid down in the busy years after the *Inquiry* appeared. The word 'vaccine' soon came into general use and 'vaccination' was coined in 1800 by Richard Dunning, a surgeon in Plymouth.[89] Also, the first references were made to the possibility that vaccination could eventually be used to exterminate smallpox. In 1801, Jenner wrote in *The Origin of the Vaccine Inoculation*:[90]

> It now becomes too manifest to admit of controversy that the annihilation of the Small Pox, the most dreadful scourge of the human species, must be the final result of this practice.

Jenner also wrote in the same work:

> The distrust and scepticism which naturally arose in the minds of medical men, on my first announcing so unexpected a discovery, is now nearly disappeared.

As will be seen, his first comment came true but his second turned out to be hopelessly naïve.

10 Crusaders and Infidels

While the bickering continued in London, vaccination set off to conquer the rest of the world. The main message to break out of England was one of excitement and hope: here was a powerful new weapon in the eternal battle against smallpox, purged of the dangers of variolation. Outside England, the natural response was gratitude for Jenner's discovery and the will to exploit it without delay.

Vaccination spread out rapidly from England and within a few years had all but encircled the globe. Vaccinators were active in Austria and Switzerland by 1799, in France, Spain, Germany and North America in 1800, followed by Sweden, Denmark and Russia in 1801, and India a year later. By 1805, vaccination was taking hold in South America, the Philippines and China.[1]

As with variolation half a century earlier, the advance of vaccination was spearheaded by passionate and persuasive evangelists, some of whom were prepared to risk their reputation, health and life for the cause. The forces helping them along included royal and presidential patronage, religious coercion and some ingenious fraud. And as before, their main obstacle was human nature.

Global ambitions

Jenner, Lettsom and Pearson all helped to kickstart the spread of vaccination worldwide. Soon after the *Inquiry* was published, Jenner sent a copy and some dried lymph to John Clinch, a medical missionary in Newfoundland who had been a friend since their schooldays together in Cirencester. It was a good time to test vaccination there as smallpox was rife. Clinch followed Jenner's instructions and vaccinated a reluctant teenage nephew and then, in the tradition of Sloane and Newgate, confined the boy to bed with "one of the worst cases of small-pox at that time under his attention". The boy remained unmarked, local excitement took off and Clinch went on to perform several hundred vaccinations before his supply temporarily ran out.[2]

Lettsom also fired up interest in North America by sending a copy of the *Inquiry* in 1799 to Benjamin Waterhouse, Professor of Medicine at Harvard University. Waterhouse immediately abstracted Jenner's work into an article in the local *Colombian Sentinel* of 12th March 1799 entitled, *Something Curious in the Medical Line.* [3] He had to wait to receive lymph from England, as cowpox (he called it "kine-pox") was then unknown in America. Waterhouse was a fortunate man, with enough children (13) to conduct a small-scale clinical trial. He vaccinated seven of them (one of whom failed to take), then made them all play and sleep together in order to check that inoculated cowpox was not contagious. Protection against smallpox was later tested by a twelve-year-old Daniel Waterhouse in a grisly visit to Boston's Smallpox Hospital. There, Daniel was inoculated three times with fresh smallpox pus and left in a room with a seriously ill smallpox patient. Worryingly, his vaccinated arm became red and inflamed but this soon settled and smallpox did not break out. Waterhouse was instantly converted: "One fact is worth a thousand arguments."[4]

It was also worth a lot of money, even though this was against the open-access spirit of Jenner and the *Inquiry*. Waterhouse, the only person then receiving lymph from England, quickly built up a highly profitable vaccination business that initially made money on a Suttonian scale. He was only persuaded to share his supply of lymph after a desperate colleague resorted to vaccination from a pustule that he believed to be cowpox; it turned out to be smallpox and killed 68 people.[5]

Philanthropy eventually got the better of Waterhouse. In spring 1801, he wrote to inform the newly-elected President Thomas Jefferson and sent him some lymph. Jefferson was excited, vaccinated his family and slaves and distributed the lymph and the new wisdom among his neighbours, including some Native American Indians.[6] Thanks to Jefferson's patronage, vaccination was rapidly accepted in New England and the Southern states – except for some pockets of resistance, such as the Anti-Vaccination Society founded in Boston in 1798 by physicians and clergy.[7]

Back to London and Pearson who, for all his underhand motives, sent ripples far around the world with his multiple mailings of threads impregnated with lymph. He triggered one spectacular chain-reaction that carried vaccination across thousands of miles to three countries and another continent. The recipients of Pearson's lymph in spring 1799 included the equally energetic Jean de Carro, a Geneva physician working in Vienna. De Carro immediately began vaccinating there and also sent lymph to Thomas Bruce, Earl of Elgin, who had not yet begun plundering marble friezes from the Parthenon and was British Ambassador in Constantinople. Lady Mary Wortley Montagu would have approved as Bruce promptly vaccinated his son and spread the word – not just in Turkey but also to India.[8] The lymph that Bruce sent to India survived the long trip (which had earlier defeated attempts by Jenner to introduce vaccination to that country) and was used to vaccinate a girl in Bombay in 1802. This single case began a chain-reaction of arm-to-arm vaccination which ultimately spread across India. The country was later divided into four areas, each administered by an English doctor who headed a network of local teams of Indian vaccinators. Progress was checked by detailed monthly returns. This pioneering hierarchical system proved highly effective and was later adopted in other countries.[9]

Some unexpected difficulties were encountered that required imaginative solutions. Cows were sacred to the Brahmins and vaccination was therefore sacrilege. Fortunately, a historic Sanskrit manuscript was unearthed which showed that cowpox had been known to the ancient Brahmins and that they had even practised vaccination using the fluid from cowpox blisters. The Brahmins were reassured and accepted vaccination. They might not have done had they known that the venerable text had been concocted for that purpose by a Mr. Ellis, an expert in Sanskrit literature, and written out on artificially aged paper in a hotel room in Madras.[10] In Ceylon (Sri Lanka), many people revered the smallpox god Patina and her victims and so refused vaccination; others were suspicious of any intervention which left a permanent scar that could mark individuals down for some sinister British purpose.[11]

Most of Europe had yielded to vaccination by 1801. In 1800 France was back at war with England, but this did not prevent Woodville from being granted exceptional leave to bring in his knowledge and a supply of lymph. Woodville, sent as Jenner's representative, was accompanied by Dr. Antoine Aubert of the newly formed Comité Central de Vaccine in Paris.[12] Apart from Woodville's temporary detention at passport control in Boulogne and the failure of some of his lymph to work, the trip was a great success. The Comité Central oversaw the introduction of vaccination throughout the country, although the armed forces were not treated until after the Battle of Trafalgar in 1805.

The Mediterranean theatre of war provided the backdrop for an extraordinary mission involving two maverick doctors and HMS *Endymion* (whose career would be completed as one of the three smallpox hospital ships moored on a forlorn stretch of the Thames Estuary). Joseph Marshall, physician of Eastington near Berkeley, and John Walker, ship's surgeon, sailed from Plymouth on 1 July 1800 to vaccinate British garrisons in the Mediterranean.[13] This they did in Gibraltar, Menorca and Malta, where Marshall stayed on to coordinate vaccination of the local population. His return to England was delayed by a series of humanitarian adventures. Smallpox had broken out at Palermo and in response to an appeal from the Viceroy of Sicily, Marshall diverted there. He received a rapturous reception and was detained for some months because of the wish of most of the population to be vaccinated. Word spread to Naples, the next stop in Marshall's crusade, where with royal approval he set up a Jennerian Institution to train Italian vaccinators. His pupils were surgeons from all the provinces of Italy, obliged by royal command to learn and spread the knowledge at home. Marshall's final diversion in 1801 was with Sir Ralph Abercromby's expedition to chase the French out of Egypt.[14]

In northern Italy, it was the Church rather than wandering Englishmen which helped to convince the population of the benefits of vaccination. The phenomenal success of Dr. Luigi Sacco in bringing vaccination to Lombardy was due both to the energy of himself and his team and a mass propaganda campaign, with proclamations "from every pulpit" that urged the devout to follow God's guidance and get themselves and their families vaccinated. The result: 70,000 vaccinations in three years and, according to Sacco, the elimination of smallpox from Lombardy – perhaps the first example of the deliberate and successful eradication of smallpox from an entire region.[15]

To the north, vaccination infiltrated Sweden, Denmark and Russia in 1801. In contrast to England, its reception was enthusiastic, at least at the top of society. The Swedish Medical Board soon proposed that it should be universally applied – but with strong encouragement rather than the legally-enforced compulsion that later created havoc in England and some other countries. In Russia, the Dowager Empress followed the example of her ancestor Catherine the Great and took a personal interest, organising a high-profile celebration of the first Russian child to be vaccinated. This abandoned baby girl had her life transformed: a ride to a new and better orphanage in an Imperial coach and a pension for life, which probably compensated for the name which the Empress chose for her: Vaccinoff. The Empress later expressed her admiration and gratitude to Jenner in a cordial letter in French (signed "Marie") accompanied by a flashy diamond ring.[16]

Large gaps persisted in vaccination's map of world conquest. In China, vaccination spread slowly inland from two coastal foci – Canton and the former Portuguese possession of Macau – but these were tiny nibbles out of the vastness of the country. Parts of the interior had to wait for over a century and vaccination only reached Tibet in 1943. [17]

Japan had a long-standing academic interest in smallpox and had established the world's first chair dedicated to the study of the disease in Tokyo in 1798. However, the advent of vaccination caused no obvious reaction. European evangelists set up local vaccination networks – in Deshima by 1810 and then in Nagasaki 50 years later – but interest collapsed when these individuals moved on. This was blamed on "universal apathy" in these pre-Reformation times rather than antagonism to vaccination. It was only in 1874 that vaccination was finally brought in. Japan then made up for lost time, setting up a Vaccine Institute in Tokyo headed by a Minister of Public Health and annexed to a dedicated lymph production farm which distributed supplies throughout the country. In 1876, vaccination was made compulsory at the ages of twelve months and again at seven years.[18]

Another nearly blank canvas was Africa, one of the Angel's favourite stamping grounds. The continent was omitted completely from the *British Medical Journal*'s triumphant review of a century of vaccination in 1896. In fact, small and isolated patches of the canvas had been blocked in since the 1820s: the Cape, Egypt, Sudan and Madagascar.[19] The main obstacles were the dispersal of populations across huge distances and the scarcity of vaccine lymph. Cowpox did not occur naturally in Africa and supplies sent from Europe often perished in the heat during shipment.

Spreading salvation to the Colonies

The first vaccinations in Spain, performed in late 1800, triggered one of the most spectacular vaccination missions of the time. King Carlos (Charles) IV was already sensitised to smallpox. His brother and sister in law had died of it and several close relatives had been badly scarred by variolation. Also, Spain was still living with the collateral damage it had inflicted during its conquest of South America nearly three centuries earlier. Smallpox was still rampant in Spanish possessions in the Caribbean and South America, where epidemics flared up periodically, hacking down the population and badly denting their contributions to the Treasury. The King had read Jenner's *Inquiry* and other texts on vaccination and commanded that a Royal Philanthropic Expedition be set up to take the miracle of vaccination to the Colonies.[20]

The motive may have been financial as well as philanthropic but the concept was ground-breaking. Detailed plans were drawn up by Francisco Javier de Balmis, who like John Hunter carried the title of Surgeon-Extraordinary to the King. De Balmis appointed Joseph Salvany as his deputy and assembled a team of physicians and nurses. The Expedition set sail in three frigates from La Coruna on 20 November 1803, with 500 copies of a French tract on vaccination which de Balmis had translated into Spanish, and a vaccine production facility that would guarantee a fresh supply throughout the trip. This facility consisted of 22 orphaned boys from La Coruna aged eight to ten, under the care of the Director of the orphanage, Isabel de Zendala y Gomez, who was the only woman on the voyage. Every ten days, a fresh pair of boys were vaccinated using lymph from their predecessors, forming a chain that carried vaccination successfully across the Atlantic. Lymph was also collected and preserved between glass slides that were sealed with paraffin and stored under a partial vacuum.[21]

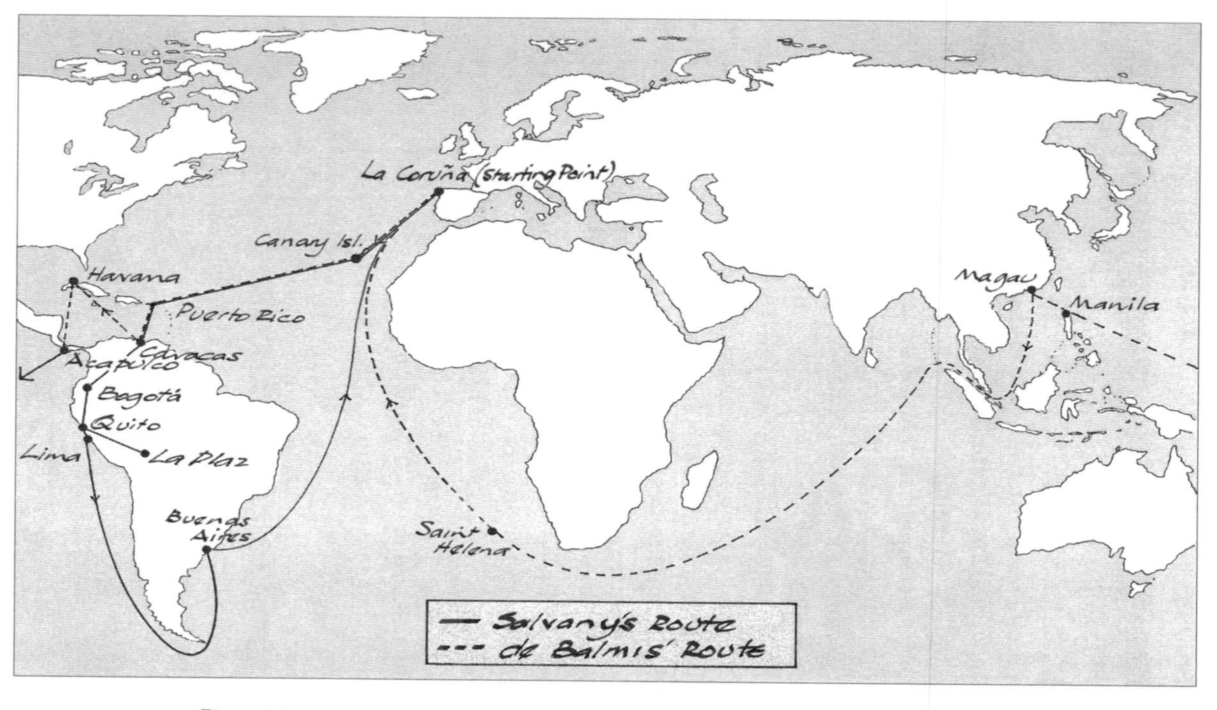

Figure 10.1 Routes taken by the Royal Spanish Philanthropic Expedition of 1803–06.

The Expedition lasted almost three years, circumnavigated the world and covered a total distance of nearly 50,000 miles.[22] It first sailed to the Canaries, then to Puerto Rico and Caracas in Venezuela, where the Expedition divided into two. Between them, these forces retraced the routes of the original Conquistadores. De Balmis followed Cortéz's trail to Mexico, crossing to the Pacific coast where an updated brief from the King told him to continue sailing west to the Philippines and then to the Spanish outposts in China at Canton (Guangzhou) and Macau. Taking in St. Helena on the way, de Balmis eventually reached home in early September 1806, and on the 7th of that month had the privilege of kissing His Majesty's hand in recognition of his Expedition's achievement. Meanwhile, Salvany's half of the Expedition had followed in Pizarro's footsteps, heading south through Colombia, down the coast of Peru to Argentina and Tierra del Fuego, then home across the Atlantic.

Both groups operated in the same way, vaccinating as many people as they could at each stopover and, crucially, establishing a self-sustaining vaccination centre in major cities, staffed by "the most zealous practitioners" and watched over by "the highest authorities" in government and Church. Continuing supplies of vaccine were ensured by teaching the locals arm-to-arm vaccination and, when the opportunity presented itself to Salvany in Colombia, by arm-to-cow inoculation. The Expedition was astonishingly productive. Several thousand people were generally vaccinated at each port of call, with over 200,000 in Peru. By the time the Expedition returned to Spain, its legacy of stand-alone vaccination networks had treated millions more.

The Expedition also had a huge emotional impact. It was a winning combination: the magical gift of vaccination, the inspirational voyage that had brought it and the royal patronage which proved that Spain really cared about its people in the Colonies. De Balmis and Salvany created a sensation almost everywhere they went, with emotional welcomes for the heroes and especially the young boys who had risked their lives by sailing halfway around the world to bring salvation. Even the *British Medical Journal* was moved (after its own fashion) and reported the "many affecting scenes ... and thanks to God for having been the witnesses to so happy an event".[23] Naturally, the warm welcome was also extended to vaccination. De Balmis' vaccination chain was renewed in Mexico where the Spanish boys resumed their childhood under the care of a Bishop, and 25 Mexican orphans were taken on board for the trip across the Pacific. They were later repatriated to be educated at the expense of the Spanish Court.

There were also surprises and disappointments. The speed with which Jenner's invention was spreading across the world was demonstrated at their first port of call in the Canaries and later in Lima, Peru: vaccination had already been brought there by other Spanish travellers. In Lima, their reception was cold because vaccination had already been taken over by a money-making cartel of doctors who were hostile to any threat to their monopoly. The trip also took its toll. Salvany's ship was wrecked on the coast of Colombia in early 1804 and he later lost the sight in one eye and caught tuberculosis. De Balmis lost 20 members of his crew during a storm on the crossing to Macau.[24]

The Royal Philanthropic Expedition deserves to be recognised as a masterpiece of ambition, planning and execution. It was also partial recompense for the curse of smallpox which the Spanish had originally brought to South America. The work that it began was completed a century and a half later when South America was finally declared free of smallpox (see Chapter 15). This was the result of the World Health Organisation's global eradication campaign, for which the Expedition was the original template.

De Balmis and Salvany also deserve their place in history. When de Balmis returned to Spain in 1806, the world had moved on and Nelson had defeated the combined French and Spanish fleet at Trafalgar the previous autumn. De Balmis may have kissed the King's hand but his courage and brilliance were soon forgotten. He died in obscurity a few years later. Salvany never even made it home; his tuberculosis killed him before he reached Argentina.

Cash cows

The rapid spread of vaccination across the world highlighted some practical difficulties – especially with the supply and preservation of lymph – but also recruited many more minds to find solutions.

Natural cowpox was always in short and unreliable supply and apparently absent from large areas of the world, including North America and Africa. Even in dairy areas, its appearance was sporadic. Jenner's sources in Gloucestershire dried up four months after the *Inquiry* was published and he had to use Woodville's.[25] (Cowpox remains a rare disease today and is more likely to be caught from a cat than from a cow.)

Good preservation helped to eke out stocks of vaccine and was essential if it needed to be transported for weeks or months and in the heat. The virus survived well if dried – like the powdered smallpox scabs which remained viable for variolation during many months. Jenner collected lymph from Hannah Excell's blisters into a feather quill cut obliquely like a toothpick and allowed it to dry; Henry Cline used this material successfully in London several weeks later.[26] Other vehicles for dried lymph included lancets, 'points' like the teeth of a comb made of ivory, and impregnated threads as sent out by Pearson. Threads were inserted into a cut in the recipient's skin and covered with a dressing, while the other instruments were rubbed into scratches and often used to vaccinate more than one subject. More sophisticated storage methods were needed for overseas distribution, such as drying the lymph between glass plates, sealed around the edges with paraffin or wax, as used by de Balmis. Liquid vaccine was easier to handle and 1891 saw the welcome discovery that mixing lymph with glycerine markedly extended its shelf-life and also killed off many of the skin and farmyard bacteria that contaminated it. Glycerinised lymph remained viable for months even in extremes of temperature, especially if kept in the dark and ideally sealed to exclude air.[27]

Much effort went into growing up new sources of vaccine. The arm-to-arm propagation which Jenner first tried out (his son received lymph from Hannah Excell's arm) was useful to carry fresh vaccine over long distances or to start up vaccination programs, as used in de Balmis' Expedition. However, there were risks that human diseases could be spread from donor to recipient, and the later realisation that syphilis could be transmitted in this way was a propaganda gift for the anti-vaccinationists.

The cow (replaced by water buffaloes in the Philippines) was an obvious target for exploitation and could be inoculated either from an infected cow or a human vaccination blister. From the early 1800s, the Italians led on the large-scale production of calf vaccine and this was taken up across the world after the method was demonstrated at an international medical conference in 1846.[28] Rather than having to rely on a few lesions inconveniently placed around the udder, the animal's flank was shaved and lymph rubbed into an array of vertical cuts. When the blisters appeared, lymph or the crusts into which it dried were collected and either used on the spot or cleaned up (crudely) and preserved by drying or mixing with glycerine. Heifers under one year old were generally used, because bovine tuberculosis (which could be spread by vaccination) was rare before that age. A good calf could produce enough lymph for 200–300 effective doses, or up to ten times that number if the lymph was mixed and diluted with glycerine.[29]

Vaccination calves were soon at the heart of a booming new business. Before long, they were being led through the streets of Paris and other European cities as mobile vaccination centres or raised in special farms (for example in America, Japan and England) run either as independent commercial concerns or government institutes that supplied vaccine nationally.[30] Some vaccine producers looked after their animals well and checked them at post-mortem to make sure that they were free of tuberculosis and other illnesses before releasing the vaccine. Others badly neglected their animals and sold them on for meat after their vaccine-producing days had ended. Bacteriological sterility and quality control came later; until then, various infections could be readily spread by vaccination.

The act of vaccination was so simple that a medical qualification was superfluous – as illustrated by the vast numbers of unskilled local people who were quickly trained up to be vaccinators in places as far apart as Italy, India and South America (and much later, during the World Health Organisation's global eradication campaign of the 1970s). In England, the clergy took an increasingly active role in promoting and even performing vaccinations. They took their cue from leading churchmen such as the Reverend Rowland Hill, a strong defender of Jenner who gamely traded insulting pamphlets with Moseley.[31] Hill also practised what he preached and became an expert vaccinator, as did Jenner's friend the Reverend Robert Ferryman. Some clerics even set aside a day each week to vaccinate.

The fact that almost anyone could vaccinate and that it did not need any specialised preparation or aftercare meant that it was difficult for doctors to monopolise. Vaccination would never bring in the absurd income that variolation could generate, although some doctors and vaccine-producers were able to corner lucrative markets with particular 'refinements' to the vaccine recipe or the instruments for administering it. The cartels that grew up to spread vaccination and protect the financial interests of doctors and vaccine-producers later attracted the anger of the anti-vaccinationists in America and Europe, especially when it seemed that the risks of vaccination were being deliberately covered up (see Chapter 12).

The simplicity of vaccination was also deceptive. Going through the motions was often not good enough, as protection against smallpox depended on the operator's skill as well as the viability of the vaccine. Experts like Jenner, Pearson and Sacco took care to ensure that the vaccine made good contact with the inner surfaces of the cuts in the skin. If possible, they also checked a few days later for the characteristic nodule and redness that indicated a successful 'take', and repeated the vaccination in cases of doubt. Because of poor technique or inactive lymph, vaccination frequently failed. Vaccinated people who went on to catch smallpox provided much useful ammunition for the enemies of vaccination. It was only later that systematic attempts were made to gauge the success or failure of a vaccination from the scars it left behind. Large or multiple scars were mementos of a vigorous local reaction that almost always indicated full immunity, whereas a small or absent mark often meant that the 'vaccinated' individual was not protected at all.[32]

An even more important reason that vaccination sometimes failed was the unfortunate fact that it had a limited duration of action. Jenner had assumed, by over-extrapolating from his few cases of natural cowpox, that inoculated cowpox would protect for life.[33] He clung stubbornly to this belief even after convincing evidence emerged that immunity often wore off after several years. His refusal to accept that he was wrong helped to fuel opposition to vaccination and undoubtedly cost a proportion of the lives that he set out to save.

Doing good?

"As for the fools and fanatics who deny Jenner, and seek to undo his work, it is only charitable to believe that they know not what they do."
British Medical Journal, Jenner Centenary Issue, 5 May 1896

The *British Medical Journal*, in its 1896 Centenary issue to commemorate Jenner's first vaccination, attempted to "gather into a few lines the teaching of the century".[34] The headline statistics were overwhelmingly convincing and the bottom-line message just as clear: vaccination was now widely practised and was causing dramatic falls in deaths from smallpox. Also, national differences in the uptake of vaccination provided a natural test-bed to see how effectively vaccination could chase away the Angel of Death.

In some countries such as Spain and Russia, uptake remained very low despite the best efforts of government. Here, mortality remained close to the historical pre-vaccination level of about 2,000 deaths per year per million of the living population. Mortality rates were over 2,000 in the Spanish provinces of Almería and Murcia and a disappointing 1,000 per million in de Balmis' native La Coruna.

Compare with the dramatic stepwise reductions in mortality in countries where vaccination was made 'permissive' (encouraged but not legally enforced) or compulsory. Those with permissive vaccination included Belgium and Austria and showed smallpox mortality rates of around 500 per million, a quarter of that before vaccination. With compulsory vaccination, the death rate tumbled to below 100 per million and was close to zero in Sweden by the time that vaccination celebrated its hundredth birthday. Indeed, the *British Medical Journal* claimed that the effect was so obvious that it was possible to tell whether compulsory vaccination was in place just by looking at the smallpox mortality rate.[35]

However, all things become clearer in retrospect and the first frenetic decade of vaccination was driven by expectation rather than hard evidence that it was doing good. This is the atmosphere of optimism but uncertainty which takes us back to England in the early years of vaccination and to Jenner, still defending his corner against the nastiness of Moseley and Pearson – and his own shortcomings as a scientist.

A prophet unrecognised in his own land

Jenner had been badly bruised by the enemy. Pearson's campaign to supplant him had picked up momentum when the Duke of York became the patron of his Vaccine-Pock Institute, which began operating near Golden Square in early 1800.[36] Around this time, Moseley and his associates were trying to raise the spectre of the "bestial humours" that could turn people into cows, and other dangers of vaccination. The thin-skinned Jenner was shocked and hurt by the intensity and breadth of the attacks against him. Benjamin Travers, the President of the Royal College of Surgeons, tried to put this into perspective for him: "Your liberality and disinterestedness everyone must admire and extol, but you are sadly deficient in worldly wisdom."[37]

Lettsom and others fought back to get Jenner the recognition they felt he deserved. In 1802, he was encouraged to apply to Parliament for an honorarium to reward him for his discovery.[38] Jenner's medical practice and income had undoubtedly suffered because of all the time and energy he was having to put into defending vaccination, but his main interest was probably the signal which the award of a government grant would send out. An unequivocal statement that he had invented vaccination, backed up by the full authority of Parliament, would help to silence his critics and kill off Pearson's attempts to undermine him.

There were precedents in the grants which Parliament had made to British inventors who had lifted the nation's wealth or health. John Harrison had received £8,000 in 1773 for his highly accurate chronometer which at last made it possible to calculate longitude and therefore pinpoint a ship's position at sea.[39] There was little relationship between the amount awarded and the value of the inventions. Notable duds included payments of several thousand pounds for a patent medicine to dissolve bladder stones and a chemical fumigant, both of which turned out to be useless.[40]

Jenner presented his petition to the House of Commons in March 1802. Thanks to the vindictiveness of his enemies, consideration of the application took several weeks and was an unpleasant process. Pearson tried hard to prove that others had discovered vaccination before Jenner, dragging up various anecdotal and poorly-documented accounts.[41] His best example would have been Benjamin Jesty, but Pearson got his name wrong and could not gather enough information in time. Pearson also pointed out that Jenner's paltry number of vaccinations had been far outstripped by the hundreds done by himself and Woodville. It followed that they, not Jenner, had built up the clinical experience needed to broadcast the technique to the world. Woodville, although on cool terms with Jenner, distanced himself from Pearson's assault. Others who attacked Jenner included John Birch, a successful London surgeon who hated everything about vaccination from the outset. Birch believed that vaccination was unnatural and damaging, because if it did work (which he doubted) it could undermine the valuable role that smallpox played in killing off the children of the lowest classes, as "a merciful provision on the part of Providence to lessen the burden of a poor man's family".[42]

Jenner's petition was mostly aspirational, as no solid data had yet emerged to show that smallpox deaths were falling following the introduction of vaccination. He again claimed, without evidence, the "singularly beneficial effect of rendering through life the persons so inoculated perfectly secure from ... smallpox".[43] Those who spoke for Jenner included William Wilberforce, later famous for leading the movement to abolish slavery, and the King's Physician, Sir Walter Farquhar, who declared: "I think it the greatest discovery that has been for many years."[44] Dr. Matthew Baillie, a physician at St. George's Hospital, went further: "It is the most important discovery that has ever been made in medicine."[45] Other supporters invoked mathematics and money. Vaccination had already saved an estimated 40,000 British men's lives. Assuming that each life was worth £5 to the Treasury (roughly Dr. Jenner's postal expenses for a week), Jenner had already spared the country £200,000, and so an award of £10,000 or even £20,000 was trifling by comparison.[46]

In the end, the House of Commons Committee voted to award Jenner an honorarium of £10,000 (an amendment to give him

£20,000 was only narrowly defeated). This was a large sum, but close to the annual earnings that he had been promised if he moved his practice to London, and perhaps one-tenth of the income that he could have earned from vaccination if he had protected it with Suttonian-style secrecy. The Chancellor of the Exchequer also pointed out that Jenner had already received the greatest reward of all – the unanimous approval of the House for the value of his work. Jenner was delighted and his enemies disgusted.[47]

Over the next few years, evidence that vaccination worked and prevented deaths from smallpox flowed in from around the world and, more importantly for the warring factions in London, from the capital itself. According to the London Bills of Mortality, the death rate from smallpox was plummeting. Before vaccination, it had hovered at around 3,000 deaths per year per million but had fallen steadily to 1,173 in 1803 and only 622 in 1804.[48]

Emboldened by the positive reports and still desperate to quash the enemy, Jenner's supporters applied to Parliament for a further honorarium for him in July 1807.[49] Pearson again mobilised the opposition and picked up where he left off. By then, he had chased up the story of Benjamin Jesty and had invited Jesty to the Vaccine-Pock Institute for the joint purposes of having his ceremonial portrait painted and of angering Jenner by publishing the evidence that the first vaccinator was actually a farmer from Dorset. Pearson's arguments were duly noted, but so was all the other evidence and the Committee voted a further award of £20,000 to Jenner. This second honorarium and was paid up front and tax-free (by the time the first grant was paid, two years late, taxes had stripped it down to just £9,000).[50]

Jenner's enemies reacted predictably to this poke in the eye. Moseley wrote, "It will not be credited by future generations that both these large sums were granted by Parliament without even a symptom of controversial discussion."[51] His bitterness evidently made him forget that he and Pearson had thrown all their energy into "controversial discussion", not to mention character assassination, and had failed.

Jenner's supporters also set about out-flanking Pearson and his Vaccine-Pock Institute. Lettsom and others were behind a proposal to set up a rival vaccination establishment, the Jennerian Society. From its first formal meeting in February 1803, with Jenner in the chair, this initially went from strength to strength. Approval from the King gave it the suffix 'Royal', and patrons included the Prime Minister, several peers, the Lord Mayor of London, a brace of Archbishops and numerous top doctors. The Society established a chain of 13 vaccination stations across London which performed over 12,000 vaccinations in its first 18 months. It was soon recognised as the national distribution centre for vaccine, sending out over 20,000 doses free of charge to doctors by the end of 1805.[52]

Membership of the Royal Jennerian Society brought a posh certificate adorned with the Royal Coat of Arms, the motto "Kings shall be thy nursing Fathers, and their Queens thy nursing Mothers." This also featured a badly-drawn Jenner on a pedestal (ornamented with cows and children), standing on the head of the defunct spotted serpent of smallpox whose flaccid coils were draped over his arm. Each year, Jenner's birthday was celebrated with an Anniversary Dinner, often too lavish for Jenner to want to attend. Everything went well until it dawned on Jenner and his colleagues that they had made a disastrous choice in appointing John Walker as their Secretary and Resident Vaccinator.[53] Walker was self-opinionated and arrogant, especially towards women and children whom he terrorised by waving his scalpel about when he got angry with them. The grim necessity of getting rid of Walker eventually tore the guts out of the Royal Jennerian Society, and Jenner resigned in 1806. The Society tottered on with its Anniversary Dinners (mostly without Jenner) but folded a couple of years later. The vacuum was filled by the National Vaccine Establishment (basically pro-Jenner, although he had no formal role in it) and the rival London Vaccine Institution, which was headed by Walker and virulently anti-Jenner.[54]

The exasperated Jenner, ground down by these incessant squabbles, contrasts sharply with the image of brilliance and generosity that was projected to the rest of the world. While the "blockheads" and "snakes" tried to do him down in England,

Figure 10.2 Membership certificate, Royal Jennerian Society. Reproduced by kind permission of the Edward Jenner Museum, Berkeley, Gloucestershire.

grateful nations across the planet were heaping recognition and adulation on him. He received honorary degrees and fellowships from Harvard, Göttingen and Paris, several thousand pounds in cash from three Indian states (instigated by Lettsom who thought that the Parliamentary awards had been "niggardly"), a ceremonial belt and string of precious cowrie shells from the Five Nations of Indians in North America and Canada, and of course the diamond ring from 'Marie', Empress Dowager of Russia.[55]

Despite ongoing hostilities with England, France held Jenner in particular esteem and affection. During the temporary peace in 1802, an official dinner to honour him was held in Paris. As

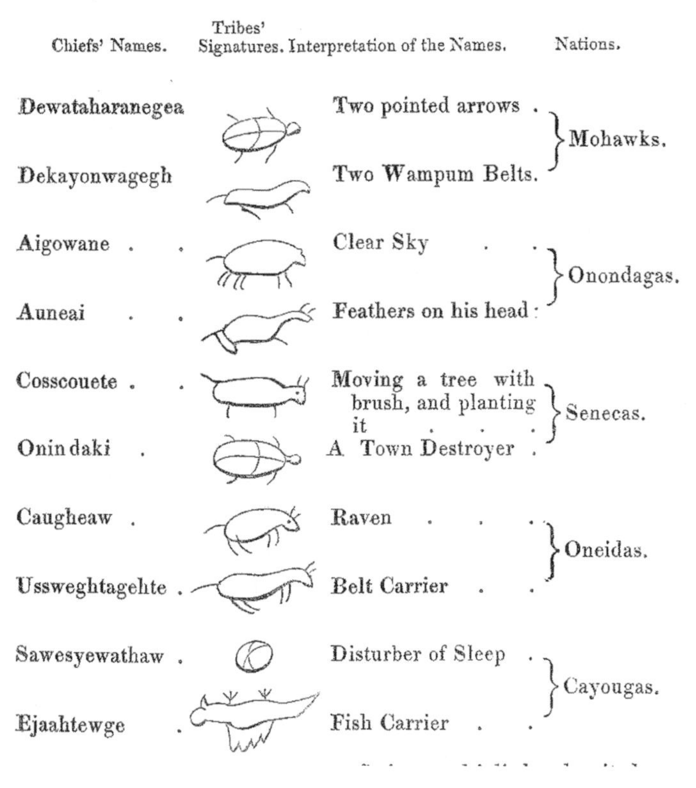

Chiefs' Names.	Tribes' Signatures.	Interpretation of the Names.	Nations.
Dewataharanegea		Two pointed arrows .	Mohawks.
Dekayonwagegh		Two Wampum Belts.	
Aigowane . .		Clear Sky . .	Onondagas.
Auneai . .		Feathers on his head :	
Cosscouete . .		Moving a tree with brush, and planting it . . .	Senecas.
Onin daki .		A Town Destroyer .	
Caugheaw .		Raven . . .	Oneidas.
Ussweghtagehte .		Belt Carrier . .	
Sawesyewathaw .		Disturber of Sleep .	Cayougas.
Ejaahtewge .		Fish Carrier . .	

Figure 10.3 Citation for Jenner from the Five Nations of North American and Canadian Indians.

usual, Jenner did not attend but was represented by his flower-wreathed portrait placed at the head of the table and by his nephew George.[56] Later, when war resumed, two Englishmen were trapped in France. Jenner wrote directly to Napoleon, whose ear was also bent by his Empress Josephine, and agreed that he could refuse nothing to Jenner.[57] Both prisoners were promptly freed to return to England. Safe passage was also guaranteed to a young man sailing to Madeira to recover from illness. Instead of an English passport, which would have had him interned as a prisoner of war, he carried a letter from Jenner, signed in his capacity as an honorary member of the French Institut National.[58]

At home, he was plied with some honours while others were conspicuously withheld.[59] He was awarded an honorary Doctor of Medicine degree from Oxford, given the freedom of London and Dublin, presented with numerous gold medals and in 1821 won an accolade that would have made John Hunter smile with satisfaction: Physician-Extraordinary to His Majesty King George IV. The omissions in Jenner's recognition are also telling. He was never knighted nor awarded the Fellowship of the Royal College of Physicians. Both probably reflect the pervasive influence of his enemies, although the latter was also due to Jenner's refusal to sit the College's statutory examination in Latin and Greek. He had been allergic to classics since school at Cirencester but it is sad that neither party felt able to bend its principles in this instance.

Jenner was easily hurt by criticism but also revelled in praise. He took particular pride in the letter that President Thomas Jefferson wrote to him in 1806:[60]

> You have erased from the calendar of human afflictions one of its greatest. Yours is a comfortable reflection that mankind can never forget that you have lived. Future nations will know by history only that smallpox has existed.

Actions spoke even louder than words as vaccination began to take hold, changing medical practice and eventually pushing the Angel of Death into retreat.

In England, this was a slow process of incremental drift rather than the big-bang conversions in other countries. In its best year, 1805, the Royal Jennerian Society carried out almost 7,000 vaccinations; during this same period, over 20 times as many (146,000) were performed in Madras.[61] The highest medical authorities in Sweden, Denmark and France were immediately convinced of the value of vaccination, while the Royal College of Physicians in London dithered for years before concluding in 1807 that vaccination was safer than variolation and generally protected against smallpox.[62] In many other countries, vaccination was so obviously better than variolation that the old procedure was declared redundant or even illegal. Meanwhile, in Jenner's own country, variolation continued alongside vaccination in an uncomfortable mixed economy that often played to the Angel's advantage. A fatal smallpox outbreak in Norwich occurred because someone decided that variolation was as good as vaccination, when smallpox rapidly spread among the unvaccinated susceptible population.[63] Even in Woodville's Smallpox Hospital at St. Pancras – where they should have known better – patients were allowed to choose between smallpox and cowpox until the late 1810s and the two procedures were conducted side-by-side. Jenner pointed out the stupidity of this practice and Woodville succeeded in tipping the balance in favour of vaccination – but after his death in 1805 the Hospital's preference swung back to variolation.[64]

Gardening leave

As always, Jenner was most comfortable out of London and he spent the last quarter of his life at the Chantry in Berkeley with spells in Cheltenham. He settled into a routine that was dominated by his own vaccination practice and the struggle to keep up with correspondence from around the world (see Plate 9).[65] His clerical friend Robert Ferryman paid for a stone outhouse to be built in the Chantry's garden near the wall of Berkeley Castle. The building was faced with rough-cut slabs of wood to give it a rustic look and Jenner nicknamed it 'The Temple of Vaccinia'.[66] Here, Jenner installed himself for a day each week to vaccinate the poor. On busy days, the waiting queue snaked over the lawn, across the gravel sweep in front of the Chantry and down to the High Street in Berkeley. Jenner treated the poor free of charge. He presumably had his tongue in his cheek when he wrote:

> In London my practice is tied to the higher order of Society – in the Country, I can always find little Cottagers on which I can introduce vaccine Virus in any form.[67]

Answering correspondence blocked off much of his remaining time:

> On the average I am at least six hours daily with my pen in my hand bending over writing paper till I am grown as crooked as a cow's horn and tawny as whey butter.[68]

He complained of being "Vaccine Clerk to the World" but seems to have enjoyed the distinction.[69]

Medicine threw many personal challenges at Jenner. Recurrent attacks of typhus fever laid him low, together with his wife and son Edward. In 1806, Jenner employed a bright young tutor for Edward, the 17-year-old John Worgan from Bristol.[70] Worgan was already an accomplished poet and brought new energy and fun into the Jenner's house. Unfortunately, he also brought tuberculosis. This killed Worgan in July 1808 and infected Edward, who wasted away over two distressing years and died in 1810.[71] His parents were devastated and moved to the house that they kept in Cheltenham. Soon after, they would have realised that the perpetually sickly Mrs J. had also caught tuberculosis. Despite Cheltenham's reputation as a first-class healing spa, Catherine followed her son's downhill course and died on 13 September 1815.[72]

Jenner returned to the Chantry. The last eight years of his life were lonely, especially after his daughter Catherine married in 1822 and moved to Birmingham. However, he kept his brain and his body active. He revisited some old research ideas, including (yet again) tartar emetic and the migration of birds, first mentioned to Joseph Banks back in 1787 with the promise that it would soon be ready.[73] This was a well-argued piece of fieldwork which suggested that birds such as swallows and hobbies disappeared from English skies in autumn because they flew off to warmer countries where food was more abundant than in an English winter. This sounds blindingly obvious today, but the prevailing view then – only slightly less daft than Cotton Mather's invisible moon – was that all these birds spent the winter under water or buried in mud. Jenner toyed with the paper but didn't quite finish it. He also maintained his interest in geology and in his seventieth year rode several rough miles up to a quarry gouged out of the flank of Stinchcombe Hill where he discovered a plesiosaur's skeleton, one of the first to be found in Britain.[74]

John Baron, a young doctor from Gloucester, became Jenner's confidant and close friend – too close for the biography which he later wrote to take a balanced view of Jenner's achievements and his faults. Baron owed Jenner more than friendship and advice. In 1816, he developed raging tonsillitis which nearly killed him. Jenner, then aged 67, was called urgently to attend Baron and found him panic-stricken and gasping for breath with his airway blocked by massively enlarged tonsils. Jenner immediately took out his knife and opened Baron's windpipe. The emergency tracheotomy that he had learned in London nearly half a century earlier saved Baron's life.[75]

It was a favour that Baron could not repay. On the morning of 25 January 1823, Jenner was late down to breakfast. He was found unconscious in his bedroom, having had a massive stroke during the night. His physician nephew Henry immediately administered the usual first aid treatment of a large bleed while someone rode to Gloucester to summon Baron. The distraught Baron was badly shaken to see his mentor and hero reduced to this pitiful state. He later wrote that "the symptoms were most formidable" and that "every effort which skill could suggest was employed". These efforts terminated in the desperate measure of cutting into Jenner's face just in front of his ear to sever the temporal artery.

Edward Jenner died at two o'clock the following morning, Sunday 26 January 1823.[76] His death was marked that morning by the cancellation of the service at Rockhampton Church and, over the following days and weeks, by a massive outpouring of grief and tributes from across the world.

In other countries, his death would have been marked by national mourning and a state funeral. His family were offered the option of burial in Westminster Abbey, but at their expense. They chose the alternative which Jenner himself would certainly have preferred, of staying with his roots. On 3 February 1823, he was buried in the chancel of Berkeley Church where his father had been vicar, beside his parents, wife and son. His pallbearers included James Phipps. According to the obituary which Baron wrote for the *Gloucester Journal*,[77] the turnout was good:

> The concourse of persons was immense; the indications of respect, reverence and regret, were unequivocally conspicuous; every eye was moistened; and every heart oppressed.

A flowery epitaph was composed for "Immortal JENNER! Whose gigantic mind brought life and health to more than half of mankind." Just as fitting, and more in keeping with the man, were his own words spoken to Baron a couple of days before he died:[78]

> I do not marvel that men are not grateful to me, but I am surprised that they do not feel gratitude to God for making me the medium of good.

11 An Affront to the Rights of Man

What do a believer in natural salt therapy, George Bernard Shaw, the professor of statistics in Bern and a distraught mother about to drown her baby all have in common? Answer: they all hated vaccination and cursed Jenner for having brought it into the world.

These were not lone voices but part of a worldwide movement that was mobilised during the nineteenth century as the vaccination juggernaut (or bandwagon, according to taste) gained momentum. The call of the anti-vaccinationists rallied people from all walks of life and cut across groups that should logically have been united in support of Jenner's discovery – doctors, preachers, politicians and reformers.

Opposition to vaccination sprang up almost as soon as the *Inquiry* appeared and triggered one of the most heated and longest running debates about medicine and society. Indeed, antagonism to Jenner and his invention has never gone away and is reawakened whenever ammunition is required for battles over immunisation against other diseases such as polio and measles-mumps-rubella (MMR). This is a 200-year war that shows no sign of ending.

So welcome to the great vaccination debate – a tangled mess of big egos and self-interest, bent statistics, lies and fraud that crescendoed through the second half of the nineteenth century. The confrontation caused damage by holding up the spread of vaccination but also did good by highlighting the flaws in Jenner's original concepts and forcing solutions to be found.

As with vaccination itself, England led the movement to oppose it. This began as the legacy of the campaign against Jenner by Pearson and his associates but the defining event that made so many rise up in anger was the introduction of laws to make vaccination compulsory in England and Wales.

An abomination against the health of the nation

In an ideal world, those entrusted with public health would have an easy job. Each new treatment would be tested to define its benefits and risks and people would then apply logic to make their own choice about whether or not to use it. When the evidence finally came in, the benefits of vaccination so clearly outweighed its risks that it should have been an obvious choice for parents who wished their children to remain healthy and happy. However, logic lost out in the first decades of vaccination because of many factors including suspicion, the lack of reliable information about its efficacy and safety, and foul play.

Different countries handled the introduction of vaccination in their own ways.[1] In some, such as Belgium and Austria, the benefits of vaccination and the risks of remaining unvaccinated were simply explained in terms that people could understand. Vaccination was strongly encouraged and voluntary uptake was judged high enough for compulsion to be thought unnecessary. Other governments were so convinced that vaccination would benefit the vast majority that it was made compulsory. Those leading the way included Bavaria (1807) and Denmark (1810), followed by Sweden (1815), Germany (1818) and Egypt (1819). Relative latecomers included much of Australia (1872–84), Hungary (1876) and Italy (1891). By 1896, vaccination remained voluntary in Russia, Spain, Belgium and in upper- and middle-class France (the poor had to be vaccinated in order to claim state benefits and for their children to attend school). In America, many states enforced vaccination, while others, like Canada, left it to individual choice.

Compulsion often stirred up protests, mostly from those who saw it as a violation of their freedom and right to choose. Some miasmatists and natural healers also pitched in to tilt at their usual targets, notably Jenner, doctors and the unnatural act of cowpoxing. In most countries, dissent was managed sensibly and did not derail the spread of vaccination.

In England and Wales, however, it was a different story, thanks mainly to the clumsy and overbearing way in which vaccination was enforced; typical British antagonism to being told what to do also contributed. A series of Vaccination Acts brought in between 1840 and 1871[2] progressively ramped up the pressure to be vaccinated and, in the view of civil libertarians, steadily tightened the noose around the neck of the individual's freedom to choose.

The first Act to Extend the Practice of Vaccination in 1840 was a hasty reaction to a severe smallpox outbreak in 1837–39. This Act simply encouraged the vaccination of children and those who had not had smallpox (and simultaneously killed off variolation by making it illegal). Progress was not thought fast enough. A revised Act in 1853 immediately made hackles rise with its uncompromising title, 'An Act to Extend and Make Compulsory the Practice of Vaccination'. Children had to be taken to a public vaccinator within three months of birth, with a repeat visit eight days later and revaccination if the first inoculation had not taken. Parents who refused or otherwise did not comply were fined 20 shillings, to be paid into the fund for relief of the poor.

The State's grip on the individual tightened with further revisions of the Act in 1858, 1867 and 1871. These brought in the right of vaccinators to collect lymph from any vaccinated child (with another 20-shilling fine for anyone who resisted) and the appointment of the euphemistically named 'Guardians' in every town to chase up and prosecute those who had evaded vaccination. These draconian measures and the authorities' refusal to listen to reasonable concerns stuck in the craw of libertarians. William Tebb, then rising to prominence in the anti-vaccination campaign, wrote in 1884 that those arguing against the Acts were denied rights of expression "like the Negro race in America during the reign of the Democratic slavocracy".[3]

Once set in motion, the machinery of the Acts ground on inexorably and was enforced to the letter and with a heavy hand. Parents could be prosecuted repeatedly for not having a child vaccinated. A man in Farringdon, London, was hauled up 32 times despite explaining that he was afraid of vaccination because a neighbour's child had been left disabled by it.[4] Another father was imprisoned for refusing to have his child vaccinated and then swearing that he would not pay the fine. The authorities were unmoved even though he had made his oath on the body of the child, who had died of other causes.[5] Desperate parents resorted to avoidance tactics, including not reporting their child's birth and, in Tebb's words, "flitting about from one part of the country to another to avoid the surgeons and the police". [6] Some even left the country. Mr. F. Scrimshaw, a "highly intelligent resident of Nottingham", emigrated to the United States in 1883 after being penalised for refusing to have his child vaccinated. As a farewell gesture of disgust, Scrimshaw stirred up local dissent with "many small indignation meetings". Unfortunately, he had not done his homework and was on board ship before he discovered that proof of vaccination was a condition of entry to America. Furious, he had to submit to vaccination by a pompous "Medical Ignoramus" in a gold-braided cap, who, like most doctors, was "without capacity to understand our concerns".[7]

The Acts became a socially divisive force, as their perverse consequences rebounded especially on the lower classes. Children born in the workhouse were vaccinated almost immediately after birth (against medical advice) before their mother could run away with them. The 20-shilling fine for each breach of the Act was easily affordable by the wealthy, but not the poor. Parents who refused or could not pay were often sent to prison, breaking up and potentially ruining the family. The power of the Act to terrorise the isolated and vulnerable was demonstrated by the tragedy of Mary Clarke, whose suicide was reported by the London papers in August 1882. Rather than have her youngest child vaccinated, she had killed the infant and then herself. Her action was desperate but determined; she had to tear up the floorboards with her bare hands to expose the water storage tank in which they both drowned.[8]

As the Acts were rolled out, foci of resistance appeared and the map of England gradually became a patchwork of areas that accepted, tolerated or opposed vaccination. By 1880, active resistance had swelled into civil disobedience in several provincial cities such as Leicester, Manchester and Jenner's own county town of Gloucester.[9] Often supported by their Member of Parliament, the civic authorities refused to enforce the Act and dissenters were not prosecuted. The Guardians therefore had an impossible job – except those who were carefully appointed because of their anti-vaccination sympathies, as in Leicester.[10] Popular antagonism to vaccination was easly stirred up among the poor, especially in the industrial cities and slums. Dr. John Scott, a physician in Manchester, reported: "There is no getting over the fact that vaccination is hated, amongst the working classes in Lancashire at least."[11]

Vaccination came to pervade other aspects of daily life. Members of the British armed forces and police had to be vaccinated on recruitment and at each outbreak, as did boys at public schools (their parents' permission was not required, but payment of half a guinea was).[12] Some of these policies did not seem to be driven by purely medical imperatives: from 1864, new recruits for the Navy had to be vaccinated "except natives joining abroad" and it was only in 1873 that it became mandatory for all.[13] Vaccination came of age when it became a criterion for life insurance. By 1880, two-thirds of the London-based assurance companies required proof of vaccination, and some refused insurance to the unvaccinated, or demanded an extra premium.[14]

Vaccination was attacked across a broad front that united an unlikely assortment of followers: doctors with sincere (or insincere) doubts, doctor-haters, inflamed clerics, slippery politicians and what can reasonably be called the lunatic fringe. Dissenters coalesced into pressure groups such as the Anti-Vaccination League and the London Society for the Abolition of Compulsory Vaccination. These spread the word in many ways. Meetings ranged from small parish events to massive protests that attracted hundreds and sometimes thousands, such as the 20,000-strong crowd that publicly burned the Vaccination Acts in Leicester in 1885.[15] Agitators produced a stream of pamphlets in the 1870s that summarised their main arguments, such as *Vaccination in the Light of History* (1878), an attack on Jenner and his disciples which also picked up some legitimate concerns about vaccination; and *Vaccination, Opposed to Science and a Disgrace to English Law* (1879) which dug deeper into the evidence that vaccination was ineffective and dangerous as well as immoral and illegal.[16] The anti-vaccinators made much of the 'martyrs' who fell foul of the law, organising high-profile welcome celebrations for those released from imprisonment for having refused to have their children vaccinated. These well-orchestrated events, with rabble-rousing orators drafted in to pull a good crowd, helped to swell the ranks of those opposed to vaccination.

Leadership of the anti-vaccination campaign in England was mostly a middle-class affair that soon linked up with counterparts in America and on the continent, especially in France and Belgium. The anti-vaccinationists were always on the prowl for propaganda coups and triumphantly paraded big names in medicine, science and society who were prepared to support the cause. These included international smallpox experts – Creighton and Crookshank in England, Boëns in Belgium – scientists like Alfred Russel Wallace, the writer George Bernard Shaw and high society figures, ideally foreign for extra impact. The anti-vaccinationists made the most of these personalities, rolling them out at meetings and recycling their words of wisdom wherever possible.

Damned lies and disrelished statistics

The medical arguments against vaccination were that it was dangerous and useless – and superfluous because any falls in smallpox were due to general improvements in sanitation and living conditions. The perceived risks of vaccination had evolved beyond being turned into a cow and now ranged from syphilis to cancer. This formed an entire flank of the anti-vaccinationists' attack, as described in Chapter 12.

A recurrent theme was the evidence that vaccination did not protect against smallpox. Various studies supported anecdotal reports of people who had been vaccinated but later caught smallpox. Professor A. Vogt of the Hygiene Institute in Bern ("probably the largest collector of statistical information in the world", according to P. A. Taylor, Member of Parliament for Leicester) studied over 40,000 smallpox deaths and found that most had apparently been vaccinated.[17] Until then a fervent believer in Jenner, Vogt was so shocked that he lost his faith and became a hardened opponent. The same question was examined by Dr. Leander Joseph Keller, the chief doctor of the Austrian Railway Company, who had access to the medical records of all the Company's employees and their families. Following a smallpox outbreak in 1872–73, Keller looked at the 2,600 cases of smallpox (of whom 470 had died) that occurred among this 55,000-strong population and compared the death rates between vaccinated and unvaccinated cases in various age groups. He reported a stark and unexpected difference across all ages, with the vaccinated showing a significantly *higher* mortality from smallpox than the unvaccinated.[18]

Both studies caused heat and confusion when they were published, with both sides hurling insults and statistics at each other with little regard for objectivity or scientific integrity. Vogt's 40,000 cases undoubtedly included some who had never been properly vaccinated, either because of technical error or dishonesty (unvaccinated people wishing to avoid the procedure often claimed that they had been vaccinated). However, the residue – and probably a large majority – could only be explained by the protective effect of vaccination wearing off. Jenner's disciples immediately rejected this notion. Jenner's assumption that vaccination protected for life was dogma that had to be defended at all costs, even though a dispassionate observer would have seen this belief overturned by the weight of evidence that emerged within a decade of Jenner's death.

There was a simpler explanation for Keller's surprising finding that vaccination actually increased the risk of dying from smallpox at all ages. The data looked too clear to be true to Dr. Körösi, whose suspicions were confirmed when he went back to the original records. He found that Keller had invented his findings, and in fact smallpox deaths were *less* frequent among vaccinated subjects. Körösi's damning rebuttal was published in 1887,[19] but Keller never retracted his paper.

Neither side learned the lessons of these experiences. The Jennerians continued to insist that vaccination conferred lifelong protection and their stance delayed the introduction of a fully protective vaccination schedule in England. Defending this increasingly untenable line also dragged them into fabrication and deceit. A study from Birkenhead, across the Mersey from Liverpool, reported that none of a group of 115 smallpox victims had been vaccinated.[20] Just as with Körösi, this neat conclusion raised suspicions and when the subjects were traced and properly examined, most of them turned out to have the telltale vaccination scars. For their part, the anti-vaccinationists persisted in quoting the groundbreaking study by the good Dr. Keller long after it was proved to be fraudulent.

These two skirmishes were part of a much larger battle of statistics that ran for decades. P. A. Taylor noted that "the public disrelish statistics",[21] but that did not stop either side from using them to convince and confuse. Both sides were guilty of concealing and manipulating data, although the anti-vaccinationists did more to confirm Disraeli's famous complaint that "There are three kinds of lies: lies, damned lies and statistics."[22]

By 1870, there were convincing falls in smallpox cases and deaths in the countries that had embraced vaccination and especially those that had made it compulsory.[23] The anti-vaccinationists put much effort into finding examples that seemed to buck this trend. Vaccination had been introduced into Prussia in 1835, with little obvious effect: the smallpox mortality rate continued to fluctuate around 2,000 deaths per year per million people (the same as before vaccination) and shot up to over ten times that level during the devastating epidemic that swept Europe in 1871–72. [24] Closer to home, there had been three large smallpox outbreaks in England since compulsory vaccination had been brought in, and these had claimed progressively *more* lives: over 14,000 in 1857–59, 20,000 in 1863–65 and a shocking 45,000 in 1871–72. Of the 8,000 who died in London in 1871–72, one-third had been vaccinated.[25] These disasters were good news for the anti-vaccination campaigners, who gloated over the fact that – thanks to the brutal enforcement of the Act – these had occurred in a population that was "vaccinated to the hilt".[26] At first sight, these data seem to support the anti-vaccinationists' claim that vaccination did not work. However, smallpox was never a predictable disease, especially during outbreaks. The deaths in London during this period were consistently dwarfed by the toll of smallpox before the vaccination era began at the turn of the nineteenth century, and even the shocking blip of the 1871–72 outbreak killed only half as many as the epidemic of 1836–37.[27]

At the same time, these and other data also brought out the uncomfortable truth that vaccination did not necessarily equal immunity. It was now undeniable that many older smallpox victims had been vaccinated in childhood. In reality, populations that were vaccinated only at birth were nowhere near "vaccinated to the hilt". By 1870, the overall proportion of babies vaccinated in the counties of England and Wales varied between 90 per cent and 97 per cent (with much lower rates of 15 per cent or less in centres with the fiercest opposition).[28] However, there was a large susceptible population of older adults who had either never met smallpox or had never been vaccinated – and which was steadily expanded by others as they lost the protection of vaccination given 30, 20 or even ten years earlier. When it was finally acknowledged that the effect of vaccination wore off, it was topped up by revaccination, generally at around the age of ten and, for added security, again if smallpox broke out. Revaccination had been adopted by several other European countries by 1860 but, thanks to resistance by the pro-Jenner lobby, the general British public were denied this crucial safety net until 1871. Prussia followed suit by bringing in compulsory revaccination in 1874, and saw a prompt and gratifying fall in smallpox mortality rates to fewer than 20 deaths per million within five years.[29]

The anti-vaccinationist propaganda machine used other ingenious tricks, including misreporting or ignoring the opposition's responses to their challenges. One pamphlet reprinted a spirited attack in the House of Commons on 11 June 1880, with two Members of Parliament, P. A. Taylor (Leicester) and C. H. Hopwood (Leeds), laying into the cynical evil of vaccination with great energy and eloquence.[30] Both their speeches were reported verbatim, together with the reactions of their supporters ("Cheers and laughter", "Ironical cheers" and "Hear, hear!") but with no mention of the opposing side of the debate or its outcome (a resounding defeat of Taylor's motion by 634 votes to 18). Unfortunately, the anti-vaccinationists were able to seize on various gaffes made by their opponents. Charles Cameron, a committed pro-vaccinationist, must have cursed the day that he wrote an informal note to friends following the 1871–72 outbreak, in which he remarked:

> The recurrence ... of a mortality almost as high as that experienced prior to the Vaccination Act, shows either that the protective virtues

of vaccination are mythical, or that there is something fundamentally wrong with our natural system of vaccination.[31]

Cameron had realised that vaccination in infancy did not protect for life, but his words came back to haunt him, recycled as proof that even the most ardent of Jenner's disciples no longer believed in the 'myth' of vaccination.

Prominent doctors who came out against vaccination were among the anti-vaccinationists' most treasured resources. In England, they included two authorities who, for scientific and personal reasons, took against vaccination and Jenner. Charles Creighton, physician and epidemiologist, wrote the section on smallpox for the *Encyclopaedia Britannica* with a lengthy diatribe about the dangers of vaccination (see Chapter 12). Edgar Crookshank, a professor at King's College Hospital in London, despised Jenner and set out on a Pearson-style personal vendetta to discredit him. Crookshank disparaged not only Jenner's reputation as a scientist but also his claim to be the inventor of vaccination – and drove home this point by having an engraving of Benjamin Jesty as the frontispiece to his influential book, *History and Pathology of Vaccination* (1889).[32] Crookshank's hatred of Jenner was so intense that it eventually destroyed his own career. Nonetheless, he remained a valuable asset to the anti-Jennerians, as did Creighton, later vice-president of the Anti-Vaccination League.

As well as Creighton and the unhinged Crookshank, the anti-vaccinationists had other darlings further afield. The head of the Vaccination Service in Paris was disgusted by the introduction of compulsory vaccination in France in 1881 under the Loi de Liouville, the counterpart of the Vaccination Acts: "vexatious, ineffective and impracticable. I cannot put out of my mind the offensiveness of such a law to free men." He conceded that he could accept enforced vaccination if it had a substantial effect on smallpox, "but as it is not so, I cannot assent to this compulsion".[33] Neither he nor his employer seemed to attach enough importance to his job description for him to leave his post.

Another valuable figurehead was Dr. Hubert Boëns from Charleroi, a prominent member of the Belgian Academy of Science. Boëns hated Jenner, vaccination and its descendant, the '*école vaccinatrice*' of Pasteur which, despite early success with a vaccine against anthrax, came under attack during the 1870s from a noisy faction in the medical establishment in Paris. Enraged by the Loi de Liouville, Boëns founded the impressively-named Ligue Universelle des Antivaccinateurs in 1880.[34] The Ligue got off to a flying start with a convention in Paris later that year, a four-day bonanza of well-staged lectures by carefully chosen doctors and scientists on the non-benefits and risks of vaccination. The public's attention was skilfully seized with showcase presentations and well targeted propaganda – and with results that Boëns must have dreamt of. Public outrage at the horrors of vaccination grew across France, forcing the government to review the wisdom of compulsion, and the hateful Loi went in the bin. Even better, the French decisiveness was noted by Belgium, Spain, Italy and Poland, all of which scrapped their own plans for making vaccination compulsory.[35] For the moment, though, Britain resisted.

Further Congresses of the Ligue followed in Cologne (1882) and Bern (1883). These were valuable opportunities to share knowledge about how best to bring down vaccination, especially in Britain. The English activist William Tebb attacked "scientists blinded enough by national fetishism to approve these eccentric doctrines, which now lacked the merit of novelty".[36] The Bern Convention also celebrated another own-goal by a prominent supporter of vaccination, Sir John Simon, who in 1858 had become England's first Chief Medical Officer. Simon presented a detailed report to Parliament showing that the incidence of smallpox infection, scarring and death had fallen dramatically in England since the introduction of vaccination. This was embarrassing for the anti-vaccinationists but, fortunately for them, Simon had also written to a friend, admitting that vaccination could obviously not be responsible for similarly rapid falls in other countries where it had not yet been introduced. Somehow, Simon's note found its way into the hands of Dr. H. Oidtman, chief physician at Verdun Hospital and a virulent opponent of vaccination.[37] Simon was widely quoted as a man of stature and integrity who had been forced to conclude that improved sanitation was responsible for the defeat of smallpox and that vaccination could not claim any of the credit.

The anti-vaccinationists attacked their enemies and their reputations with great energy but somehow seemed to avoid that fate themselves. Boëns was quick to lambast the "brainless assertions on which the whole edifice of vaccination was founded"[38] but his own credentials were dubious. The reader may remember him as the proponent of anal leech therapy in smallpox. His other therapeutic coups for treating the disease included emptying the bowel with emetics in children or drastic purgatives in adults (presumably with reapplication of any leeches unfortunate enough to be dislodged by the diarrhoea). He also revived a recipe from over a century earlier, namely lemonade fortified with sulphuric acid. If this all sounds passive, it was in line with his own philosophy about the nature of the disease. Smallpox, he said, was caused by "sharp moral shocks" such as fear and anger, as he sought to prove with a series of 18 "likely cases" of "spontaneous" smallpox which he published in 1884.[39] Above all, it was a mild disorder, which is why the risks of vaccination were entirely unacceptable:

> Many people and even doctors consider smallpox to be a terrible disease which carries off hordes of victims or which stamps indelible stigmata on their faces for the whole of their life. Such exaggeration needs to be tempered. We are determined to show that smallpox is not such a murderous or disfiguring disorder, as some are happy to say in order to scare people and make them resort to vaccination.

These comforting words would have come as a surprise to the inhabitants of Montreal, where a smallpox outbreak was then gaining momentum and by the following year would kill over 3,000 people. Boëns must have had a powerful and persuasive personality, or perhaps nobody thought to dig behind the façade of the Ligue Universelle. His views were never exposed or tested against reality, let alone attacked with the ferocity that he and his associates reserved for the work of Jenner and Pasteur.

All in a good cause

Vaccination and compulsion were powerful magnets that dragged many other people and causes on board the anti-vaccination bandwagon. Compulsion was the main motive force for the Personal Rights Association, founded in 1871 in response to the Vaccination Act. The Association was determined to defend (in surprisingly modern terms) "the equality of all citizens ... without regard to health, birth, sex, culture, religious belief or any other circumstances".[40] It was implacably opposed to the "tyranny of the Few over the Many" – exemplified by the Act – and also covered other options with its resistance to "the tyranny of the Many over the Few".

The abrogation of the individual's rights pulled in the writer George Bernard Shaw, who was also disgusted at the whole notion of vaccination – "A particularly filthy bit of witchcraft."[41] Alfred Russel Wallace, scientific colleague of Darwin and spiritualist, also came on board. Wallace found vaccination scientifically weak as well as morally repellent and later became fixed on its dangers. With impressive precision he declared that "An average of 52 children are officially murdered every year", although how he derived this number is not clear. Also at variance with the source data was his analysis of annual mortality returns, from which he concluded that vaccination "caused many deaths ... and is probably the cause of greater mortality than smallpox itself, but which cannot be proved to have ever saved a single life".[42] Wallace's formidable reputation presumably out-trumped any impression that he might be a loose cannon, and his name was widely used by the opponents of vaccination to further their cause. In a circular letter dated 9 February 1911, the Secretary of the National Anti-Vaccination League wrote:

> Dear Sir,
> I am directed by my Council to send you the enclosed copy of a pamphlet which Dr. Alfred Russel Wallace, one of the greatest scientists of our day, pronounced to be the best work of its kind on the subject which he had seen. The importance of the Vaccination question is a sufficient reason for bringing the pamphlet to your notice, and when, in addition to this, my Council does so at the suggestion of such a noted man as Dr. Wallace, you will not grudge the 50 or 60 minutes which are all that are required for its perusal.[43]

Great men such as Shaw and Wallace added kudos to the cause but cannot be said to have advanced the quality of the scientific or social debates.

Men of the cloth also rode out against vaccination as they had done before to express '*odium theologicum*' over variolation and Jenner. They might have been more subtle than John Birch and his view that vaccination prevented the Godsend of smallpox from containing the numbers of the poor, but they still bristled at the invasion by medicine of God's right to determine life and death. In 1896, the centenary year of vaccination, the Reverend Andrew Dick White preached a wide-ranging anti-medical sermon on "Theological opposition to inoculation, vaccination, and the use of anaesthetics".[44] Anaesthetics were included because the first use of chloroform (by Simpson in Edinburgh, 1847) had been to relieve the pain of childbirth – another clear breach of Divine directives, as the Bible stated, "in sorrow thou shalt bring forth children". (Simpson had retaliated by quoting back the Bible: "And the Lord caused a deep sleep to fall on Adam; and he slept; and He took one of his ribs and closed up the flesh thereof.")

Other dissident causes were also swept up. Dr. H. Valentine Knaggs, an anti-vaccinationist physician who claimed that cowpox was identical to foot-and-mouth disease, was incensed by the gravy train of vaccination – "soft jobs" paying £1,000 a year but "available only to vaccinationist doctors".[45] It is not clear whether the salary would have made him bend his principles if he had been eligible to apply. Many other people were also disgusted by doctors. The contempt that Mr. Scrimshaw of Nottingham felt for the "Medical Ignoramus" on his steamer bound for New York was widely felt and the refusal of Jenner's devotees to acknowledge his errors also helped to discredit the medical profession.

People were not allowed to forget the dirty farmyard origins of cowpox. P. A. Taylor MP, never lost for purple prose, picked up Jenner's theory of the origin of cowpox (by then long since jettisoned) and described the transition from "greasy-heeled horses ... stroked down by rugged horse-boys' hands" and thence to cows' dirty nipples and the milkmaid's hand.[46] Public revulsion was stirred up by photographs of calves, their flanks shaved and engraved with vertical cuts to receive the lymph, strapped to a tilting table with suitable containers to catch the "discharges" that might pour out during inoculation or lymph harvesting. In 1909, Ernest McCormack tried to resurrect the ominous warning delivered by Immanuel Kant in his *Fragments Littéraires* (1843), about "the dangerous consequences from the absorption of a brutal miasma into the human blood".[47] By now the "dangerous consequences" were more sophisticated than Rowley's bovine transformation and included syphilis, cancer and madness (see Chapter 12).

Which brings us to the lunatic fringe. This was the age of the pamphleteer, and anyone could spread their views (no matter how crazy or outrageous) far and wide at low cost and without the inconvenience of peer review or editorial interference. Diseases and treatments were common targets. Sensible reflections about the dangers of smoking (*Tobacco-taking is Poisoning!*), drinking and drug-taking appeared alongside contributions from fanatics, such as a disarmingly candid treatise about the mistreatment of the mentally ill by "A Certified Lunatic" and a pamphlet whose title made it hard to ignore, *The Perils of Premature Burial*.[48]

Smallpox and vaccination attracted the attention of many pamphleteers, most of whom were vehemently opposed. As well as 'official' publications by the Anti-Vaccination League, the Ligue Universelle and the Personal Liberty Association, hundreds of others were written, paid for and distributed by people with views to share, spleens to vent and money to make. 'Dr. Sly' (translated from the French) lined Jenner up for ridicule alongside Pasteur and his "scientific method for rearing and floating scientific canards [hoaxes] at will".[49] Natural healers sided with Boëns' view that smallpox was mild and broadcast news of the horrible complications (real and imagined) of vaccination that they believed were being hushed up by the medical establishment.

And 1902 saw the publication by C. Godfrey Gümpel of a long-winded pamphlet: *Smallpox. An Enquiry into its Real Nature and its Possible Prevention showing the Extent and Duration of the Protection Afforded by Vaccination. An Attempt to Solve this Vexed Question Without Statistics*.[50] Gümpel, an engineer with no medical background, had already written books about plague ("One of the most useful books this Century has produced", according to his own blurb) and cholera (dedicated to the recently deceased von Pettenkofer, the "greatest expert on cholera"). Noting that "fear and panic drive the general public to seek protection in a poisoned lancet", he offered a safe alternative, already proven effective in plague, cholera, etc.: common salt. Statistics were not needed to prove that it worked, although he used mathematics to show that smallpox outbreaks correlated with sunspots and the Northern Lights. Smallpox was due to "adverse meteorological influences ... the presence or absence of ozone, or the electrical activity of the air we breathe". The pamphlet ended with another shameless plug for his books and an application form for membership of the Association for Promoting the Prevention of Smallpox by Simple Physiological Means, to be sent to a temporary address in Beckenham, Kent.

To the rescue

The counter-offensive to defend vaccination also spread across several fronts but was less strident, probably because it was proceeding anyway and backed up in Britain and other countries by the full force of the law. As well as set-piece exchanges in the House of Commons and pro-vaccination articles in the medical press, there were determined grass-roots campaigns to counter some of the anti-vaccinationist misinformation. Some of the toughest crusades were those directed at the working-class populations in the heart of the anti-vaccinationist strongholds in England's industrial cities. Spirited examples were those led by two middle-class ladies of Leeds, Mrs. Emily C. Kitson and Mrs. Catherine M. Buckton, who in 1873 produced instructive pamphlets about their experiences so that others could follow their example: *Sanitary Lessons to Working Women in Leeds*[51] and *Two Winters' Experience in Giving Lectures to my Fellow Townswomen on Physiology and Hygiene.*[52]

These pamphlets paint an incongruous and seemingly unpromising picture. At the front of a cold hall in midwinter stood the well-to-do, smartly dressed lecturer, determined to engage the interest of 80 or more "working men's wives, mill-girls, dressmakers &c." Many of the audience were weighed down with the fight to survive in poverty and squalor, and resigned to the fact that their children had an even chance of dying before the age of five. Hanging above their heads were large-print banners explaining that 1,538 inhabitants of Leeds had died the previous year from preventable causes. Luckily, there were also uplifting messages:

TO ENSURE GOOD HEALTH, We must have fresh air. PURE WATER. WHOLESOME FOOD. CLEAN HOUSES. GOOD DRAINS. TEMPERANCE. CLEANLY HABITS.

The nice cups of tea that began and ended the proceedings were probably an unusually safe drink for many in the audience, who had open sewers and the ever-present risk of faeces-borne infections running past their doors.

The course of lectures covered the ladies' concepts of healthy living – the dangers of smallpox and scarlet fever and how to prevent them, "diseases induced in children by impure air", how general health had improved since "the time when the use of baths was unknown", the dangers of drink and the "evil effects of tight lacing". As well as practical advice about covering drains, drinking only clean water and first-aid treatment for burns and drowning, attendees learned why it was unhealthy to live in cellars and back-to-back housing (an unhelpful message for all those trapped in those conditions) and how to keep their husbands faithful and out of public houses by cooking good meals and dressing unfussily but with a "harmony of colours".

The workings of the body in health and disease were explained with the help of props such as a glass tubing mockup of the circulation perfused by red and blue solutions, drawings of the internal organs (natural and squashed by over-tight lacing) and papier-mâché models of the heart, the eye and a tooth. Everyday health hazards were revealed by microscopes trained on drops of dirty water and the cysts of tapeworms in 'measly' pork, while the ability of the demon drink to pickle its victims was demonstrated by a blood clot (kindly lent by the Medical School) perfectly preserved in a glass jar full of alcohol.

Despite the vast gulf between them, Mrs. Kitson and Mrs. Buckton succeeded in locking wavelengths with their "fellow Townswomen" – at least those who turned up. Attendance was good throughout the winter lecture courses and almost all the husbands joined the last session. Mrs. Buckton also distributed question papers at the end of each lecture. Levels of literacy were higher than might be expected as over a dozen women generally stayed behind; the prize for the best answers went to a 16-year-old mill-girl, praised by Mrs. Buckton for the originality of her written responses.

This was also a voyage of discovery for the good ladies. Mrs Buckton made several visits to the slums where she admitted that "the dreadful air ... would create a desire to drink". The smell was due to dumping of the 'night soil' (human excrement) on open ground nearby. She also learned that it was customary for newborn babies to have their heads tightly pressed to bring together the bones of the skull, and that babies' heads were never washed because this might bring on hydrocephalus.

In the middle of her course, Mrs. Kitson devoted a whole evening to dispelling "erroneous notions about vaccination" and particularly the strong prejudice that the Vaccination Act was "a gross tyranny". Beginning with the story of Jenner, she showed how the devastation of smallpox had been "wonderfully modified" by vaccination and built up to her uncompromising message:

> The fact that small-pox is still so general arises, not from the failure of the protective power of vaccination, but from large masses of the population, either from carelessness or ignorance, rejecting its benefits. Such persons not only voluntarily sacrifice their own lives, but endanger the lives of those about them as, when seized, they form the centre of a circle for the propagation of the disease.

This information, she said, was gratefully received and "removed much misapprehension on the subject". Conversion indeed: many of the women told her "how much they wished they could have had this knowledge at the beginning of their married life, as they could now see what care and misery it could have saved them".

Mrs. Kitson (who died shortly after giving her lecture course in winter 1873) and Mrs. Buckton did not choose an easy mission.

We cannot tell whether they could claim any lasting success, but they deserve praise for having tried.

Their down-to-earth foray into the slums of the anti-vaccinationist enclaves of England stands in sharp contrast to the ethereal ramblings of Dr. William W. Parker, President of the Medical Society of Virginia. Parker's devotion to Jenner was total and uncritical and would have cut no ice with the mill-girls of Leeds. His paper *The Ancient and Modern Physician – St. Luke and Jenner* (1891) is embarrassing in its adoration of the saintly doctor and his invention.[53] Parker himself had refined vaccination to the point where his failure rate was "one in 10,000" and having vaccinated himself, he "would not be afraid to sleep between two patients with confluent smallpox" – while admitting that "it would not be very pleasant however".

To show that he was a serious scientist, Parker described his new invention to protect against smallpox, "the most intense and deadly of all contagions". This was a face-mask of Chinese hair, soaked in thymol, turpentine spirits and glycerine, which he planned to test on non-immune subjects shut in a room with smallpox victims. He did not expect failures, but could rescue any with vaccination at the first sign of trouble. The results were not apparently reported.

Dr. Parker's naïveté could well have helped rather than hindered the anti-vaccinationist cause. However, he got one thing right, even though it was to take another 90 years:

> The time is coming when smallpox will be banished from the face of the globe.

12 More Fatal than Smallpox

"I have told this long story, not because my case is sadder than that of other bereaved mothers, but because this is what I have felt, what I have known, and I can tell it. The same heartbreak, the same sense of irrecoverable loss has been the lot of other mothers of murdered children. First stunned, then wild with grief. And the long years stretch out afterward, with that brightness gone forever."

Lora C. Little *Crimes of the Cowpox Ring*, 1907

Burying a child is one of the greatest agonies that any parent has to endure, and a cataclysm that can wreck the lives of those left behind.

Kenneth Marion Little was the only child of Elijah and Lora C. Little, then residents of Yonkers in New York. His photograph shows a composed little boy, smartly turned out and with a fresh haircut, very much his mother's "reserved and dignified child". His expression also seems to hold a resigned sadness, almost as though he has somehow seen into his own short future. Kenneth died in early April 1896, just three months after his seventh birthday.

His mother was devastated. Her account of Kenneth's death,[1] from which the above quotation is taken, was not published until a decade later but grief and anger still cut through clearly. Her anger was directed mainly against the injustice of her son's death, and perhaps against herself for having failed to prevent it.

Kenneth died after being vaccinated against smallpox. His mother protested that vaccination was unreliable and potentially dangerous but the authorities told her it was mandatory for children starting primary school, and that was that. One morning, Kenneth was taken out of class and forcibly vaccinated. His mother wrote that "it was as though he had died the day his arm was punctured". Powerless, she watched the happy little boy turn into an invalid who fell victim to one illness after another. He struggled to reach his seventh birthday and then went steadily downhill.

KENNETH MARION LITTLE.
Only child of Mrs. Lora C. Little. Vaccinated Sept., 1895. Died from the effects, April 10, 1896. Age 7 yrs. 3 mos. (No. 30.)

Figure 12.1 Kenneth Little. Reproduced by kind permission of St. Deiniol's Library, Hawarden, North Wales.

Afterwards, the authorities denied any responsibility for Kenneth's death and carried on regardless with their enforced vaccinations – even though they were unable to give Mrs. Little the basic evidence to justify vaccination. And now that she was tuned in, she found that hers was just one of many dissenting voices: not just grieving parents whose children had been killed or maimed by vaccination, but also doctors, scientists, politicians and clerics, and from England and Europe as well as across the United States.

The more she found out, the more convinced she became that vaccination was useless, dangerous and only kept alive by a money-making cartel of doctors and vaccine producers. This cartel had influence and was at the centre of a conspiracy to silence doubts about vaccination. Her anger focused into a growing hatred of vaccination – "If expletives were deadly, Vaccination and Vaccinators would both have died of violence and moldered to dust a long time ago" – and the determination to rid the world of it.

Two years after Kenneth's death, the Littles moved to Minneapolis, following Elijah's work as an engineer.[2] There, Lora Little began a crusade and founded *The Liberator*, an uncompromising monthly 'Journal of Health and Freedom' that railed against the tyranny of conventional medicine, drugs and vaccination. The fearless energy of *The Liberator* and its formidable one-woman editorial-production team soon won the admiration of anti-vaccination activists across the United States. However, success carried a hefty price tag that included her marriage as well as the costs of producing and distributing *The Liberator*. After Elijah drifted away, she continued to use her married name, but referred to Kenneth simply as "the only son of Mrs. Lora C. Little".

In 1906, Lora Little self-published a book which had been in gestation for several years. *Crimes of the Cowpox Ring*, subtitled *Some Moving Pictures Thrown on the Dead Wall of Official Silence*, was a 75-page exposé of the dangers of vaccination and how these were covered up by the "Vampires fattening themselves on the life-blood of children".[3] The core of the book was a series of over 300 medical disasters following vaccination, collected by herself and the network of contacts that she built up during her travels across the United States. The cases range from three-line vignettes to graphic clinical reports, some illustrated with shocking clinical images. Case No. 22 was a boy with a deep-seated infection in the shin which eroded through to the skin, exposing several inches of bone. Case No. 275 was a young man who was dead by the time his photograph was published: his right shoulder and upper chest were engulfed by a huge tumour, four times the size of his head, mushrooming out from his vaccination site. The legend to his image reads, "Saved from smallpox by Vaccination" (see Fig. 13.2, page 296). The catalogue of misery was punctuated by affidavits from medical authorities, confirming that vaccination was to blame. For example, Dr. Dennis Turnbull (a cancer specialist) wrote: "I have no hesitation in saying that in my judgment the most frequent disposing condition for cancerous development is infused into the blood by vaccination and revaccination."

Case No. 30 is Kenneth Marion Little, described in nine terse lines as one of the 100-plus fatalities following vaccination. He is accorded some special consideration: a full-page photograph and a five-page appendix to the book, containing a moving dialogue between the bereaved mother and her dead son.

Crimes of the Cowpox Ring paints a disturbing picture of vaccination as practised in the United States around the turn of the twentieth century. Kenneth Little's forced vaccination turns out to be a tame example of the authorities' dirty tricks. Six-year-old Lucille Sturdevant (Case No. 21) was vaccinated "against her own protest" at school while a policeman stood over her and "playfully flourished his club". She also died, of blood poisoning, Sixty members of a Negro religious community in Sunnyside, Georgia, were rounded up by police en route to a mass baptism and forcibly vaccinated. Other dissenters were tricked into unlocking their doors to vaccinators who lied to them, claiming that a friend was seriously ill with smallpox.

Intimidating vaccination propaganda was everywhere, such as the huge posters outside Owensboro, Kentucky – "Be Vaccinated or Leave the City" – which had scared an impressionable ten-year-old boy into vaccinating 50 of his friends, using the pus from his own inoculation blister.

Lora Little travelled widely across the United States, preaching a doctrine of 'natural health' which slotted in nicely with networks of activists who were variously opposed to vaccination, doctors, drugs, carnivorism, or subjugation by authority. 'Natural health' brought all these strands together. All diseases, including cancers as well as infections, were temporary disturbances that could be avoided altogether by a healthy diet and lifestyle. If they occurred, they must be left to run their natural course. Any attempt to treat them with drugs, or to prevent them with measures such as vaccination, was futile and dangerous. What doctors in their naïveté saw as 'diseases' were actually signs that the body was curing itself. Take smallpox, for example: the rash was nothing more than the body eliminating dangerous waste through the skin and was therefore life-saving. This doctrine sat uncomfortably with certain inconvenient facts (such as the 20 per cent death rate from smallpox), but Lora Little found receptive and welcoming audiences wherever she went.[4]

In 1907–08, she visited Britain, still the major arena for the vaccination debate. She travelled from London as far north as Glasgow and Edinburgh, spreading the word about natural health and injecting New World insights and energy into the battle against vaccination. The highlight of her tour may well have been in the industrial sprawl of Leicester, the shining example of how a whole community could unite to fight back against the curse of vaccination. Here, the civic authorities had sided with the largely working-class population to oppose vaccination, tearing up the Vaccination Act and scrapping punishments for parents who refused to have their children vaccinated.[5] In Britain, Lora Little enjoyed celebrity status as a visiting speaker and made a lasting impact: material from *Crimes of the Cowpox Ring* was still being recycled by British anti-vaccinationists in 1957,[6] half a century after her tour. Her travels may have been more than an evangelical mission to spread her version of the gospel; they may also have helped her to live with bereavement and perhaps to avenge her son's death.

In late 1908, Lora Little returned to the United States and eventually settled in Portland, Oregon – just about as far from New York as she could go within the bounds of the country. There, she again built a new life for herself, based around the lucrative Little School of Health and her self-care philosophy which appealed to many in middle-class Portland: "Saves money, saves suffering, saves life. All diseases reached by my natural methods."[7]

And she continued to campaign against vaccination and the conspiracy of silence which covered up all the harm that it did.

Primum non nocere

"Vaccination propagates a variety of other diseases more fatal than smallpox – such as scarlet fever, croup, typhoid fever, scrofula, consumption, syphilis, tuberculous formations, diphtheria, & Co."

Dr. Schiefferdeck, New York, quoted by P. A. Taylor MP, in a speech to the House of Commons, 1880

Primum non nocere – 'Above all, do no harm' – is the time-honoured guiding principle for all medical treatments. For Lora Little and her associates, vaccination fell at this first hurdle. It caused a long list of serious and sometimes fatal complications, thanks to the toxins introduced in the vaccine lymph. Her own collection in *Crimes of the Cowpox Ring* was headed by infections: tuberculosis, massive ulcers, the runaway infection of skin and soft tissues called erysipelas, blood poisoning and diphtheria. Also featured were cancers, leukaemia, epilepsy, heart disease, stroke, blindness and insanity.

When she published *Crimes of the Cowpox Ring* in 1906, there was already a substantial literature about the dangers of vaccination. Some 30 years earlier, an article in the influential *Glasgow Herald* had listed 15 diseases caused by vaccination; the top four were syphilis, bronchitis, blood poisoning and skin disease, with diphtheria, cholera and whooping cough not far behind.[8] The ninth edition of *Encyclopaedia Britannica* (1888) – the fount of all knowledge for Britain and its Empire – contained a lengthy entry about vaccination, which was claimed to spread five major diseases: erysipelas, jaundice, various skin disorders, vaccination-site ulcers and syphilis.[9] The article was written by Dr. Charles Creighton, who was revered by the anti-vaccinationists because of his blunt criticisms of Jenner and his invention.

Problems were also reported from mainland Europe. In 1884, Dr. Hubert Boëns, a member of the Belgian Academy of Science and a major international player in the Universal League of Anti-Vaccinationists, published his own gruesome compendium of medical disasters following vaccination.[10] The 27 cases included a large tumour in the neck of a six-year-old boy, and lupus (destructive ulceration) that had steadily gnawed away the face of a man who had been compulsorily vaccinated in New York. This victim had been further mutilated by botched surgery that had cost him an eye and needed the weekly application of leeches to suck away the residual swelling. On the brink of suicide, he had contacted Boëns, asking to be variolated in the hope of expelling the cowpox virus that he believed had poisoned him. The outcome was not reported.

The anti-vaccinationists also picked through annual mortality returns and found that the rise in vaccination had been paralleled by increases in deaths from over a dozen diseases. A widely-circulated pamphlet, published in 1883 by William Scott Tebb,[11] quoted statistics that had an impeccable pedigree – the Annual Reports on deaths in England, issued by the Registrar-General. Concurrently with tighter enforcement of vaccination between 1850 and 1880, the death rates for syphilis (mostly in infants) and bronchitis had more than doubled, while cancer deaths had increased by over 60 per cent. To Tebb, these were not coincidences. Vaccination had caused the upswing in all these diseases – but not, of course, the remarkable decline in smallpox, which he attributed instead to general improvements in hygiene and sanitation.

In all, the anti-vaccinationists claimed that vaccination spread over 30 diseases, of which syphilis and erysipelas caused particular concern. During the nineteenth and early twentieth centuries, both these infections had a powerful grip on people from all social strata. Both targeted children and routinely carried off thousands of infants every year. These deaths were concentrated in the first few months of life, the period which the Act stipulated for vaccination, so any suggestion that this could be responsible instantly touched a raw nerve with parents.

Syphilis, with its judgemental overtones of divine punishment for the sin of illicit sex, aroused an almost primeval paranoia, and this was skilfully exploited by the anti-vaccinationists.

An even greater pox

"You see a little coffin and realise that the lymph was tainted."

P. A. Taylor MP, speech to the House of Commons, 1880

It is difficult to imagine a disease nastier than smallpox, but syphilis – the 'Great Pox' – is a strong contender. Its initial stages are less dramatic but, once its claws have sunk in, untreated syphilis never relaxes its grip. Ulcers initially chew away at the genitals, later transferring their attention to the face, skull or bones. The brain eventually takes the brunt, ending in the pitiful *sans everything* state termed 'general paralysis of the insane'. The corkscrew-shaped spirochaete bacterium that causes syphilis can also drill its way through the placenta and into the unborn child of an infected woman unlucky enough to fall pregnant. Congenital syphilis (Plate 12) makes no concessions for the blamelessness of the child: it stunts growth and development and causes telltale facial and other deformities, and until twentieth-century antibiotics arrived on the scene, it was often fatal during the first year of life.[12] Treatment of a sort became available in the early 1800s: mercurial salts, almost as toxic to human tissues as to the spirochaete. Their side-effects were often worse than the non-cure – hence the ironic question which many pondered in miserable retrospect: "Is one night with Venus worth a lifetime with Mercury?"

The anti-vaccinationists exploited the dread and revulsion generated by syphilis by claiming that vaccination could spread the infection. The first seeds of suspicion had been sown soon after the birth of vaccination, in Benjamin Moseley's caustic Latin nickname for cowpox, *Lues bovilla*.[13] *Lues* originally meant any kind of plague, but had come to be shorthand for *lues venerea*, or syphilis. Others pointed out that cowpox lesions resembled some syphilitic ulcers, although the similarity was superficial and syphilis was known as 'the great imitator' because it could mimic so many other lesions. Dr. Charles Creighton, of *Encyclopaedia Britannica* fame, and Edgar Crookshank, the eminent Professor of Bacteriology and Pathology at King's College in London, maintained that "smallpox and cowpox are distinct, and the nearest analogue to cowpox is syphilis".[14] Their opinion carried much weight, even though they were completely wrong – syphilis is a bacterial infection whereas cowpox is caused by a virus. Dr. J. W. Hodge, a reformed public vaccinator from Niagara Falls who publicly recanted his sins, went further in his book *Vaccination Superstition*. He stated that "cowpox and venereal pox have much in common" and that cowpox was nothing more than the Great Pox passed on to the cow's teats by syphilitic milkmaids.[15]

More worryingly, arm-to-arm vaccination stood accused of spreading the disease via lymph collected from donors (vaccinifers) with unsuspected syphilis: the early stages, although highly infectious, were often asymptomatic. Fear of being given vaccinal syphilis was widespread. In 1885, this stirred up the French Canadians in Montreal to march in protest on the City Hall; the resulting violence rapidly persuaded the authorities to scrap their proposal to make vaccination compulsory.[16] The pro-vaccinationists should have recognised this risk, but many simply dismissed the notion as scaremongering. In 1870, 500 English doctors signed a declaration stating categorically that vaccination could not transmit syphilis.[17] Unfortunately, they were wrong.

In 1861, an outbreak of syphilis in Rivalta, near Turin, had been traced to lymph obtained by vaccinating an apparently healthy eleven-month-old boy, Giovanni Chiabresa, which was then used to inoculate another 40 infants.[18] In retrospect, more care should have been taken. The lymph collected from the baby was obviously bloodstained and it turned out that he had been breast-fed by a wet nurse who had syphilitic ulcers around both nipples. But it was too late by then – a total of 50 children and 21 mothers and nurses had been given syphilis, and several children, including baby Giovanni, later died of the disease. Other well-documented outbreaks of vaccination-borne syphilis occurred between 1856 and 1866 at Lupara and Bergamo in Italy, and Auray in Brittany.[19] Each claimed up to 60 victims, but had been hushed up by doctors fearful for their own reputations as vaccinators and for the future of vaccination in general.

The Rivalta outbreak occurred nine years before the 500 English doctors declared that this could not possibly happen. Rivalta had been well publicised at the time, and all four outbreaks were described in exhaustive detail in 1868 by Dr. Edward Ballard, the Medical Officer in Islington in London. Ballard had won the coveted £100 annual prize put up by the Ladies' Sanitary Association for an essay on vaccination, with a 400-page dissertation entitled *On Vaccination: Benefits and Alleged Dangers* which devoted 84 pages to the hazards of vaccinal syphilis. The medical signatories included alleged experts on vaccination and syphilis; if they did not know about

Rivalta, they should have done. Whatever the excuses, signing up to the declaration was an act of idiotic negligence.

Also idiotic and negligent was a bold experiment which a Dr. Robert Cory, a "popular and much esteemed" government medical officer, conducted on himself in 1881. Cory was at least brave enough to put his beliefs to the test and, in the spirit of von Pettenkofer knocking back Koch's flask of cholera, he deliberately inoculated himself with lymph from several syphilitic children. This *grand geste* turned into a tragic *grande folie*. Cory nearly died of acute syphilis and was left partially paralysed.[20]

If doubters needed more evidence, it continued to accumulate. In 1870, the year of the 500 doctors' declaration, 18 twelve-year-old girls vaccinated in Frankfurt with 'animalised' lymph (i.e. transferred from a human vaccinifer into calves) all developed syphilis.[21] The fact that the lymph was approved and certificated by the National Vaccine Institute of Germany was of no comfort to the several girls who died, or to the survivors who were consigned to the scrapheap of spinsterhood and unemployability. Then in 1880, all 58 recruits in the Fourth Regiment of the French Zouaves in Algiers went down with the disease after being vaccinated with lymph from a supposedly healthy Spanish child. All survived but, on recovery from the acute illness, found themselves discharged from the army as well as the hospital, with "their health and future prospects fatally blighted".[22]

The anti-vaccinationists milked syphilis as hard as they could. They blamed vaccination for the six-fold rise in infant deaths from syphilis in Britain between 1847 and 1875, even though their argument was readily demolished. Post-vaccination syphilis did occur, but was rare (perhaps 1 per cent of the total cases of childhood syphilis) and the time-course did not fit. Most infant deaths from syphilis occurred in the first three months of life, whereas vaccination was not given in various countries (such as Scotland) until after that time.[23] Nevertheless, the threat of the Great Pox helped to fuel fears about vaccination. As late as the 1950s, heart-rending photographs of babies killed by vaccination-borne syphilis still figured in British anti-vaccination literature, together with the often-quoted epitaph by P. A. Taylor, Member of Parliament for Leicester, which heads this section.[24]

Vaccination was not cleared of the taint of syphilis for many years, mainly because the there were no diagnostic tests to guarantee that lymph donors were free from the disease. The only quality control of vaccine lymph was a visual check to make sure it was not obviously bloodstained. Unfortunately, the blood of subjects with early syphilis teemed with spirochaetes and an infective dose could be transmitted in lymph that appeared crystal-clear to the naked eye. Eventually, the threat of syphilis was abolished only by barring humans from all stages of vaccine production and using lymph generated exclusively in calves or other animals. In England, arm-to-arm vaccination was not banned until 1898.[25]

For the anti-vaccinationists, the doctors' stubborness and their conspicuous failure to apologise when the truth about syphilis eventually came out were a propaganda gift on a plate. If doctors could not be trusted over syphilis, then what else were they covering up? The anti-vaccinationists had opened up a can of worms and, on peering inside, thought they could see many other horrors wriggling away with the spirochaetes. Infections such as erysipelas, blood poisoning, tetanus, tuberculosis and diphtheria either lurked in the "rotten fluid" of the vaccine or were triggered by the "constitutional weakening" caused by toxins in the lymph – and the same toxins could also cause cancer and other diseases.

Many fears about vaccination tracked back to its bestial origins, which the anti-vaccinationists continued to play on, even if they had retracted the claim that vaccination would turn people into cows (see Chapter 11). In 1909, Ernest McCormack reiterated the ominous but evidence-free warning by Immanuel Kant – from 66 years earlier, of the "dangerous consequences from the absorption of a brutal miasma into the human blood".[26] Some pseudo-science was offered, presumably to combat all the serious-looking data pushed out by the germ-theorists. A fine example was McCormack's explanation of how vaccination had doubled the number of cancer deaths in England during the previous 20 years. Noting that humans need 21 years to reach maturity, whereas cows only take five years, he concluded that a cow's cells must grow much faster than a human's. Therefore, he declared, introducing a cow's protoplasm into a human "must upset the constitutional balance and promote the general condition in which cancer finds birth".[27]

More of the same came from Dr. W. A. Farr, who proposed that the fatal infectious (zymotic) diseases operated to a quota. When deaths from one infection declined for whatever reason, others compensated by killing more people. For example, a fall in smallpox deaths in Glasgow had been balanced by a concomitant rise in measles. Farr's notion was based on an embarrassingly simplistic look at selected mortality data but acquired some credibility because it was aired in the Registrar-General's Annual Report.[28]

William Tebb made inventive use of the Registrar-General's statistics. He calculated the excess mortality from six selected diseases that had increased while vaccination gained ground between 1850 and 1880. The additional deaths from these disorders totted up to 1,852 per million people, equivalent to 48,000 deaths for the whole population of England. Tebb blamed all these deaths on vaccination, on the grounds that it diminished "constitutional vitality" and so made "recovery from any disease more precarious". And what had vaccination achieved during this period? Certainly not the 85 per cent fall in smallpox, which Tebb maintained was due entirely to improved sanitation and effective isolation of cases. Hence Tebb's conclusion that "48,000 lives are now *annually* sacrificed" through the "vain attempt" of vaccination against smallpox.[29]

The poisoned lancet

"With due care in the performance of the operation, no risk of any injurious effects from it may be feared."

Facts concerning Vaccination for Heads of Families
The National Health Society, 1884

There is no doubt that various bacterial infections were commonly spread by vaccination during the eighteenth and nineteenth centuries. The medical profession was kind to germs throughout this period. For example, nobody saw anything wrong with medical students going straight from dissecting fresh corpses to attend mothers giving birth; their professors were not much better, sometimes pocketing organs cut out of the recently deceased, to teach students in the delivery room. It was only in 1847 that the Hungarian physician Ignasz Semmelweis wondered whether the "cadaveric poison" smeared over his students' unwashed hands could be connected with the shocking 10 per cent of mothers who died from fever after giving birth in his hospital in Vienna. Semmelweis made the students wash their hands in a calcium salt solution, and the number of deaths from puerperal fever immediately fell by 80 per cent.[30]

This is the distinctly grubby context in which we have to consider vaccination at the peak of the debate. Lister had established the principles of antisepsis by 1870, but the meticulous washing of hands and instruments was thought superfluous for the trivial procedure of vaccination. Until antisepsis was recommended in the early twentieth century, vaccination provided many opportunities for bacteria to hitch a lift, at all stages from lymph production to the patient's arm.

The lymph itself was often of dubious quality. A bewildering variety of vaccines was available around the turn of the twentieth century.[31] These were derived by inoculating the blister fluid from various infections, including horsepox and 'grease' as well as cowpox and smallpox, into human vaccinifers and/or other species, mostly calves. Vaccinators could choose from a menu that included 'horse-grease pure', 'horse-grease cowpox', 'vaccinised smallpox', 'spontaneous cowpox', 'arm-to-arm lymph', and even 'monkey lymph' and the ominously named 'corpse lymph'. Vaccines were sold by independent producers, such as 'Mulford's Pure Lymph' in the United States and 'Badcock's Lymph' in England. Numerous state-approved centres also distributed lymph, often free of charge; one such as was the Jennerian Institute in England (whose lymph was ranked twelfth for quality and reliability by the *Lancet* in 1890).[32]

Commercial vaccine production was big business – a 20-million-dollar-a year industry in the United States in 1906 – and Lora C. Little's assertion that financial imperatives ruled supreme may be close to the truth.[33] Proprietary names often emphasised purity and quality but covered up what was essentially a farmyard-based industry. Calves, often kept in squalid conditions, had their flanks shaved and inoculated with lymph from whichever source. After ten days, the blisters were cut open and the fluid allowed to run into a vessel pressed against the animal's skin. Some centres also used the juice squeezed out of scabs that were picked off the inoculation wounds. Looking back now, we might find some sympathy with the revulsion of the anti-vaccinationist E. McCormack, when he described vaccine lymph as "the virus obtained by squeezing the filth from sores raised on a calf's belly".[34]

There were no routine checks for bacterial contamination. This is unfortunate, as microscopy of commercial lymph samples readily revealed their origins from Blossom's kind, with bits of straw and manure bobbing about in a lively soup of bacteria.[35] The only purification step between the newly-harvested fluid and the patient's skin was filtration to remove obvious particles. A famous outbreak of erysipelas that killed several newly-vaccinated children in Norwich in 1882 was apparently caused by bacterial contamination of the lymph.[36] The introduction of glycerine in the 1880s to extend the vaccine's shelf-life was also found to inhibit the growth of some (but not all) bacteria but it was only decades later, with the addition of phenol, that vaccine lymph could be declared bacteriologically sterile.[37]

The act of vaccination provided a ready-made portal of entry into the body for bacteria in the lymph or on the skin, especially with aggressive inoculation techniques such as cross-hatching a square inch of skin with cuts that often drew blood. The various vaccination instruments – needles, 'points' made of metal or ivory carrying dried lymph, traditional lancets or even the elaborate 'Badcock's double lancet' – were no safer than the rusty needle with which the old Greek woman had terrorised Lady Mary's son Edward in Constantinople in 1718. For most vaccinators, the only nod to hygiene was the occasional wipe to keep their weapon socially clean. The 1892 edition of the physician's bible, *The Principles and Practice of Medicine* by William Osler, contained step-by-step instructions on vaccination but made no mention of sterile precautions.[38] In the same year, Queen Victoria's Royal Commission on Vaccination heard harrowing evidence about the infections caused by vaccination (see the case history below). In their final report, the Commission strongly advised that instruments should be sterilised by boiling between inoculations and that vaccination should be deferred if the child had erysipelas or scarlet fever – but this was only in 1896.[39]

Vaccination was very much at the mercy of the cleanliness of the environment. The leisurely, chocolate-box pictures of Jenner at work are in stark contrast to the grim reality of everyday vaccination practice – which at its worst could even constitute a war crime. In 1864, during the American Civil War, all 30,000 Union prisoners held in Camp Sumpter outside Andersonsville, Georgia, were vaccinated. Many developed severe vaccination-site infections, with gangrenous ulcers that bored deep into the flesh; 100 men lost the use of their arms, many needed amputation, and 200 died. A later casualty was the camp's Commandant, hanged after the war in November 1865 by the victorious Union forces for having deliberately introduced poison into the prisoners. The generally atrocious conditions in the camp were more likely responsible: the 200 deaths blamed directly on vaccination were eclipsed by the 13,000 fatalities from starvation, neglect and other infections.[40]

Another harsh glimpse of the working conditions of some vaccinators was provided by Dr. G. H. Merkel, who sailed from Bremen in Germany to New York in November 1882, on board the steamship *Neckar*.[41] The Atlantic crossing was rough in every sense, especially for the 800 steerage passengers crammed in below decks. Many were desperate to be vaccinated, by order of US Immigration, to avoid "much trouble and serious detention" in the notorious quarantine camp on Blackwell's Island in New York Harbour. Dr. Merkel may have felt some sneaking sympathy for his harassed colleague, working flat out (276 vaccinations on the first day) and facing a fine if he failed to inoculate all those who needed it. But Merkel pulls no punches in capturing the squalor of the scene: a store-room deep in the airless bowels of the ship, stinking of sweat and vomit, crammed to its walls with people jostling to get their entry ticket to the Land of the Free. At the centre of the pandemonium was the ship's doctor, perched on an upturned packing case, head down and grimly getting on with the job. The doctor evidently had "no fear of inoculating disease": he blew the lymph from its storage tubes into a cup, into which he dipped his one and only vaccination lancet. Dr. Merkel noted that he often drew blood, yet made no attempt to clean the lancet between patients. Jenner would have been appalled by this "pitiful sight", which Merkel blamed on the compulsion to vaccinate rather than the procedure itself: "Could the gentlemen through whose instrumentality the law was enacted, see what I saw of the manner in which it is carried into effect, they would have been as zealous in seeking its repeal."

Erysipelas, named for the fiery redness of the skin as the infection ploughs through the underlying tissues, topped Creighton's list of complications in the *Encyclopaedia Britannica*.[42] It is caused by streptococci bacteria which readily invade the bloodstream; until the advent of antibiotics after the Second World War, death from blood poisoning (septicaemia) was a depressingly common ending to this and other streptococcal infections. During the nineteenth and early twentieth centuries, erysipelas was ever-present and a major killer of infants. The tragic account of a 15-year-old girl from Liverpool, who died from erysipelas after being vaccinated by Dr. Thomas Skinner, illustrates the unpredictable and violent course of the disease and its horrific aftermath (see page 373).[43] This must have been an intensely disturbing Damascene moment for the shaken Dr. Skinner, who until then had been an unquestioning supporter of vaccination.

The medical profession took decades to recognise that vaccination could cause erysipelas. In Britain, deaths due to 'erysipelas after vaccination' were eventually listed separately in the Registrar-General's Annual Report, between 1856 and 1882 (after which they were lumped together with all vaccine-related deaths). With proper records, it became clear that erysipelas accounted for half of all deaths attributed to vaccination – but also that these numbers were swamped by all the other cases of fatal erysipelas. Each year in England, vaccination caused only 20–40 recorded deaths from erysipelas, a few percent of the 600-plus total.[44]

The slowness of the medical profession to acknowledge the risk, and evidence that they had actively covered it up, was highlighted by several high-profile *causes célèbres* in England during the 1870s and 1880s. Public vaccinators in Norwich and St. Pancras in London found themselves in the Coroner's Court and the unwelcome glare of publicity, having to explain why some of their cases had died from erysipelas and blood poisoning.[45] All were highly experienced – the Londoner had vaccinated over 40,000 children – but none seemed aware that they could introduce infection. In their defence, vaccinators were required to follow up their cases only to ensure that the inoculation had taken successfully; others filled in death certificates, and simply recorded erysipelas with no mention of vaccination. The evidence incriminating vaccination was compelling but the Coroners sat on the fence and recorded open verdicts. Elsewhere, judgments on similar cases were bizarre and evasive. In Leeds, the blame for a child's death from sepsis was successfully dumped on the Almighty. The Coroner directed the jury that there was no such legal entity as 'death from vaccination'; they therefore acquitted the two surgeons accused of manslaughter and pronounced that the child had died "from a visitation of God".[46]

Other infections commonly linked to vaccination included abscesses (carbuncles) and ulcers at vaccination sites, often caused by staphylococci. These bacteria tend to produce thick pus but can break out into the bloodstream, settling to form secondary abscesses in brain, liver or bones (osteomyelitis). Returning to *Crimes of the Cowpox Ring*, many of the cases of carbuncles, osteomyelitis (including Case No. 22) and blood poisoning could reasonably be blamed on staphylococci scratched into the skin at vaccination, or colonising an open vaccination sore afterwards. Fortunately much rarer was tetanus, with agonising muscle spasms leading to lockjaw and finally death by suffocation. The tetanus bacterium thrives in the colon of cattle – hence the connection with farmyard-contaminated lymph. Lora Little reported 33 cases of tetanus, all fatal, some of which were undoubtedly introduced by vaccination.

Occasionally, the vaccinia virus itself could cause a local or general infection, sometimes severe and life-threatening.[47] A vaccinal ulcer at the inoculation site could be large, agonisingly painful and take many weeks to heal. In the unlucky, including some of Lora Little's cases, the common name of 'bad arm' turned out to be a gross understatement and the limb had to be amputated. Fortunately much rarer (perhaps 40 vaccinations per million) was 'generalised vaccinia', when the whole of the skin broke out in blisters that resembled the usual vaccination-site pustules. Here, the vaccinia virus invaded the body, much as smallpox did, because of failure of the immune defences that normally kept the vaccinia virus corralled within the vaccination site. Generalised vaccinia carried a mortality rate of about 50 per cent. Another all-over eruption sometimes occurred when people with severe eczema were vaccinated – so-called 'eczema vaccinatum'. Initially, it was thought that vaccination had triggered the ezcema; in fact, the rash arose from virus replicating in previously damaged skin. As well as disfiguring, this could be fatal, especially in children. It was only during the 1930s that the risk was acknowledged and eczema became an absolute contraindication for vaccination. During the nineteenth century there were also scattered reports of 'sleepy sickness' which probably represented the inflammation of the brain and spinal cord (encephalomyelitis) that later emerged as a very rare allergic complication of vaccination.

The anti-vaccinationists also claimed that vaccination caused many other diseases, for which there is no convincing evidence of any causative link. Some diseases were incriminated by implication, as they had become commoner more or less in step with vaccination, while others were noted in particular patients who had been vaccinated. The table from Wheeler's pamphlet shows the death rates – the annual mortality in England and Wales and also the chances of dying from an attack – that were typical in the early 1890s for some of these diseases. These should be contrasted with death-rates for vaccination and smallpox (which by now, had sunk to 72nd place in the British mortality league table; see Fig. 13.1, page 284).[48]

Tuberculosis – conspicuous, widely feared and deadly – was often seized upon by the anti-vaccinationists. The lungs filled up with 'tubercles' of cheesy debris, which broke down into cavities that ate into blood vessels – hence the common symptom of coughing up blood. End-stage tuberculosis was called 'consumption' because the victims wasted away, their flesh seemingly eaten up by the infection. During the nineteenth century, pulmonary tuberculosis killed one person in five, including Jenner's wife Catherine and his son Edward.

Tuberculosis could also infect the lymph glands in the neck, which liquefied and chronically discharged pus through a sinus that broke out through the skin. This unpleasant condition – scrofula – had been claimed to be a complication of vaccination by 'Dr. Squirrel', soon after the publication of Jenner's *Inquiry* (see Chapter 9).

Tuberculosis had a ready-made link with cows, which the anti-vaccinationists tried to exploit. Cattle also suffered from tuberculosis and it could spread to humans who drank infected milk – so why could it not be transmitted by vaccination? Tuberculosis deaths had risen steadily during the enforced vaccination campaign of 1850–80, and McCormack and others held vaccination entirely responsible. Rare cases were documented of bovine tuberculosis being spread by vaccination, but the overall risk was extremely low – if only because the calves used to raise lymph were killed (often for the table) before they were old enough to become heavily infected. Globally, tuberculosis tracks closely with social deprivation and overcrowding, and the rise in deaths in England and Wales reflected the failure to improve living conditions among their lower classes.[49] Eventually, the weight of evidence suffocated the notion that vaccination spread tuberculosis, but anti-vaccinationists continued to claim a link well into the 1950s.[50]

Vaccination was also held responsible for numerous other disorders, including diphtheria, bronchitis, whooping cough, cancers, heart disease, stroke and diabetes. It was often difficult to rebut the accusation that vaccination caused a particular disease, especially when this was backed up by official data (even if these had been massaged into absurdity) or by emotionally-charged case reports. In some cases, it took decades to debunk the mythology of causation, but by the start of the twentieth century, the weight of evidence was already tipping against any convincing link with these diseases. For the hard-core opponents of vaccination, however, this was simply more defensive posturing by all those who had a vested interest in keeping alive the "medical Frankenstein" of vaccination.[51]

Those injured or bereaved following vaccination could be forgiven for attaching blame to the procedure. Human nature has always pushed us to find a cause for our misfortunes and there is a long tradition of sensible people convincing themselves that their illnesses are due to specific events, be these comets in the night sky, a sin or some otherwise random event that coincidence has thrown into prominence. Vaccination was bound to attract blame, especially in a time of imperfect knowledge, deliberate disinformation and a high childhood mortality rate (which even in the 1890s exceeded 40 per cent for the under-fives in Britain's slums). High-profile and widely resented, vaccination had built up a sinister reputation for causing harm in many ways – and there was a vocal community keen to welcome in anyone who had a grievance. The personalities and attitudes of some pro-vaccinationists also helped the anti-vaccination cause. It was much more satisfying to blame people, rather than chance or an undeserved act of God – especially when the people concerned were easy to hate because they were arrogant, evasive or dishonest.

Dead reckoning

"25,000 babies are yearly sacrificed by diseases excited by Vaccination."
Anti-Vaccination League, 1880

"6,000 infants are killed every year by Vaccination."
Anti-Vaccination League, 1921

The vaccination debate saw volleys of imaginative statistics fired off repeatedly by both sides. The anti-vaccinationists routinely inflated the risks of illness and death, while their opponents were guilty of crimes of omission and obfuscation. Without any evidence, the Anti-Vaccination League made the shocking claim in 1880 that vaccination killed 25,000 infants each year, while William Tebb at least made some attempt (even if spurious) to justify his cobbled-together total of 48,000 deaths annually in England and Wales.[52] These numbers could not be challenged while doctors were unaware of the risks of vaccination – or denied them. Moreover, official records such as the London Bills of Mortality and the Registrar-General's returns were so full of holes and errors that doubters could reasonably suspect a cover-up.

Eventually, and against resistance from the medical establishment, systematic monitoring and reporting of the complications of vaccination were brought in. The possibility that vaccination could kill was only conceded on death certificates and annual mortality returns during the 1850s. The data proved that vaccination did indeed kill, but also showed that these deaths were rare events and a tiny fraction of the numbers bandied around by the anti-vaccinationists. In the decade from 1881–91, there were 6,739,902 vaccinations (including revaccinations in adults), at a cost of 476 deaths reported as a direct result of the procedure. This represented an overall mortality rate of 1 in 14,159, or 0.007 per cent. As already mentioned, about half the deaths were caused by erysipelas.[53]

These broadly reassuring statistics had little impact on the opponents of vaccination. For some, including the two dissenting members of the Royal Commission (Collins and Picton) and Darwin's celebrated colleague Alfred Russel Wallace, these enemy data proved that vaccination was hazardous and therefore unacceptable because it brought no benefits to offset its dangers. Others, convinced that all official statistics were fudged to defend vaccination, continued to pluck much more impressive numbers out of the air. In 1907 the Anti-Vaccination League claimed that 6,000 children died each year from vaccination in England and Wales. This figure, although still about 100 times the actual risk, was more temperate than the 800-fold exaggeration of 30 years earlier.[54]

End of a campaign trail

"The cry of the anti-vaccinator ... is that thousands of children are crying for the infantine health which nature offers, but which professional interest does not permit them to enjoy. They may be born of healthy parents, yet they must be exposed to suffering and possible death, through this system of universal State blood-poisoning: and Rachels are weeping throughout the land because their hearths are made desolate."

International Anti-Vaccination League, 1880

This brings us back to Lora C. Little, now settled in Portland, Oregon, and making a good income from her school of natural health.[55] She was a versatile campaigner who took on many causes and stirred up frequent confrontations with the city and state authorities. She locked horns with sugar, millionaires, white flour, germ theory, high heels and enemas and, as Vice-President of the Anti-Sterilization League in Portland, led the fight against the compulsory castration of male "habitual criminals, moral degenerates and sexual perverts". Above all, she remained faithful to the causes of stamping out vaccination. In 1916, the Medical Education Freedom Committee of Battle Creek, which enjoyed a national reputation for militancy, sent her on a pan-American crusade against vaccination. She spread the word so effectively that she fell foul of the Espionage Act and landed up in prison in North Dakota, for inciting mutiny by telling servicemen to resist vaccination.[56]

For the rest of her life, Lora Little continued to practise what she preached and was a perpetual thorn in the side of the authorities. According to viewpoint, she was either a shining beacon in the fight against racketeering by the medical mafia, or the dangerously deluded irritant who required lifelong monitoring by the Propaganda Department of the American Medical Association.[57] She died, aged 75, in Chicago in October 1931. Kenneth would have been 42 years old.

Her photograph, taken in 1913, shows a no-nonsense, rather spinsterly lady with a determined lift to her chin, very much

Figure 12.2 Lora C. Little. Reproduced by kind permission of St. Deiniol's Library, Hawarden.

the seasoned campaigner and the founder-editor-publisher of *The Liberator.* We can well believe that this was a woman of strongly-held beliefs, voiced without mincing words or fearing consequences. Little's ideas about health and disease were both odd and extreme but seem to have gone largely unchallenged. Her force of character probably bulldozed aside criticisms, but she also found a ready audience in Middle America where many were revolted by the "ghoulish work" of Jenner and Pasteur or enraged by the authorities' presumption that they could enforce vaccination against the will of the individual. Her views about smallpox were entirely in line with the doctrine of leading anti-vaccinationists such as Hubert Boëns, the evangelist determined to prove that smallpox "was not such a murderous or disfiguring disorder" and the inventor of anal leech therapy to cure the disease (see Chapter 11). A few minutes spent on the internet will confirm that there are still people today who would give credence to her views.

As with many of the combatants, we are left wondering what really drove Lora C. Little into the heat of the vaccination battle. Was she was above all a loving mother, devastated by her son's death and fighting to spare others her agony? Or a calculating troublemaker, cynically exploiting the deaths of children – including her own – to make a living out of natural health and topple the authority figures for whom she had a pathological hatred? Or simply a crackpot, unhinged by her bizarre beliefs? There may be a hint in *Crimes of the Cowpox Ring*, where she takes a gratuitously cruel swipe at parents whose children had died after vaccination but who were "unwilling to serve humanity by allowing their names to be used here" and who "therefore ... seem to have deserved some punishment".[58] Could it be that this unspecified punishment was the death of their child?

The fate of Kenneth Marion Little, her index case, may also provide some insights. The legend to his photograph reads simply, "Vaccinated Sept., 1895. Died from the effects, April 10, 1896", but his mother's brief report on Case No. 30 reveals a more complicated sequence of events. Kenneth's last seven months were beset with medical challenges, although not exceptional for a time when only three out of five American children lived to count ten candles on their birthday cake. He suffered from recurrent ear and throat infections, then ran up against two of the most accomplished slayers of children: measles, from which he recovered, then diphtheria, which killed him. To Lora Little, it was an open and shut case. Vaccination, the "artificial pollution of the blood", had fatally weakened his constitution and left him at the mercy of the subsequent infections. Measles behaved as expected, with the rash confirming that the little boy's body was expelling its toxins. But diphtheria, with its leathery grey membranes choking her terrified son to death before her eyes, could have shaken her belief in natural health.

Later in her life, Kenneth's death could have come back to haunt her in a different way. Her reputation as a natural healer rested on her guarantee that she could teach how "to keep well, to get well quickly when sick, to know how to take care of your family" and that her methods covered "all diseases".[59] Her failure to save her own son could have dealt a fatal blow to her credibility – but luckily for the Little School of Health, any flak was safely deflected on to vaccination.

The year in which Kenneth Little died, 1896, was a turning point in the treatment of diphtheria. Until then, the disease was incurable and killed half of its victims, but things were about to change dramatically for the better. A year earlier, the work of the recently-deceased Louis Pasteur had come to fruition when diphtheria antitoxin was successfully raised in horses. Within five years, the antitoxin was in clinical use, snatching diphtheria victims back from death and pushing the mortality rate down to just 12 per cent.[60]

Had the life-saving antitoxin been available at the time, Lora Little would have faced a terrible dilemma. If she remained true to her principles, the hardened campaigner against the bestial products of the medical cartel would have had to refuse the treatment; but a loving mother might have pulled back from sacrificing her only son on the altar of her beliefs.

13 The Most Beautiful Discovery or a Disastrous Illusion?

"Out of the crooked timber of humanity no straight thing can be made."
Immanuel Kant (1724–1804)

If smallpox had been a sentient predator rather than a chunk of biological machine code, it might have been puzzled by the spectacle of its victims fighting among themselves rather than strengthening their defences. The commanders of the human army – the leading lights of medicine, science and politics – should have buried their differences, solved the problems of vaccination and then put it to the best possible use. Instead, they poured their energy into posturing, trying to defend the indefensible and picking holes in the opposition. Both sides were crippled by tunnel vision which blinded them to the valid claims of their opponents and to their own faults.

Allegiance to either cause bulldozed aside the duty and conscience of the doctor as well as the objectivity of the scientist. The supporters of vaccination were rightly disgusted to discover that Dr. Keller of the Austrian Railways had falsified his paper showing that vaccination increased the risk of dying of smallpox. Yet they remained strangely silent when it emerged that doctors on their own side had consistently lied about the vaccination status of smallpox victims – and even when God, rather than the complications of vaccination, was blamed for killing children in Norwich.

This was the first big public debate in which statistics served as ammunition (see Fig. 13.1). Both sides were guilty of using numbers to numb and confuse rather than to inform, and of massaging data. P. A. Taylor conceded that determined statisticians could make numbers dance to whichever tune they

VACCINATION RESULTS.

MEAN ANNUAL RATE OF MORTALITY IN ENGLAND from SMALL-POX, and from six directly or indirectly inoculable causes, during each Quinquenniad of the 33 years, 1850–80. (P. lxxix., Table 34, of the 43rd Annual Report of the Registrar-General, 1882.) N.B.—Vaccination made compulsory, 1853; more stringently so, 1867.

Causes of Death.	Annual Deaths per Million Living.							Increase of Mortality per cent. 1880 over 1850.
	1850–4	1855–9	1860–4	1865–9	1870–4	1875–9	1878–80	
Small-pox *	279	199	190	147	433‡	81	40	—
Syphilis......	37	50	63	82	81	85	84	127 p. cent.
Cancer	302	327	368	404	442	492	510	70 p. cent.
Tabes Mesenterica..	264	261	272	315	298	330	340	29 p. cent.
Phlegmon & Pyæmia...	20	18	23	23	29	39	40	100 p. cent.
Skin Disease	11	15	15	17	18	23	23	109 p. cent.
Bronchitis †.	1,016	1,358	1,658	1,839	2,105	2,464	2,505	144 p. cent.
Total 6 causes	1,650	2,029	2,399	2,680	2,973	3,433	3,502	112 p. cent.

Increase 1,852, or, in round numbers, 48,000 annually. ‡

* Small-pox is an epidemical disease, and there has occurred but one great outbreak (1871-2), 18 years subsequent to the introduction of enforced Vaccination, and four years after the further and more stringent Act.

† Bronchitis, though not, perhaps, directly inoculable, is often observed by intelligent medical authorities to supervene upon, or soon after, Vaccination; and it is obvious that diminished constitutional vitality will render recovery from any disease more precarious; hence the benefit derived by newly-born children from improved sanitation is nearly neutralized by Vaccination.

‡ The single epidemic (1871-2) carried off 44,000 only, the general death-rate of the years being somewhat under the average of the period. There was, therefore, no loss thereby, because, had small-pox not been the epidemic, more would have died from other causes; nevertheless, we see that in the vain attempt to prevent this erroneously supposed loss in two years out of thirty, 48,000 lives are now *annually* sacrificed.

Figure 13.1 Table of death rates from smallpox and other diseases, from a pamphlet by Alexander Wheeler. Reproduced by kind permission of St. Deiniol's Library, Hawarden.

wished – "I know how easy it is to make [mortality] statistics show almost anything by selecting years and manipulating averages"[1] – but he and others kept on doing it. Eleanor McBean, a twentieth-century anti-vaccinationist, put it more succinctly: "Figures cannot lie, but liars can figure."[2] True, but it was the anti-vaccinationists who had the worse record for torturing statistics into making false confessions.

The vaccination debate permeated all levels of society and spilled out on to the streets through pamphlets, newspapers and public meetings. Before long, the media had come to confuse the message. The most inflammatory sound-bites came from the anti-vaccinationists such as Alexander Wheeler, ranting about Jenner:

> His unscientific, foolish, unsupported assertions show that it was time that he should die ... Only in oblivion should his name and doings be decently buried.[3]

However, the supporters of Jenner and vaccination were guilty of the same crimes of pigheadedness and prejudice, even if their language was more moderate. Sir John Simon epitomised the arrogance with which Jenner's disciples (and Jenner himself) brushed off their critics: "Against the vast gain of vaccination, there is no loss to count."[4]

Not surprisingly, confusion reigned.

A prolonged enquiry

The stand-off triggered by the first Vaccination Act had lasted nearly 50 years when, in May 1889, Queen Victoria ordered a Royal Commission to determine once and for all whether the country should continue to carry the burden of compulsory vaccination. Her Majesty commanded the Commission to "report to us with as little delay as possible" and to provide clear answers to simple questions. Did vaccination decrease the risk of catching smallpox and of dying from it, and was there any viable alternative to it?[5]

A few years later, Lora C. Little remarked, "It seems peculiar for an experiment that has been running for a century, that the only way to persuade people to undergo it is through compulsion."[6] To explore the outcome of that 'experiment', the Royal Commission met 136 times, interviewed 187 witnesses from both sides of the debate and picked through mountains of data, beginning with Jenner's original experiments and ending with current reviews of the successes, failures and complications of vaccination. The members of the Commission worked hard but, as the long-suffering Chairman wrote to preface their Final Report: "Our inquiry has been a prolonged one."[7] The Commission sat for seven years, long enough to be mocked by William Tebb for its "extreme mental exercise"[8] and for three of its 15 original members to die. It was also divided by disagreements. The 137-page Final Report of the Commission was a detailed analysis of the evidence about the benefits and risks of vaccination. So was the 64-page Addendum written by two dissenting members of the Commission, William Collins and Allinson Picton, who were so convinced that vaccination was useless and dangerous that they refused to sign the main Report. The majority Report concluded with a form of words that suggests that the Chairman had tried hard to keep the dissenters on board:

> On the whole, we think that the marked decline of smallpox mortality in the first quarter of this century affords substantial evidence of the protective influence of vaccination.

The majority (but not the dissenters) felt that the overall risks of vaccination were acceptable, given its likely benefits. Based on data from 1881 to 1891, the risks of dying after vaccination were only 1 in 14,149 (for comparison, the chances of being killed in a rail accident were calculated at 1 in 169 million). Were there real alternatives to vaccination? Apart from Collins and Picton, the Commission believed not. The only other possible strategy – rigidly enforced isolation of all smallpox cases – was expensive and not feasible in many parts of Britain.[9]

The answers to the Queen's questions were therefore not as clear-cut as had been hoped but were taken to indicate that vaccination had withstood close scrutiny. However, compulsion was not the solution, even though another two members of the Commission made it clear that both vaccination and revaccination should be made compulsory.[10]

As it stood, the Vaccination Act was clearly not working. The Commission had surveyed over 600 local authorities and found that one-fifth were not enforcing the Act at all – some of them apparently because they were waiting for the Commission's verdict.[11] The Commission noted the intense popular hostility to compulsory vaccination in England and acknowledged that this was not simply due to "the work of agitators". They also cast longing glances towards Scotland, where a softer-touch and more family-friendly introduction of compulsory vaccination had caused little friction.

The Final Report recommended some softening of the Act. Those imprisoned for refusing to pay the vaccination fine would no longer have a criminal record. Most significantly, those refusing vaccination because it offended their beliefs should have the opportunity to plead their case before a magistrate who could decide to accept "a sworn deposition of conscientious objection".[12]

This marked the turning-point in vaccination policy in Britain and the beginning of the twelve-year wind-down that ended in victory for the anti-vaccinationists. The first milestone was the amendment to the Act in 1898, two years after the Final Report appeared, which allowed a plea of 'conscientious objection' to be heard by a magistrate.[13] If the magistrate was not convinced, then the parent had to submit the child to vaccination or face the same penalties as before. This subjective assessment was obviously weighted against the poorly-educated and the tongue-tied and was doomed from the start to cause pain and anger. The legislation was soon shown to be unworkable but it limped along until 1907, when a further amendment broadened the definition of 'conscientious objection' so that all that was needed was the parent's signature on a declaration. This relaxation of the rules made it virtually impossible to enforce compulsory vaccination and it was only a matter of time before it was killed off completely. The Vaccination Act, one of the most divisive and unpopular pieces of legislation in British history, was eventually repealed in 1909.

Other countries followed the same cycle of compulsion–revolt–repeal at their own tempo. Permissive and then compulsory vaccination had been introduced across most American states by the turn of the twentieth century, but were repealed in seven states (including California and Utah) around the same time as in England. Other states followed, notably Arizona and Washington, and by the 1930s the balance had tipped so far that some had introduced laws that banned compulsory vaccination.[14]

Further south, opposition to compulsory vaccination was even more robust. An attempt by the Texas Rangers to force Tejano Mexicans to be vaccinated ended unhappily (and fatally for one of the Rangers' officers).[15] In Brazil, Latin American energy compressed 67 years of English debate into a few violent weeks. In 1904, the government's plans to enforce compulsory vaccination in Rio de Janeiro provoked widespread rioting, the wrecking of large parts of the city, and the rapid scrapping of the legislation.[16]

Comeuppance

The most obvious consequence of the vaccination debate was damage. Lives were lost through the efforts of both sides, by obstructing both the spread of vaccination and the introduction of revaccination. Some good also came out of the confrontation by improving vaccination practice, although this could have been achieved much more easily.

The centres of resistance to vaccination built up large susceptible populations that should, according to the pro-vaccinationists, have been asking for trouble. Whether or not they got what they deserved was sometimes hard to interpret, mainly because of the unpredictability of smallpox itself and the fact that the protective effect of a single vaccination in infancy wore off after a variable interval.

Montreal had a violent rite of passage through legislation to make vaccination compulsory.[17] The French-Canadian population had long been opposed to arm-to-arm vaccination because they believed (correctly, although this was rare) that it transmitted syphilis. By the early 1880s, most of the city's population of about 168,000 were unvaccinated and all it took was a single, well-placed case of smallpox to ignite a major outbreak. This happened at the end of February 1885, when a Pullman car conductor off a train from Chicago was admitted to a general ward in the Hôtel-Dieu hospital with 'chickenpox'. The true diagnosis only declared itself after he had been discharged, when a hospital servant died of classical smallpox. Inexplicably, the doctors forgot about the infectious prodromal period and sent home all the hospital's patients who did not have a rash. The result was entirely predictable. As William Osler wrote, "Like fire in dry grass the contagion spread"[18] – whereupon the authorities panicked and rushed through a bill to enforce compulsory vaccination.

The result: 10,000 people rioting in the streets, pitched battles with police and troops, and limited victories for both vaccination and the Angel of Death. Almost half the population were eventually vaccinated but the year-long outbreak claimed 10,000 victims and killed over 3,000. Children paid dearly for their parents' beliefs. Abandoning vaccination in infancy had restored the age-old preference of smallpox for children, and almost 90 per cent of those who died were under 10 years of age.[19]

Leicester, the shining example of resistance against vaccination, was in the spotlight for years while both sides of the debate waited to see what would happen there. The city's attitude to vaccination had hardened after the great epidemic of 1871–72, which caused 3,000 cases and nearly 350 deaths – hardly proof of efficacy for vaccination, which had been given to over 90 per cent of its infants. Following protests and anger stirred up by P. A. Taylor and others, the vaccination rates slumped.[20] In 1881, Hubert Boëns wrote to congratulate the 1,200 parents of Leicester who had refused to let their children receive "poisoned blood".[21] By 1885, 4,000 parents were awaiting prosecution for non-compliance with the Act and by 1892, only 3 per cent of infants were vaccinated. This would have been a major headache for the Guardians if they had not been selected for their anti-vaccination leanings.[22]

For over 20 years, Leicester seemed to lead a charmed existence. The disaster that was waiting to happen was largely averted by Dr. Killick Millard, appointed as Medical Officer of Health in 1901.[23] Millard put in place defences that were almost as proscriptive as the Vaccination Act: compulsory notification of all new smallpox cases and their strict confinement in an isolation hospital. Amazingly, he also managed to persuade the authorities that vaccination was justified under specific circumstances, namely to protect nurses working in the isolation hospital and to treat the immediate contacts of each new case, to immunise them and stop the infection from spreading. This strategy worked. Despite the large susceptible population, the next outbreak was swiftly contained with only 150 cases and 20 deaths. However, the detail behind these headline numbers was revealing. Two-thirds of the cases and deaths were in children – over twice the relative mortality rate for children in well-vaccinated cities such as Sheffield. Also, 22 of the 28 nurses chose revaccination and none of them caught smallpox; of the six who refused, four became infected and one died.[24] Leicester therefore survived its anti-vaccinationist stance thanks to an imaginative strategy that won over public opinion and was allowed to work.

Gloucester, another famous centre of resistance, was less fortunate.[25] Its hospital was dirty, badly run, understaffed and overcrowded, often with two (sometimes four) patients to a bed. Its local reputation as a death-trap meant that smallpox victims refused to go there. All attempts at notification and isolation, the foundations of Leicester's success, therefore failed. Dr. Millard's counterparts in Gloucester were more interested in private practice than in public health, and the anti-vaccinationists had free rein. They were headed by Walter Hadwen, a local doctor who had already won popular sympathy following his repeated prosecutions for infringing the Vaccination Act.[26]

Conditions for a major outbreak matured steadily until 1895, when smallpox flared up in a primary school and spread quickly. Again, the children took the brunt: fewer than 15 per cent of them had been vaccinated and, as in Leicester, two-thirds of the 400 deaths were in the under-tens.[27] To the disgust of the good Dr. Hadwen, compulsory vaccination was rushed in for children and adults and even worse, it was accepted as a lesser evil. Fortunately, this rescued a potentially catastrophic situation that could have cost many more lives.

Gloucester illustrated the resilience of the anti-vaccinationists and the ease with which mistrust of vaccination could be whipped up. Dr. Hadwen – venerated in the style of Francis of Assisi as 'Hadwen of Gloucester' – was well liked, charismatic and passionate.[28] He was also anti- other things. In an early recording, he can be heard hectoring people to join the Anti-Vivisection League, a theme which he tied in with the filthy ritual of vaccination in his pamphlet, *The Modern Medicine Man – How he Manufactures his Vaccines and Serum*. (Hadwen also dabbled in another popular topic, namely *Premature Burial – and How it may be Avoided*.)

Hadwen kept the anti-vaccination spirit alive in the city, which was hit by another smallpox outbreak in 1923. Fortunately, this was the milder *Variola minor*. He was later accused of manslaughter over his management of a girl who may or may not have had diphtheria, as he failed to give her diphtheria antiserum. He was found guilty at the High Court in London but an appeal in the more sympathetic setting of Gloucester Assizes (with "dramatic scenes" reported by the *Gloucester Chronicle* of 8 November 1924) led to his acquittal. When he emerged from the courtroom, the man who had done more than any other to keep a warm welcome for the Angel of Death in Gloucester was given a rapturous reception by his grateful townsfolk.[29]

Silver lining

There were also reasons to be grateful for the antagonism which vaccination stirred up. The debate cast a harsh and unforgiving light on glib assumptions and sloppy practice that the pro-vaccinationists had taken for granted. Two false idols – both raised by Jenner himself – were toppled, to the benefit of all.

The first was the disastrous notion that a single vaccination in infancy would protect for life. Jenner considered this a crucial selling point for vaccination. It was assumed that variolation gave permanent immunity, so vaccination must do the same. He had no direct evidence to support this view and systematic studies soon confirmed that vaccinated infants could not expect permanent immunity. The first suspicions were raised by Dr. Goldson of Portsea as early as 1804, although the demonstration by Dr. Herder of St. Petersburg that immunity generally ran out after 14 years of age was not published until the year of Jenner's death.[30] Jenner reacted petulantly to Goldson's report, possibly because he took it as a personal slight, and bracketed him with the other enemies of the cause: "The impudence of Pearson, the folly of Goldson, the baseness of Moseley and Squirrel ... the absurdity of Birch."[31]

McCormack certainly had grounds for criticising Jenner's blind adherence to his dogma and his "pathetic remonstrances" that the need for revaccination would "rob his discovery of more than half its virtues".[32] It is also easy to understand the anger provoked when Jenner's disciples refused to acknowledge that he and they were wrong, and by their repeated attempts to cover up the fact that vaccinated subjects could catch smallpox. As a Leeds City Councillor told an anti-vaccination meeting in 1876, he did not believe the official statistics produced by the Medical Officers of the city's Smallpox Hospital "because I know they are cooked" – as indeed they were.[33] In the end, hard evidence prevailed in England but only after eight decades of fighting against determined resistance. By 1883, revaccination was recommended at around the age of 14 and whenever smallpox broke out in the vicinity.[34]

Another Jennerian invention was arm-to-arm vaccination which he had pioneered using Hannah Excell as a vaccinifer to produce lymph for inoculating others, including his son Robert. Arm-to-arm vaccination had kept the missionary torch alight during crossings of the great oceans and had fired up the mass vaccination campaigns in many parts of the world where cowpox did not occur. However, it should have been scrapped as soon as solid evidence emerged that it could spread human diseases such as syphilis (see Chapter 12). Again, this notion was actively resisted because of misplaced loyalty to the cause. Arm-to-arm vaccination was finally abandoned in England in 1898 (when it was proscribed by the Vaccination Act) and in most countries by 1900.

The debate also focused attention on the risks of vaccination. It set in motion the formal monitoring of complications and led to much-needed improvements in vaccine preparation and vaccination technique – although the descendants of Mr. Scrimshaw's "Medical Ignoramus" could always be trusted to flaunt the rules. A more systematic approach to defining the risks and benefits of vaccination was also hinted at by P. A. Taylor in his speech to the House of Commons in June 1880.[35] He suggested investigating

> 1,000 children under the same condition and of the same constitution, half of whom are vaccinated and half not, and see what becomes of them.

This was a template for a 'controlled' clinical trial (i.e. comparing the effects of a treatment in a matched but untreated group of 'control' subjects) of the type that is now mandatory for all new vaccines and indeed all new treatments. Rigorous clinical trials were only introduced from the mid-twentieth century and vaccination against smallpox was never formally put to the test. Unfortunately, Taylor did not follow through his pioneering idea. Instead, he fell back on a study that he considered to be "the most satisfactory experiment" that he had ever seen. This turned out to be a study that we have already met – a survey comparing death rates from smallpox in vaccinated versus unvaccinated adults and children, published by a Dr. Joseph Keller of the Austrian Railways Company.

The final, oblique spin-off of the debate was the notion of 'conscientious objection'. The phrase entered the language in 1907 and the concept was soon picked up by other causes when the will of the State cut across an individual's beliefs and convictions. The most imminent and obvious example was the pacifists who refused to take up arms during the First World War.

Strange bedfellows and long shadows

The anti-vaccination campaign appealed to many factions because of its refreshing contempt of medical dogma and doctors, and its courageous stand against flawed legislation and the State's attempts to demolish personal freedom.

Various groups in 'fringe' medicine aligned themselves with the cause, including natural healers, homeopaths and chiropractors. Chiropractic, the brainchild of the father and son Palmers in the 1890s, explained all diseases as blockages of an innate life-force that normally flowed throughout the body but could be obstructed by 'luxated' vertebrae that had slipped out of line and pressed on nerves.[36] The inflamed nerves "expressed too much heat at their twig ends" causing (for example) epilepsy, fever, diphtheria and smallpox (which was "one and the same disease" as chickenpox). Diphtheria could be treated by a single manipulation to realign the displaced vertebra, which was much neater than "being poisoned" by doctors or suffering "200 movements" inflicted by an osteopath. The same applied to smallpox, in which the fifth cervical vertebra was characteristically luxated. The Palmers observed:

> When it is generally known that a large share of diseases, including small pox [sic], are caused by luxation of the vertebrae and that replacing these will reduce the temperature to normal and cure the patient, then we will give the death blow to the vaccine poison swindle.[37]

The Palmers poured further contempt on vaccination and highlighted its dangers and horrors with graphic cases. One of these, "Another Awful Death", describes the fate of a 23-year-old man. Before vaccination – "the monster which pollutes the purity of our children with the foul excretions thrown off from diseased beasts" – he had enjoyed perfect health and his skin had been "smooth, clean and beautiful." Following vaccination, a huge tumour had grown out of the vaccination site, finally killing him.[38] His name was Benjamin F. Olewine of Altoona, Pennsylvania: we have already met him as Lora C. Little's Case No. 275. Whether he liked it or not, the unfortunate Benjamin Olewine became a multipurpose martyr of vaccination who was revered by several causes. A brief internet search will show that he is still remembered and exploited today.

The anti-vaccinationists remain alive and well. With the demise of vaccination against smallpox, their focus has moved on to other targets – polio, measles-mumps-rubella (MMR), human papilloma virus (HPV) – but their rhetoric, concepts and attack strategies can all be traced back to the original great vaccination debate. The anti-vaccination crusade received a new lease of life during the 1950s, thanks to the significant side-effects of the injected Salk polio vaccine, later replaced by the safer oral Sabin vaccine.[39] The medical authorities proved again that they were slow to learn by failing to listen to or act on legitimate concerns about the vaccine's safety, or to demonstrate quickly and convincingly that its general benefits outweighed individual risk.

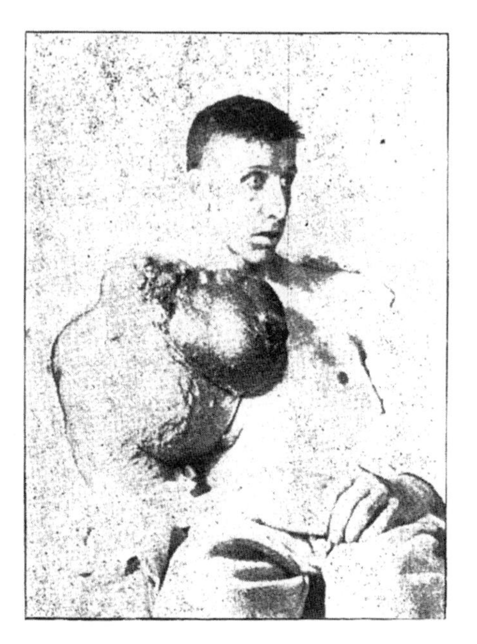

Figure 13.2 Benjamin Olewine. Case No. 275 in Lora C. Little's *Crimes of the Cowpox Ring*. Reproduced by kind permission of St. Deiniol's Library, Hawarden.

One of the responses to the controversy over smallpox vaccination was *The Poisoned Needle*, a book published by Eleanor McBean in 1953.[40] McBean recycled many of the elements of the original vaccination debate, including heart-rending pictures of babies who had died from vaccine-transmitted syphilis – and a 23-year-old man from Altoona in Pennsylvania, previously "clean and smooth of skin", killed by a massive tumour at his vaccination site. McBean also strayed into the realms of fiction, describing a meeting between Lady

Mary Wortley Montague [sic] (died 1762) and Edward Jenner (born 1749) about the relative benefits of variolation and vaccination, and the death of James Phipps at the age of 20 from the complications of Jenner's vaccination (which would have made it difficult for Phipps to carry Jenner's coffin some eleven years later). Despite its many errors, *The Poisoned Needle* remained a standard work for the anti-vaccinationists and was reprinted, apparently not just for historical interest, in 1993. The book still features on various websites, often alongside advertisements for natural health cures which show that other beliefs also die hard. Typical are 'colon-cleansing products' that would have delighted Richard Mead, fighting his duel in 1719 against the 'vomiter' John Woodward.

Mission accomplished?

By the end of the nineteenth century it was hard to deny that smallpox was being pushed into retreat across Europe, North Africa and North America.[41] In England and Wales in 1882–83, mortality statistics were topped by bronchitis (57,000 deaths), tuberculosis (48,000) and scarlet fever (17,000). In the same year, smallpox claimed only 648 lives, just a few per cent of its average toll a century earlier, and it had plummeted to 72nd place in the ranking of killer diseases (see also Fig. 13.1, page 284). By 1893, the relative mortality rate from smallpox had also tumbled to 5 per cent or less of its pre-vaccination levels in other countries and records from England, Scotland, Sweden and Copenhagen showed consistent declines that had steepened with the introduction of compulsory vaccination.

In 1895, Jenner's dream that vaccination would result in "the annihilation of smallpox, the most dreadful scourge of the human species" took its first step into the realm of reality with the declaration that Sweden was free of endemic smallpox. Following suit in 1899 came Puerto Rico, which had been one of the first countries to be devastated when the Spanish brought smallpox to the Caribbean nearly 400 years earlier.[42]

For believers, this was an obvious triumph of vaccination. For sceptics, however, vaccination could not possibly claim the credit. According to Collins and Picton, its protective effect lay "somewhere between none at all and very considerably less than that of a previous attack of smallpox".[43] Instead, other factors must be responsible, with which the spread of vaccination had happened to coincide. Sanitation had improved and this was bound to reduce diseases such as smallpox that fed off overcrowding and deprivation. Also, smallpox was being contained and stamped out as soon as it appeared, thanks to the policy of strict isolation as enforced in (nearly) vaccination-free Leicester. The banning of variolation in 1840 should also have reduced the number of outbreaks. And finally, as Boëns believed, the nature of the beast itself could have changed, downgrading smallpox into a benign and non-lethal condition.

The Royal Commission was preoccupied by the vexed question of whether vaccination or 'cosmic' (general) changes were responsible for defeating smallpox. Sanitation was clearly better than a century earlier but the timing was wrong. In England, living conditions had not obviously improved before or during the initial plunge in smallpox deaths in the first quarter of the nineteenth century, while vaccination had been gaining ground. The same asynchrony was seen in Egypt and other countries. Moreover, other diseases that tracked with social deprivation had not decreased in parallel with smallpox. Deaths from measles remained constant throughout the century, while scarlet fever showed a dramatic fall that almost certainly reflected better living conditions – but this occurred half a century after smallpox had begun to decline. Reviewing all the evidence, the Commission (except for Collins and Picton) concluded that 'cosmic' changes could not explain the retreat of smallpox.[44] Also, they felt that effective isolation was practised on too small a scale to have any significant impact.

There was some truth in the notion that smallpox was becoming milder: the more benign *Variola minor* had begun to spread through Florida in 1896 and would soon predominate across the USA. However, this remained rare in Europe, where the classical *Variola major* still operated and, even in retreat, continued to kill up to 40 per cent of its victims (see Chapter 15).

The majority of the Commission therefore came down in favour of vaccination and this view was generally corroborated by doctors, scientists and governments across the world. For Collins and Picton and their associates, vaccination would remain useless and dangerous. We can only speculate as to whether the eventual eradication of smallpox some 80 years later would have changed their minds.

Ironically, the success of vaccination ultimately helped to achieve their aims. Routine vaccination was abandoned in Europe and America decades before smallpox was finally eradicated, because its potential benefits became trivial in an essentially smallpox-free setting and were then outweighed by its risks.

Some mysteries solved

For the Royal Commission the key issue about vaccination was whether it worked. They did not need to know how it might work and argued that this gap in knowledge did not undermine the value of vaccination:

> We are unable to trace the steps by which vaccination exerts protective influence ... Our inability to accomplish this does not seem to us to be the slightest reason to regard with doubt the conclusions to which the facts lead us.[45]

Their position was reasonable as current theories about the mechanism of action were speculative and woolly. In his prize-winning essay of 1879, Ballard suggested that the vaccine somehow absorbed the "pabulum" (nourishment) that normally sustains smallpox, but without explaining what this might be.[46] The Centenary issue of the *British Medical Journal* showed that thinking had not moved on much by 1896. Immunity was "the result of the local culture during the evolution of the pock, of an organism (unknown) which causes the production of an immunising substance".[47]

They were, however, on the right track. Vaccination works because the outer coat of the cowpox virus contains proteins that are identical to those which the immune system uses to mark out the smallpox virus for destruction (see Chapter 7). Vaccination therefore generates an army of immune cells programmed to kill anything carrying these markers. This initial immune response, directed against cowpox, produces the local reaction around the vaccination site. This reaction contains the infection (except in the very rare individuals with defects in cell-mediated immunity, who may go on to develop a systemic infection and the widespread rash of generalised vaccinia). Afterwards, memory cells, ready to be reawakened by contact with the marker proteins, drift away and settle throughout the body. If the smallpox virus later tries to invade, it is recognised at its point of entry by local memory cells which, thanks to the shared marker proteins, treat smallpox in exactly the same way as cowpox. They trigger the immune cascade, multiplying themselves and calling in reinforcements to destroy the smallpox virus and the cells which it has infected. This stops the infection in its tracks.

However, the immune memory created by vaccination in infancy fades after several years and needs to be reinforced by revaccination in early adolescence. Even this may not be enough to keep the immune defences adequately primed against smallpox throughout life, which is why further vaccinations were recommended if smallpox broke out nearby. Thanks to Jenner's obstinacy, the need for revaccination was long resisted. The eventual demonstration that he was wrong laid the ground for recognising that other vaccines also needed to be topped up. Most need repeat 'booster' doses at intervals ranging from yearly (influenza, to keep up with its high mutation rate) to every three years (rabies, meningitis and typhoid) or every ten years (polio and tetanus).[48]

Modern immunology has solved the mystery of how vaccination works but one conundrum remains. 'Cowpox' was long used interchangeably with 'vaccination' because it was the source of Jenner's original material and of the lymph used to inoculate countless calves in the vaccine production lines. However, when it became possible in the mid-twentieth century to distinguish clearly between the various poxviruses, the 'vaccinia' virus present in all commercial smallpox vaccines turned out to be quite different from cowpox. Allan Downie first showed that the two viruses had distinct biological properties (such as the pocks they produced in chick embryo membranes) and they were later found to diverge in their DNA sequences and in some of their coat proteins (although not in the ones that conferred immunity against smallpox).[49]

So where did the vaccinia virus come from and how did it manage to infiltrate vaccines throughout the world? The Liverpool virology expert, Derrick Baxby, looked in detail at the similarities and differences between the poxviruses and concluded that vaccinia probably evolved from the horsepox virus rather than cowpox.[50] It seems incredible that this interloper could have come to dominate vaccines that were produced in tens of thousands of centres scattered across the planet. Jenner is probably partly responsible, through his early notion that 'grease', the blistering eruption on horses' heels, was the original source of cowpox. Around 1813, he experimented with lymph obtained from horses and maintained by arm-to-arm vaccination and reported that it produced "beautifully correct" pustules, just like those of inoculated cowpox.[51] Jenner sent this lymph to the National Vaccine Establishment, from where it was distributed around Britain. Vaccination lymph was also raised from horses in other countries, including France, Germany and Persia and undoubtedly worked its way into the commercial vaccine-production network – alongside smallpox and cowpox raised in people, sheep, monkeys, chickens and rabbits' testes, transferred randomly between species and even harvested from corpses.[52] As Collins and Picton noted rather acidly,

> Under the one name of "vaccination", matter derived from various sources and of diverse origins, has been introduced at various times. It is no longer possible to trace or distinguish these.[53]

We can only agree and have to assume that the virus that became vaccinia had some biological advantage that enabled it to drown all the others that also swam in the vaccination pool.

The afflictions of farm animals caused further confusion. Jenner was right in believing that several different diseases could cause blisters on cows' udders, of which only 'true' cowpox conferred protection against smallpox. He did not describe the other 'spurious' eruptions in any detail. These presumably included conditions that are recognised today, such as 'pseudo-cowpox' which can infect humans and cause the painful lumps known as "milkers' nodules", but does not stimulate immunity against smallpox. The virus responsible is a parapoxvirus, distinct from the main poxvirus family (orthopoxviruses) that includes cowpox, vaccinia and variola.[54]

Similarly, skin eruptions in horses have various causes, of which horsepox (producing a blistering rash) was only one. At least some of the cases of Jenner's 'grease', which caused discharging cracks in the skin rather than blisters, were probably not horsepox. This would explain why some of these inoculations failed to immunise against smallpox. The story will remain incomplete because, like cowpox, horsepox became steadily rarer after Jenner's day and is now believed to be extinct, at least in horses.[55]. In keeping with the promiscuity of most poxviruses, it could still survive in rodents – which may have been its original host species.

Halls of Fame

"I do not know what I may appear to the world, but to myself I seem to have been only like a boy playing on the sea-shore, and diverting myself in now and then finding a smoother pebble or a prettier shell than ordinary, whilst the great ocean of truth lay all undiscovered before me."

All those involved in the historical drama which took place in the Chauntry (now 'Chantry'), Berkeley on 14 May 1796 achieved immortality of a sort. Sarah Nelmes' hand, drawn by Jenner for the *Inquiry*, has been reproduced countless times. James Phipps is commemorated by a plaque outside the stone cottage, down the lane from the Chantry, which Jenner left him in his will. And there remains a tangible memorial to Blossom the cow. Her hide, donated after her happy retirement on the pastures of Breadstone, is still on display in St. George's Hospital Medical School, together with a pair of prosthetic wooden horns; in the best tradition of saints' body parts, several horns exist that all purport to be the genuine item.

Jenner's claim to fame has been fiercely defended and just as fiercely attacked. To John Baron, his biographer, Jenner was a demigod and everything that he touched became imbued with magic, like the sticks thrown for a faithful dog by its master. Baron, the only biographer with access to all Jenner's papers, epitomised the slavish adoration of the man and his discovery that many fought to perpetuate after Jenner's death. This was counterbalanced by a barrage of criticisms of Jenner's scientific credentials, personal insults, and hatred for all the perceived damage which vaccination had inflicted.

Why did people hate Jenner so much? Some of his peers were driven by professional jealousy of a man they considered their social and intellectual inferior and who had been blessed with an impossibly lucky break that they could have made much more of. Jenner also personified vaccination and became the lens that focused all the suspicion, fear and hatred that his discovery provoked. His personality did not help, especially the brittle and sulky façade that he presented to his critics. He also angered many with his stubborn dismissal of potentially uncomfortable evidence. His responses to the suggestion that vaccination provided only temporary immunity must have suggested arrogance, laziness and a tendency to bluff:

> [This] is refuted, not only by analogy with respect to the habits of diseases of a similar nature, but by incontrovertible facts which appear in great evidence against it.[56]

A good scientist would have set about collecting the "incontrovertible facts" and presenting the "great evidence" but Jenner did neither.

Jenner has also been criticised for his dilettante approach to medicine and science. Wheeler noted sarcastically that Jenner needed 25 years to do the 24 experiments in the *Inquiry*,[57] and he took even longer – 36 years – to not quite finish his paper on the migration of birds (this was completed by his nephew within a few months of Jenner's death and promptly accepted for publication in the *Philosophical Transactions*). Jenner might not have thrived in the cut-throat world of twenty-first-century science where success depends on fighting for grants to fund research and on a prolific publication record. However, this was a different age. The quotation heading this section is from Isaac Newton,[58] not Jenner. It could also apply to the breadth of interests of many other gentleman-scientists of Jenner's era; but Jenner clearly lacked Newton's gifts of focus and perseverance.

The ambivalence about Jenner's place in history can be judged by the mixed bag of accolades that he received after his death. In May 1858, a large bronze statue of Jenner by Calder Marshall, funded by public subscription, was unveiled in Trafalgar Square by Prince Albert, consort of Queen Victoria.[59] The pomp and circumstance seemed appropriate for a national hero who had saved millions of lives worldwide, but the anti-vaccinationists had friends in high places. Four years later, Jenner's statue was quietly carted off into the backwaters of Kensington Gardens, where it remains to this day – counterbalanced by the conspicuously empty plinth in Trafalgar Square.

In 1881, while opposition to both vaccination and germ theory were at fever pitch, Louis Pasteur used his top-bill appearance at an international conference in London to heap praise on Jenner. Pasteur proposed that (even though no cows were involved) all immunisations against infections should be called 'vaccination' in deference to Jenner's achievements.[60] The response from those still attacking the long-dead Jenner was swift and predictable. The anti-vaccinationist Garth Wilkinson wrote scathingly: "The doom of Jenner is upon us; the judgement on Pasteur is yet to come."[61]

Today, the town of Berkeley seems oddly coy about its most famous son, with only a couple of drab brown signs pointing the way to the excellent Jenner Museum, housed in his Chantry. At the heart of the museum is a recreation of Jenner's study, considerably tidier than Baron's picture of a jumbled chaos of fossils, natural history specimens and a heap of partially answered correspondence. As a memorial to Jenner, country doctor and gentleman collector, this seems just right.

Much grander is the magnificent library of the Wellcome Trust building on Euston Road in the West End of London. Around the gallery is a block-capital frieze of the greatest names in the history of medicine, beginning with Hippocrates and Galen and running through to Banting and Fleming. Jenner is right above your head as you climb the stairs to the Rare Books Collection where many of his letters and documents are preserved. He is flanked by Hunter and Darwin, and Pasteur is close by.

This too seems just right.

14 Lest We Forget

"The fact is we have forgotten what smallpox is like."
Professor D. Fraser Harris, Dalhousie University, Halifax, Nova Scotia, 1914

This brings us to the threshold of the twentieth century and the last 80 years of the free existence of smallpox on earth. During this time, smallpox continued its general retreat. Its terminal decline was due mostly to vaccination which administered the final coup de grace. This was assisted by a change in the nature of the beast itself, which brought a degree of reality to the wishful notion of Dr. Hubert Boëns that smallpox had lost its lethality. In 1896, the mild variant *Variola minor* broke into Florida, from where it rapidly invaded America. As described in the next chapter, the power struggle between *major* and *minor* helped to force smallpox out of North America. *Variola minor* brought lifelong immunity to both itself and its murderous twin, at the cost of a trivial illness that generally left few scars and was very unlikely to kill. Elsewhere, however, it was business as usual for classical *Variola major*. This was by no means a spent force: the Angel of Death carried off over 300 million people between 1900 and its extinction in 1979.

The twentieth century witnessed massive advances in the treatment of infections, especially those caused by bacteria. The development of antibiotics, galvanised by the discovery of penicillin just in time for the Second World War, brought in effective treatments for bacterial diseases such as erysipelas and syphilis, which both had bedevilled vaccination, and the secondary infections that often complicated the eruption of smallpox. By contrast, the prescription chart against smallpox itself remained blank. There were many false hopes for treatments that would kill the virus without harming the host but none was ever translated into therapeutic reality. Even with all the sophistication of twentieth-century intensive care medicine, the outlook for a victim of *Variola major* in 1975 was not materially better than it had been in 1875. The chances of dying still ranged between 20 per cent and 50 per cent.[1]

This same period also demonstrated how little mankind had learned from previous encounters with smallpox. Time after time, the same fundamental mistakes were made at all levels from individual to government: failing to recognise the disease, underestimating its dangers, responding slowly or inappropriately to outbreaks, and allowing vaccination and other preventive measures to be undermined. As always, human nature could be relied upon to wreck attempts to contain the disease, with everything from bloody-mindedness to suicidal indifference and blind panic.

This chapter describes four episodes from the final 80 years of smallpox's reign of terror. They show the continuing hopelessness of the patient and the helplessness of the doctor – stuck in a backwater of medicine where therapeutics stood still – as well as the unlearned lessons that so often played against us.

Middlesbrough, 1897–98

Middlesbrough, a shipbuilding town on the Tees Estuary in the north-east of England, epitomised many of the opportunities and threats of late nineteenth-century industrialisation. Feeding off the booming steel industry, its population had grown exponentially from 7,000 in 1850 to 55,000 in 1881 and 90,000 in 1897, of whom 40,000 were children under ten years of age.[2]

According to a local dignitary and Member of Parliament Sir Hugh Gilzean-Reid, this "youngest child of England's enterprise" was not only "brave and prosperous" but also "remarkably free from epidemics; there are a few healthier or more privileged communities in the country." In praising Middlesbrough, Gilzean-Reid singled out for particular mention its "large-hearted Mayor" who had made "life brighter and better for the rich and poor alike", adding that "there will long linger a sweet aroma of his unbounded and wisely-directed beneficence".[3]

Even by contemporary standards, this was excessive and cynical readers will have guessed that the Mayor's fragrance covered up the odour of rat. Dr. Charles Dingle, the town's Medical Officer of Health, had a better grasp of the facts.[4] The population's rapid growth had long since burst through all safe limits of housing and sanitation – symbolised by the flawed sewage system which flooded the low-lying slums with human excrement every time heavy rain coincided with a high tide. Thanks to all this and the "dirty habits of a portion of the inhabitants", it was only a matter of time before disaster struck. Despite Gilzean-Reid's reassuring words, there had recently been epidemics of typhoid and scarlet fever, which turned out to be the warm-up act for something even worse. Why had Gilzean-Reid lied about Middlesbrough's state of health? Answer: as a desperate act of damage limitation, to try to entice trade back into a town crippled by a major outbreak of smallpox in 1897.

The outbreak probably began with an infected sailor returning in late November from Bilbao where an epidemic had been raging. A few days later, his wife (unspotted but already contagious) joined the crowds at a visiting 'wild beast show'. When the sailor was admitted to the Sanatorium, the diagnosis was somehow missed even though he had a full-blown classical rash. He was not isolated and managed to infect several others before recovering and being sent home. Other undiagnosed cases then cropped up. A child was seriously ill for two weeks and died without a doctor being called. A woman who visited the child's home fell ill and died of "erysipelas". Being Irish, her corpse lay in state and shed viable virus throughout the two-day wake while relatives and friends paid their last respects (which traditionally included a farewell kiss for the deceased).[5]

Middlesbrough was relatively well covered for primary vaccination in infancy but very few of its inhabitants had been revaccinated as adults. The trickle of smallpox cases rapidly swelled into a torrent that soon overwhelmed the town's limited isolation facilities. Even when the Lunacy Committee were over-ruled to turf the insane out of the Asylum, there were fewer than 300 hospital beds – with 20–30 new cases appearing every day.[6]

It was Dr. Dingle rather than the aromatic Mayor who galvanised the town council into a military-style campaign to build an emergency smallpox hospital. The site was strategic: high ground, out-of-town and abutting the cemetery. Six corrugated-iron ward blocks were thrown up by "any man who could hit a nail squarely on the head", paid £4 per week and with the added incentive of "an unlimited supply of disinfectants in the shape of WHISKY AND TOBACCO DURING WORKING HOURS". The ward blocks were given "facetious" nicknames taken from popular icons of hope springing eternal, including 'Klondike', 'Dawson City' and 'Kimberley'. The 600-bedded complex, looking like "a ground laid out for a Royal Show", was finished within two months despite a strike when a sudden influx of new cases were housed temporarily in tents which were initially pitched too close to the workmen.[7]

Medical and nursing staff were recruited with some difficulty. The nurses were under the command of the daunting but efficient Matron, Miss Bell, who "flinches from nothing ... but with a degree of gentleness and consideration which does much to lighten the burden of the sufferers". The nurses, and especially the teenaged probationers, were protected by vaccination but unshielded against the horrors of smallpox in adults and children, or the reactions of their townsfolk. Outside the hospital, these "plucky little girls" were ostracised. People crossed the road to avoid them and they sat alone at dances.[8]

Figure 14.1 Miss Bell, matron of the temporary hospital during the smallpox outbreak, Middlesbrough, 1898. Reproduced by kind permission of the Middlesbrough Reference Library.

The patient's grim journey into isolation began with the arrival of the yellow 'Fever Van', which sometimes interrupted a party if the victim's friends had called in to say farewell and have a meal or a parting glass.

> Upon being reported and after the visit of the Sanitary Inspector and the doctor, the patient was allowed to walk downstairs and place himself or herself in the waiting van, or if they were too ill they were gently handled by the van men, who, wrapping them in their bed and bed-clothes just as they lay, placed them in the van and took them off to Hospital, where they were placed in various wards and labelled by the attendants.

The van often departed with small boys hanging off the back trying to peer inside at the victim. Meanwhile, the patient's house was fumigated with burning sulphur and their possessions collected for disinfection, although "many of the articles taken from some of the poorer houses were not worth returning and these were, therefore, destroyed".[9]

The wards were spartan but "bright and cheerful, being decorated internally with plants and pictures" (the same potted aspidistra appears in several contemporary publicity photographs). The patients were well looked after and "many made themselves at home ... There are many worse homes than that to be found at 'Klondike'". Two patients escaped but were quickly recaptured, "little worse for their escapade" and luckily without infecting anyone else.[10]

The patients were strictly isolated from the outside world, with no visits even for the children who were nursed alongside the adults. There was a daily ritual to distribute gifts of clothing, food and delicacies sent by friends and relatives to the Parcel Office. These items were laid on the ground, ringed by policemen to keep order, while an official read out the names. Those who could walk came forward from the crowd to claim their parcel; the bed-ridden had to trust to others' honesty.[11]

The official accounts of the outbreak focus on the efficiency of the civic response to the disaster and the miracle of a hospital that sprang out of the ground in a few weeks. There were a few insights into what it was like to be a patient:

> A VERY SAD CASE. A girl who came into the Hospital only three days before her wedding day, and died that same night. This was a case of haemorrhagic smallpox.[12]
>
> It was very hard at times to tell a sick patient of the death, perhaps, or serious illness of another member of the family in some other ward at the same hospital. In some cases, news had to be withheld for a time, which only made it harder to tell afterwards.[13]
>
> The hymn, "On the Resurrection Morning" became a great favourite with many of the patients who had lost those near and dear to them.

On that happy Easter morning
All the graves of their dead restore;
Father, sister, child and mother
Meet once more.[14]

> It was rather amusing to hear some of the patients in their delirious state. Many a ship has been built, engines have been hoisted in and out of the roofs of the wards.[15]

But there were brighter moments: "Though it was sad to see men, women and children afflicted with such a disease, there were times when it was forgotten."[16] And finally, when it was time for a survivor to go home:

> The exit from the discharge block led directly into the cemetery ... Happily, owing to the excellence of the manager and the skill of the medical and nursing staff, very few had to remain there.[17]

In fact, the outbreak claimed over 1,400 victims, of whom 201 died, 107 of them children under ten years of age.[18] These numbers were small by comparison with many epidemics, but nine months of smallpox brought the community to its knees and the town and its dependent region to the brink of economic ruin.

Shipbuilding and other industries were starved into a standstill as trade with the rest of the world ceased: the whole area was boycotted even by trusted business partners and paralysed by stringent quarantine restrictions imposed on all its exports. Despite the best efforts of Gilzean-Reid, the town council and the railway company to paint a positive picture, travellers and visitors also stayed away. The blight spread far beyond industrialised Middlesbrough and deep into the tourist destinations of the Cleveland Hills and its string of seaside resorts, even though these had been spared by smallpox. In economic terms, the British press probably did as much damage as the Angel of Death. Newspapers outside the region ignored the heroism and determination of the people of Middlesbrough and concentrated instead on the stuff of a good story: death, misery, slums drowning in their own sewage and the failure of vaccination to protect against smallpox. This sensational misreporting was "palpably exaggerated to it discreditable extent" but still "inflicted grievous harm to the town." It took several years for Middlesbrough to begin its financial recovery.[19]

Figure 14.2 The cover of the *Northern Weekly Gazette*, 'The story of the smallpox outbreak in Middlesbrough, 1898'. Reproduced by kind permission of the Middlesbrough Reference Library.

After the outbreak ended in August 1898, the town's *Northern Weekly Gazette* issued a commemorative supplement *The Story of the Smallpox Epidemic in Middlesbrough*, price one penny.

The cover shows an angelic nurse handing a cup of tea to a moustached man lying in bed. The man looks unusually well for a smallpox victim and his face is completely unmarked. The rest of the story is similarly sanitised – an understandable response to the horrors of the outbreak and the desperate struggle to bring health and prosperity back to the town.

Dr. Dingle, Miss Bell and the town councillors were all hailed as heroes of the moment: "And their names are recorded – and their work will live."[20] Dingle's career certainly prospered as a result of his vision and leadership during the crisis. The formidable Miss Bell, commander-in-chief of the "plucky little girls", was rewarded with an honorarium of £30. She probably did more than anyone else for the victims but was forgotten as soon as Middlesbrough had put the outbreak safely into the past. In 1904, a new Fever Hospital was opened; Klondike, Dawson City and Kimberley were all bulldozed and Miss Bell was sacked.[21]

South-west India, 1945

An account by Mrs. Kitty Hutchinson, then an Army Sister with the Queen Alexandra Imperial Military Service, based at a military field hospital for troops preparing to invade Japan.[22]

He was admitted one evening to the medical ward just before I came on duty. He had become feverish and ill en route from Singapore. The Captain could see he was ill and he was excused work, and having a consuming thirst and drank a lot of water and other fluids, which was fortunate. I tried to ascertain if he had a rash but in the bad light was not very successful.

He was a great strapping Liverpool lad of a mild and sanguine temperament and great morale, which led the Medical Specialist to underestimate the severity of his illness. The next morning, he did develop a rash and was put in a side-room. I had taken a pocket edition of a *Student's Aid to Medicine* and became pigheadedly convinced that he had smallpox. The Medical Specialist pooh-poohed this and said he thought it was scrub typhus, of which I and my *Student's Aid* had no knowledge.

I kept looking at and even feeling his "spots" and one night when they felt like hard ball-bearings I decided it was definitely smallpox. I wrote my report, noting that I thought his condition to be critical and that his rash was strongly suggestive of smallpox, and underlined this in red ink. The patient was immediately removed to a special isolation building apart from the normal isolation ward; his room was fumigated and all the nursing staff and doctors concerned were vaccinated.

An unfortunate auxiliary was detailed to "special" him. The poor girl was terrified, being isolated all alone in a lonely part of a hospital compound. I did not go near the poor man but could observe him through the window. I read with perturbation that the more pustules on the face, the worse the prognosis. His face was covered all over.

He remained cooperative and uncomplaining throughout. At first the Medical Specialist was quite blasé about his eventual recovery. Then it became apparent that the patient had pustules in his mouth and further down, and "probably down to his anus", the Medical Specialist said gloomily. By now it was realised he was in dire straits. The smallpox turned out to be a virulent form of confluent smallpox, undoubtedly picked up in Singapore.

The pustules coalesced and no drips could be put up without introducing pus into a vein. His mouth and throat were so swollen, water could not trickle down. He sank into a coma and slowly died, his remarkable constitution keeping death at bay longer than anyone thought possible. We were all in tears – he was such a marvellous patient. He was 28, and had had a vaccination with a swollen arm developing two months before he became ill. That was partly why the Medical Specialist held out so long against the diagnosis.

I copied his home address from his Army notes and wrote what I hoped was a consolatory letter to his wife in Liverpool. She sent me a pathetic letter – she had had twins, one of them stillborn, two weeks before receiving my letter. They were Catholics, so I had written, quite untruthfully, that he had received the last rites from the Catholic padre. It would have been the Commanding Officer's duty to write but as the patient was a civilian, I knew no one would do it if I didn't.

Bradford, 1962

On 11 January 1962, a blood sample reached Dr. Derrick Tovey, not long in post as a consultant pathologist at St. Luke's Hospital in Bradford.[23] All that he knew from the request form was that the sample came from a 50-year-old woman admitted as an emergency to the local Fever Hospital, that she was critically ill and that her doctors didn't know what was wrong with her. The working 'diagnosis' was the unhelpful catch-all of 'pyrexia (fever) of unknown origin' (PUO). The blood film was striking – full of immature cells of the sort that the bone marrow pushes out into the bloodstream *in extremis* – but also uninformative. There were no diagnostic characteristics of any disease that Dr. Tovey was familiar with. All he could do was report that the blood changes were consistent with a severe viral infection.

While he was puzzling over that, another blood sample arrived from a case in his own hospital: a 40-year-old man with PUO and severe headache, provisionally diagnosed as meningitis. To Tovey's astonishment, this second sample showed identical changes to the first. Perhaps an unusual tropical infection could be responsible? Bradford had a large immigrant population from India and Pakistan, but both PUO cases were English Bradfordians who had never left the country and so had few opportunities to pick up exotic fevers. The astute Dr. Tovey was nonetheless concerned that the two undiagnosed cases at different hospitals might herald an outbreak of some sort. He and a colleague went back to the books to try to find a condition that matched the bizarre blood films. Finally, they unearthed an out-of-print textbook in the hospital library which stated that identical appearances had been reported in 1925, in patients with fulminating haemorrhagic smallpox.[24]

By that time, events had moved on. The man in St. Luke's had died and his body had been taken to the mortuary. Tovey telephoned the consultant at the Fever Hospital, where the woman with PUO was now moribund, and "almost apologetically" told him that the peculiar blood changes in both the PUO patients had been reported in smallpox. The consultant immediately came to examine the man's body in the mortuary: not a single pustule anywhere, but hidden away in the creases around the elbow, the dark pinpoint bleeds into the skin that were the hallmark of haemorrhagic smallpox.

Laboratory tests would take two days to confirm the diagnosis. Tovey and his colleagues decided to take the diagnosis on trust and undoubtedly saved numerous lives by their prompt action. They informed the public health authorities, who immediately triggered the exercise they had hoped never to activate.

The two apparently isolated cases were rapidly joined up. The woman, who died the following day, had worked as a cook in the Children's Hospital, where the child of the man with PUO had recently been an in-patient. Six children at the Children's Hospital were found to have early smallpox, and more detective work traced the original source of the outbreak. A five-year-old Pakistani girl had arrived with her family by air from Karachi in mid-December. She had been admitted to the Children's Hospital just before Christmas with a fever, thought to be malaria as the parasites were found in her blood. She was treated for this, initially with some response, but she then relapsed and died on 30 December 1961 with a diagnosis of septicaemia. Her post-mortem was performed on New Year's Day 1962 by Dr. Norman Ainley, the duty pathologist, who reported findings consistent with fatal septicaemia.

Endemic smallpox had disappeared from Britain 30 years earlier but occasional imported cases, especially from Asia, still cropped up. Bradford, with its large immigrant population and heavy traffic to and from India and Pakistan, was well prepared for an outbreak. Once activated, the countermeasures swung into action with astonishing speed. The city's main hospitals were all infected and had to be closed to all admissions, which were handled by outlying hospitals until the smallpox cases could be transferred to a single site and the others disinfected. Within five days, over 1,000 contacts were traced (over 200 of them from the man who had died at St. Luke's) and 285,000 people were vaccinated. As a result, only three new cases occurred. The final total was twelve, and the outbreak ended in February 1962.

There were six deaths, representing a mortality rate of 50 per cent. In four of them, the diagnosis of smallpox was missed: the index case from Karachi, the woman and the man with PUO, and Dr. Norman Ainley, the pathologist called in on New Year's Day to do the post-mortem on the little girl who had apparently died of septicaemia. As soon as Tovey suspected the diagnosis, he told Ainley who was horrified because he had never been vaccinated. This was immediately rectified but it was already too late. Ainley fell ill that evening and died of confluent smallpox a few days later.[25] Ironically, the little girl from Karachi had been vaccinated as an infant and was revaccinated with her parents before flying to England; perhaps this second vaccination had not taken, or she was already incubating smallpox.

There were other lessons to be learned and relearned. Isolation rules were deliberately flouted, not only by the uninformed public but also by professionals who should have known better. A junior doctor left his hospital without permission to visit a senior colleague in Manchester, while a nurse later slipped away to see a friend in London. Neither could guarantee that they were not incubating smallpox and both had plenty of opportunities to infect others – especially the nurse, who travelled by train and stayed in the Nurses' Home in a major London hospital.

As before, the national newspapers made the most of the story and worked hard to ensure that the facts were not allowed to get in the way of sales. "Bradford – City in Fear!" (*Daily Mail*) echoed the bad press about Middlesbrough nearly seven decades earlier, while others sank even lower with overtly racist coverage: "Keep out Pakistanis" (*Sun*).

Quick thinking, effective isolation and wide-scale vaccination saved the day, but the benefits of vaccination came at a cost. Ten previously healthy people of the 285,000 vaccinated suffered serious side-effects of vaccination, mostly the rare allergic inflammation of the brain and spinal cord, encephalomyelitis, which killed three children. The vast difference in mortality rates – one in two with smallpox, barely one in 100,000 with vaccination – was little comfort to the bereaved families who had lost their children in this way.

Copenhagen, 1970

On the last day of August 1970, a 22-year-old Norwegian man was admitted to the Blegdams infectious diseases hospital in Copenhagen with fever and a cough.[26] He had already been treated with a broad-spectrum penicillin for suspected pneumonia, which was confirmed on admission. He also turned out to have *Salmonella* bacteria in his bloodstream, another potential cause for his fever, and had developed a non-specific skin rash of the sort which penicillins often cause. His recent past was complicated: he had just returned from travelling in Afghanistan and had spent several days in hospital in Kabul with diarrhoea and weight loss, which could have been explained by the *Salmonella*. He had been discharged twelve days earlier to fly back to Copenhagen where he caught up with numerous friends and stayed in two large halls of residence. Before he became ill, he had enough energy to go to a well-attended ball.

The day after admission, even though there were no features to suggest it, someone thought to include variola in the battery of screening tests sent off to the laboratory. Before the results came back, his rash had turned into the classical smallpox eruption. He was promptly isolated and a monumental exercise of tracing, isolating and treating his contacts began even as he was facing up to battle for his life.

The bacterial infections – pneumonia and *Salmonella* – responded well to conventional antibiotics but the eruption rapidly became confluent, "with almost total destruction of the skin". Fluid poured out across the raw flesh, with losses of up to 13 litres per day – as much as in massive full-thickness burns. His circulating volume was carefully maintained with intravenous fluids and he was artificially ventilated when his lungs became involved and his blood oxygen tension fell to critical levels. Meanwhile, the virus continued on its rampage. With the onset of multiple organ failure, the patient's temperature fell into the hypothermic range and he finally died three weeks after admission.

He had been vaccinated in childhood and again earlier that year in preparation for visiting a region where smallpox was still endemic, but the revaccination had not taken. It later transpired that two patients in adjacent beds in Kabul had been diagnosed with smallpox after his discharge.

This single case unleashed an urgent and far-reaching containment campaign which paralysed much of Copenhagen's health system for six weeks. From the typical incubation period of smallpox, the authorities estimated that they had barely twelve hours after making the diagnosis in which to trace and isolate the earliest contacts. The patient had travelled widely in Copenhagen as well as Afghanistan. The authorities initially guessed that there might be 50 contacts, hastily revised this to 300 and ended up with a list of 589; another 200 were eliminated after detailed questioning. The 589 included students in halls and all those at the ball, many of whom had since dispersed to a total of twelve European countries as well as Israel, Pakistan, Burma and New Zealand. Warnings were sent out via travel agencies, foreign ministries and Interpol.

Incredibly, all the contacts were traced and all those remaining in Copenhagen were rounded up within a week and isolated in Blegdams hospital, which was completely taken over for the purpose. In the spirit of Klondike and Kimberley, the hospital's wards were supplemented with tents and temporary beds provided by the Army, together with water, heat, disinfection facilities and telephones. Vaccination was given unless there was was a medical contraindication or conscientious objection (some of the contacts were members of a religious sect which opposed vaccination).

The alternatives included hyperimmune globulin, an antibody preparation that was injected intravenously and provided only temporary protection against smallpox, and an experimental antiviral drug, methisazone, which had recently been shown to have some activity as a prophylactic against *Variola major*.

The contacts were each examined twice daily for signs of fever or a rash. Amazingly, none caught smallpox. Dr. Viggo Faber, one of the infectious disease specialists who oversaw the operation, later wrote:

> Let me end by asking: Why did we not have any secondary cases in this extraordinary situation? The answer may be one or a combination of the following: (1) smallpox is not very infectious; (2) we were extremely lucky or (3) we were rather clever.[36]

On balance, (2) and (3) would seem the most likely.

This massive logistical exercise was hugely expensive. It absorbed over 3,000 person-hours to organise operations and communications, and direct costs of 3–4 million kroner (over £1 million at current rates). If any secondary cases had emerged, the World Health Organisation would have declared Copenhagen a "local infected area" – when the costs of a Middlesbrough-style commercial embargo would have escalated into hundreds of millions of kroner.

Flat learning curve

"The lesson of history is rarely learned by the actors themselves."
James A. Garfield (1831–1881)

Medicine has always been a process of trial and error. The accounts remind us that mistakes dominated many of mankind's encounters with smallpox and that the same errors were made repeatedly. These can be condensed into four basic unlearned lessons.

Lesson one: Never ignore smallpox

Smallpox was often missed, neglected or underestimated. This was a diagnosis that had to be made early to avoid the catastrophe of an outbreak, even if it made little material difference to those who were already infected. Even while smallpox and medical awareness of it were at their peak, key cases were missed at crucial moments, often with disastrous consequences.

The diagnosis was sometimes difficult, as smallpox could change its spots and occasionally covered its tracks almost completely. Chickenpox and smallpox could each present with an atypical rash that looked like the classical eruption of the other, and even experts could be confused. Fulminating smallpox could kill the victim before the skin showed anything but a few easily-missed haemorrhages. Smallpox could also hide behind other causes of fever such as malaria or pneumonia. Commonly, however, obvious cases of smallpox were missed or mistaken for other diseases, presumably because of inexperience, unfamiliarity or carelessness.

Further examples can be added. In 1962, a young nurse in Poland was diagnosed with acute leukaemia because her blood film was flooded with the same immature cell-types as the two PUO cases in Bradford. She was a popular girl and had a constant stream of visitors before and after her death from unrecognised haemorrhagic smallpox. Twenty-five of them became infected and sparked off a major outbreak.[27] In Yugoslavia 10 years later, a patient with undiagnosed confluent smallpox was shuttled between four different hospitals including an intensive care unit. Smallpox jumped first to 38 of his contacts and then galloped away in an epidemic that ultimately involved most of the country.[28] The disease was contained and stamped out by the efficient and ruthless application of authority that characterised other activities of President Tito's regime.

On numerous occasions, sluggishness and indecision held back the countermeasures against smallpox and all too often proved that it was a fatal error to underestimate the disease. Copenhagen was outstanding for the speed and efficiency with which it traced, isolated and treated contacts. It was also unique in this regard. The task was made easier there by having to deal with just one index case, and a bigger outbreak might not have been handled with the same panache or success.

Lesson two: Smallpox can be prevented but not cured

Vaccination of a patient incubating smallpox could abort the infection if given within a few days of exposure but not, as demonstrated by Dr. Ainley of Bradford, later in the prodromal period. Once smallpox had taken grip, the only treatment was supportive (intravenous fluids, artificial ventilation) or palliative.

Numerous therapies were proposed to treat or even cure smallpox but none have succeeded. In 1900, Dr. Monkton Copeman was on the point of perfecting an antitoxin which, by analogy with diphtheria antitoxin, would snatch smallpox victims back from certain death.[29] Unfortunately, smallpox was a virus that killed cells directly rather than the bacteria which produced a circulating toxin, and Copeman's miracle cure came to nothing.

The turn of the twentieth century saw another therapeutic non-breakthrough by Nils Finsen, a Danish professor of dermatology who had found that tuberculosis of the skin (lupus) improved when irradiated with ultraviolet light. Finsen decided that there must be substance to the oddly tenacious notion of 'red treatment' and experimented with shining red light on smallpox patients.[30] He claimed that this 'erythrotherapy' prevented pustules and therefore scarring. The effect was weak, however, and could be wiped out by even brief exposure to sunlight. Patients therefore lived in a dim red twilight, with heavy curtains nailed over the windows; sometimes the entire sick-room was painted deep red for maximum effect. Unfortunately, others could not reproduce his findings and formal trials eventually showed that erythrotherapy had no effect. Even though useless, red-light treatment was still practised in parts of Europe until the 1920s. Finsen was more fortunate at the other end of the electromagnetic spectrum. The effect of ultraviolet 'actinotherapy' on lupus was genuine and won him the Nobel Prize for Medicine in 1903, the year before he died.

Interestingly, the unsolicited advice about 'cures' for smallpox that bombarded Bradford's Medical Officer of Health after the 1962 outbreak included actinotherapy, building up to 1-minute treatments to all sides of the body. Other suggested treatments in his mailbag[31] included a rhubarb/cream of tartar purgative and a tartar emetic ointment (which would have had the approval of Richard Mead and Edward Jenner, respectively), the mind-boggling and possibly brain-frying "diathermy to the pineal gland" to keep life-preserving energy circulating in the seat of the soul, and cryptic nutritional advice delivered by telegram:

> ESSENTIAL KEEP ALL SMALLPOX CONTACTS ON STRICT VEGETARIAN DIET + WILL WRITE

More conventional drug discovery strategies came up with some initially promising leads. Methisazone (known in early trials as 33T57) was introduced in the 1960s and was found to prevent the multiplication of DNA viruses, including poxviruses (it works by inhibiting the replication of DNA). Preliminary animal work showed potentially useful effects in mice infected with poxviruses but the first human study, with high doses given for ten days, failed to prevent the death of a patient with confluent smallpox.[32] Later clinical trials found that methisazone could be used prophylactically to prevent smallpox from becoming established in contacts exposed to the disease, although the effectiveness of the drug remained controversial.[33] Methisazone was used alongside vaccination to treat contacts in Copenhagen in 1970, and none of these subjects developed smallpox.

A newer inhibitor of DNA replication, cidofovir, is more potent but has to be given intravenously and has potentially serious side-effects, especially kidney damage; an orally active form with less marked side-effects has recently been developed. Cidofovir has been used to treat various other DNA virus infections such as severe herpes. As it was introduced after the eradication of smallpox, its activity against variola in real life is unknown. However, it is effective against cowpox and vaccinia infections in mice and, like methisazone, has been earmarked as a potential prophylactic for smallpox contacts.[34] It could also be useful to treat the rare generalised vaccinia infection which can follow vaccination in subjects with defective cell-mediated immunity.

At the time of the death of smallpox, there was no 'magic bullet' capable of destroying it. This situation has not changed, despite another 30 years of medical progress and the introduction of powerful computer-driven drug design and high-throughput screening that can detect useful therapeutic activity with breathtaking speed in hundreds of thousands of compounds.

This leaves us with vaccination, which now stands revealed as falling short of Jenner's ideals. One vaccination in infancy was simply not enough to protect for life, and a routine revaccination in adulthood could not guarantee immunity. Experiences such as Middlesbrough helped to shape a rational vaccination policy (with immunisations given in infancy, adolescence and then if an outbreak occurred), while Bradford and Copenhagen proved that a well-coordinated vaccination strategy could be carried out rapidly and effectively. Bradford also highlighted the low but distinct risk of serious damage and death following vaccination – unavoidable complications that would be thrown into stark relief should large-scale vaccination against smallpox ever be needed in the future.

Lesson three: People will be people

Benjamin Franklin, one-time printers' apprentice on the *New England Courant* in Boston, later wrote to a friend, "But in this world nothing can be said to be certain, except death and taxes."[35] To this should be added, "and the illogicality of human behaviour".

Living under threat nurtures strange behaviour, whether provoked by fear of the Angel of Death or of being turned into a cow by vaccination. The accounts above show again how effectively human nature could wreck logical and well-planned strategies to contain smallpox – and that this was not confined to the uneducated and uninformed, but also involved doctors, nurses, the press and those in authority.

Many of these aberrant behaviours can still be seen in the reactions to other health threats such as AIDS, SARS and swine flu, and to other vaccinations including measles-mumps-rubella (MMR), and more recently human papilloma virus (HPV). Another Franklinesque certainty is that the emotional responses to smallpox and vaccination would be recapitulated if the disease were ever to reappear.

Lesson four: Count the full cost

Smallpox did much more damage than simply killing and disfiguring. The Middlesbrough outbreak showed clearly how it tore the heart out of a whole community, paralysed its medical services and infrastructure and brought several years of destitution to the whole region.

Even on the relatively small scale of Copenhagen, tracing and isolating contacts was a logistical nightmare which, in a less organised setting or a bigger outbreak, would not have had such a happy ending. In the instances above, medical services managed to continue functioning, but it was a close-run thing in Middlesbrough and the beginnings of an outbreak on that scale could have caused catastrophe if transposed to Copenhagen in 1970.

Our ability to cope with smallpox undoubtedly improved during its last few decades but arguably was never really put to the test. Since then, we have lost the skills needed to recognise and deal with smallpox and our defences have fallen into disuse.

Should smallpox ever return, we would not be able to rely on history to have taught us the lessons for survival.

15 Death Throes of a Clinical Curiosity

"Dr. Jenner's sanguine hope has not been realised ... Smallpox has not been eradicated. Let me add that scientific observation and reasoning give no countenance to the belief that it will ever be eradicated, even from civilised communities."

Edward Ballard *On Vaccination: its Value and Alleged Dangers*, 1868

The 80 years between the start of the twentieth century and the declaration in May 1980 that smallpox had been eradicated probably represented the last 1 per cent or so of the Angel's reign on earth. For the first half-century of that period, the area on the world map still gripped by endemic smallpox continued to shrink, although progress was interrupted by resurgences, often reinvading territory which had previously been cleared. The casualties of these rearguard actions included confidence in the belief that smallpox could be wiped out, as well as lives and health-care budgets.

Even into the 1970s, with the 'Target Zero' of total eradication less than a decade away, many saw Jenner's naïve dream turning into a logistical nightmare that would come to nothing. The doubters were not just the traditional enemies of vaccination, but also included influential people who should have been fighting for the cause. The Director-General of the World Health Organisation (WHO) during the crucial marshalling phase of the eradication programme repeatedly voiced his conviction that smallpox could never be wiped out, while one of his aides promised to eat a tyre if it was.[1] Fortunately, this corrosive pessimism was countered and eventually overcome by crusaders who could think on their feet and were determined to defeat smallpox against resistance from both friend and foe.

Vaccination should claim most of the credit for the eradication of smallpox but, for once, Nature helped. Around the turn of the twentieth century, the mild and generally non-lethal variant, *Variola minor*, took hold firmly in North America and Brazil and steadily supplanted *major*. This shift greatly reduced the total death toll from smallpox, although it created some problems of its own. Elsewhere, *major* continued to rule supreme and *minor* never penetrated significantly into the historical heartland of smallpox, the densely-packed and mobile populations of India and Pakistan.

Key change

"But how strange the change
From major to minor,
Every time we say goodbye."
Cole Porter, 'Every Time We Say Goodbye', 1934

The severity of classical smallpox (*Variola major*) was long recognised to be variable, in that some outbreaks killed 50 per cent or more of victims while others claimed 20 per cent or fewer. *Variola minor* clearly fell outside this spectrum, being markedly gentler than classical smallpox at its most benign.[2] Its clinical course was so mild that most sufferers felt well and continued their normal activities. Very few required hospitalisation and the mortality rate was generally 2 per cent or less. The rash was sparse and more superficial in the skin, so that the risk of permanent scarring was much lower than with *major* (only 7 per cent, rather than 75 per cent, had five or more facial pocks). Even the characteristic stench of suppuration that accompanied classical smallpox was said to be absent.

Variola minor showed distinct properties in other biological test-beds – for example, producing smaller pocks on the embryonic chick membrane – and its DNA sequence also diverged in places from that of *major*. Even within *minor*, there was variability: the strain prevalent in Eastern Africa tended to be more severe, and it was genetically closer to *major*. Interestingly, a third variant of smallpox – *Variola intermedius* – was identified in Africa during the 1950s. As its name implies, its severity fell between those of *major* and *minor* and it carried a mortality of about 10 per cent. However, later research raised doubts about whether a distinct entity really existed in the grey zone between *major* and *minor*.[3]

Variola minor had been noted in various parts of the world, presumably having spread from its original focus, tentatively in West Africa when its taterapox precursor underwent the defining mutation perhaps six millennia ago. It was known in Africa as *amaas* or pejoratively as 'kaffir-pox' and in the Caribbean as *alastrim* (from the Portuguese for 'spreading like fire'). It was always a diagnostic conundrum, especially where it existed alongside *Variola major* and chickenpox, with which it was persistently confused until specific laboratory tests were developed. *Variola minor* may have been the strikingly lenient form of smallpox that caught the attention of Jenner in 1798: "So mild a nature that a fatal instance was scarcely ever heard of."[4]

The wave of *Variola minor* that engulfed North America within a few years around the start of the twentieth century was another example of a re-energised poxvirus that seemed to spring from nowhere.[5] The most likely sources were long-standing foci in the Caribbean islands or ships coming from South Africa, another stronghold. The portal of entry was Pensacola, a busy port on the southern coast of Florida, where a cluster of cases with peculiarly trivial smallpox emerged in 1896. From this first toehold, *Variola minor* fanned out steadily to cover the North American landmass: the Midwest by 1898 and right to its limits at the Pacific and Atlantic seaboards and the frontiers with Canada and Mexico just two years later. Canada was also invaded, but America's southern border was less permeable and Mexico remained a bastion of *Variola major*. Further south,

Variola minor also broke into Brazil, presumably through traffic with the Caribbean, but *major* continued to be endemic throughout the rest of Latin America.

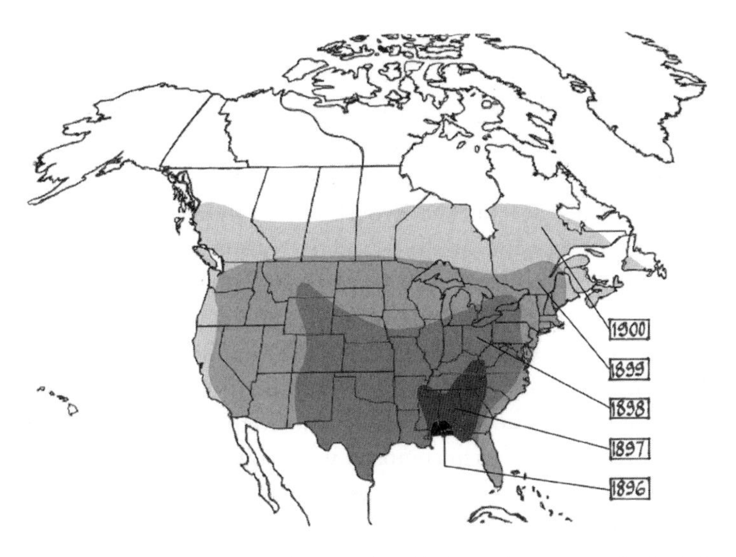

Figure 15.1 Map showing the spread of *Variola minor* across the United States around the turn of the 20th century. Modified after reference no. 5.

The spread of *Variola minor* across North America was phenomenally rapid – even faster than when *major* had cut down the virgin populations of the Aztecs and Incas nearly 400 years earlier. The mildness of the infection contributed. Classical smallpox tended to limit its own infectivity by making its victims ill enough to withdraw from company and be shunned by society. As with the 'artificial' smallpox of variolation, infection with *Variola minor* was mostly so trivial that patients remained in circulation and isolation was thought superfluous. Another formality commonly omitted was the notification of cases to the public health authorities. As a result, even the most explosive epidemics of *Variola minor* were under-reported, and the true numbers of cases must have been substantially higher.

Even though underestimated, the official figures show an exponential increase in total smallpox cases in the US between 1895 and 1900, rising successively from 2,000 in 1895–96, to 11,000 in 1899, 20,000 in 1900 and then 48,000 in 1901. The mixed nature of the beast is revealed by the risk of smallpox killing its victims, which plummeted during the same period. There were 400 deaths (a 20 per cent mortality rate) in 1895–96, 660 deaths (6 per cent mortality) in 1899, 800 (4 per cent) in 1900 and just 3 deaths (0.6 per cent) in 1901. *Variola major* held sway in 1896 but was progressively diluted by *minor* with its markedly lower death toll – and the true percentage mortality rate would have been even lower, because of the likely large numbers of unreported cases of *minor*.[6]

The *modus operandi* of the 'new' smallpox was neatly demonstrated by the impressive trails of cases left by two travelling troupes of actors as they toured the Mid- and Eastern USA.[7] Members of each company went down with 'chickenpox' but not badly enough to interrupt a hectic performance schedule or to repel audiences. Each tour was marked by an infected corridor of several hundred cases which soon spread into the surrounding areas. Fortunately, the risk of being killed by *Uncle Tom's Cabin* was a tiny fraction of that associated with the 'wild beast show' in Middlesbrough a year earlier, which had unwittingly featured *Variola major*.

For the victim, *Variola minor* was much more blessing than curse. The survivors (virtually all cases) were left with lasting protection against smallpox, both *major* and *minor*, because both variants shared the same proteins that fired up the immune response.[8] Scaling up from the individual, this meant that the advancing front of *Variola minor* left behind many previously susceptible populations which were now at least partially immune to smallpox. Thus, new outbreaks of *major* spread less rapidly and could be more easily contained by isolation of fresh cases and vaccination. Thanks to the biological advantage conferred by its mildness, *minor* gained ground at the expense of its more dangerous sibling and steadily replaced *major* as the endemic form of smallpox across North America.[9] This did not stop *minor* from rising up periodically to pick off cases it had missed during its first sweep, and 200,000 cases were recorded in an epidemic between 1920 and 1925. This coincided with *Variola major* outbreaks that affected several thousand cases. At the peak, the US had the unenviable distinction of being the country with the second highest number of cases worldwide (after India).[10]

Vaccination, effective against both *major* and *minor*, was shunned by many Americans, especially as its inconvenience and hazards seemed to pose a greater risk than a trivial attack of something no worse than chickenpox. Following an outbreak in a factory in Niagara Falls, only seven of the employees agreed to be vaccinated so that they could continue working and earning. The other 90 chose two weeks of unpaid suspension.[11]

In North America, the balance seemed to tip against smallpox in the late 1920s. This could have been due to a spontaneous decline in the infectivity of *Variola minor* as well as the tightening grip of vaccination. The numbers of cases fell dramatically through the early 1930s and suddenly, in 1935, an event occurred that had seemed almost unthinkable a decade earlier: endemic smallpox had disappeared from the United States. Canada followed in 1944. However, the subcontinent was still at risk of outbreaks of imported smallpox, especially from Mexico. *Variola major* had the last word, causing the final three outbreaks in the USA, in 1945 (Seattle), 1947 (New York) and 1949 (Rio Grande, Texas), and the last Canadian epidemic in Toronto, 1962.[12] All were rapidly stamped out by isolation and mass vaccination but all showed the Angel of Death behaving as usual – and mankind still confused and incapacitated by learning difficulties.

The penultimate American outbreak had depressingly familiar ingredients spiced with mass panic, and was rounded off with a convincing demonstration of how rapidly large-scale public health measures could be mobilised.[13] The source was a

47-year-old American man, resident in Mexico City for six years. In late February 1947, he set off with his wife on a Greyhound Bus, crossing the Southern states to visit New York. He had been vaccinated in childhood and revaccinated in Mexico, where *Variola major* was still endemic; the revaccination had not taken but nobody bothered to have it repeated. Falling ill with a fever soon after reaching New York, he was admitted to a specialised infectious diseases hospital where his rash (probably modified by his original vaccination) was mistaken for an allergic reaction. He died a few days later. The correct diagnosis was made two weeks after his death, when two fellow in-patients developed classical smallpox. By then, several other contacts were infected. In all, there were twelve cases and two deaths. An emergency vaccination campaign swung into action in New York, and treated over 6.3 million people in under a month. Luckily, the index case had not become infectious during his 2,000-mile pilgrimage by bus – or at least, no cases were identified along the route. His wife, who had been successfully revaccinated, remained healthy. Between them, the couple illustrated both the basic truth and the fatal flaw in Jenner's original dogma: vaccination protects, but not for ever, and must be topped up periodically.

Variola minor made inroads elsewhere in the world. A focus appeared in Nottingham, England in 1901 following a Mormon conference attended by a delegation from Salt Lake City, which had been overtaken by the advancing edge of the American epidemic a couple of years earlier.[14] It spread more slowly than in the US but, together with vaccination, helped to squeeze out *Variola major* – which ceased to be endemic in England in 1905. Having moved in, *Variola minor* remained endemic for another three decades and, as in America, it swelled into an epidemic during the early 1920s.[15] There were over 80,000 cases but only 200 deaths. The cities affected included Gloucester, still in Dr. Hadwen's thrall and holding out stubbornly against vaccination. With just a handful of deaths, this outbreak was light relief compared with the attack by *Variola major* 25 years earlier – and this was seized upon by the anti-vaccinationists as further evidence that vaccination had achieved nothing. Between 1928 and 1934, Dr. J. Pickford Marsden collected nearly 14,000 cases of *Variola minor*.[16] His systematic survey confirmed the very low mortality – just 19 deaths directly attributable to smallpox and its complications, a rate of 0.13 per cent – and, as far as he could tell, lasting protection against *Variola major* as well as further attacks of *minor*.

Marsden's review turned out to be a retrospective. Mirroring events in North America, both *major* and *minor* lost ground quickly after 1930 and the last case of endemic smallpox (*minor*) in Britain was recorded in 1935. However, as in America, the absence of endogenous smallpox did not guarantee freedom from imports of the disease. Britain, which has effectively defended its frontiers against rabid pets and the Colorado beetle, never attempted to bar human carriers of infection. Accordingly, smallpox slipped in and caused several outbreaks of *Variola major* during the next 30 years, including Glasgow (1950), Brighton (1950–51) and Bradford and South Wales (both 1962). *Variola minor* caused an epidemic around Birmingham in May 1966, beginning with a medical photographer misdiagnosed as having chickenpox who captured his own rash on film. The outbreak was traced to an escape of the virus from a research laboratory in the Medical School at Birmingham University. Seventy-three people were infected, but fortunately with only one death.[17] There were two subsequent outbreaks of *Variola major* from English research laboratories, in 1973 (London) and 1978 (again in Birmingham), which are described in the next chapter. The latter took place after naturally-occurring smallpox had been eradicated from the world, and caused the last ever known death from smallpox.

Switzerland became another European haven for *Variola minor* during the 1920s. The gentleness of the disease apparently seduced the Swiss representative at the League of Nations in 1926 into arguing that there was no need to make smallpox an internationally notifiable disease.[18] Echoing the words of Dr. Boëns, he said "This is no pestilence."

Variola minor showed another side to its nature when it was introduced in 1913 on an American ship to Australia and New Zealand. These countries had always enjoyed relative freedom from smallpox, due to vigorously enforced quarantine and vaccination strategies as well as their geographical remoteness. Among European victims, it behaved true to type, causing just four deaths among the 2,400 cases in Australia. However, it had a substantially greater impact on native populations. In New Zealand, only 114 Europeans were infected and all survived, whereas 2,000 Maoris fell victim, of whom 55 died.[19] This supported the long-held impression that particular ethnic groups were more susceptible to catching smallpox and to dying from it.

The last great strongholds of *Variola minor* were Brazil, hemmed in until the 1960s by *major* in the surrounding countries of Latin America, and parts of Africa, especially in the Cape region and the Horn of Africa to the south of the Arabian Gulf. In the Horn countries of Ethiopia and Somalia, *minor* coexisted with *major* and then gradually displaced it. It was in Somalia in 1977 that *Variola minor*, representing the final vestige of natural smallpox, would make its last stand.

The march of *Variola minor* helped to seal the fate of smallpox but was an irrelevance to the vast majority of those living in territory occupied by the disease; they still had to face the full horror of *Variola major*, killer and mutilator.

The long goodbye

"It was difficult to know how to tackle Africa."

D.A. Henderson, 'Smallpox – the Death of a Disease', 2009

Donald A. Henderson (known universally as 'D.A.') was the mastermind behind the WHO global eradication campaign. The quotation above shows that he was also a master of understatement. When the eradication programme began in 1966, Africa was a vast problem. Smallpox had been cleared from the north of the continent but still reigned in most other countries south of the Sahara – a total area the size of China and with a population of 278 million.[20] The imponderables of scale were compounded by difficulties with geography, people and money. The terrain ranged from rainforest to desert and could have been designed to frustrate systematic efforts to reach out to everyone – such as the nomadic wanderers who were readily swallowed up in huge tracts of desert scrubland in Somalia. Problems with people included simple bloodymindedness (which afflicted WHO operatives as well as politicians) and major armed conflicts, notably the Biafran Civil War which broke out a few months after the programme got underway. The entire campaign was cash-starved throughout, because smallpox was not seen as a high enough priority. One smallpox vaccination cost only a few cents – less than a round of ammunition for an AK-47 and one-tenth of a dose of measles vaccine – but large-scale vaccination was unaffordable in some countries, which were among the most poverty-racked in the world. The funding which trickled in from the WHO was pitifully inadequate.

And Africa, although big and difficult, was not the worst problem to confront Henderson and his newly-formed team in 1966. The number of countries with endemic smallpox had fallen from 59 in 1959 to 31, and the sash that the Angel of Death had draped across the world now looked distinctly moth-eaten (see Figure 15.2, page 338). Also, even though half of Latin America was still occupied territory, this was Brazil and the exclusive property of *Variola minor*. Elsewhere, however, *Variola major* clung on and over 1 billion people still lived in its shadow. Its most impregnable stronghold was not in Africa, but among the half-billion inhabitants of India and Pakistan.

Henderson led the eradication campaign throughout the most active phase of its life and had also smoothed the way to its conception and nursed it through a difficult gestation beset by

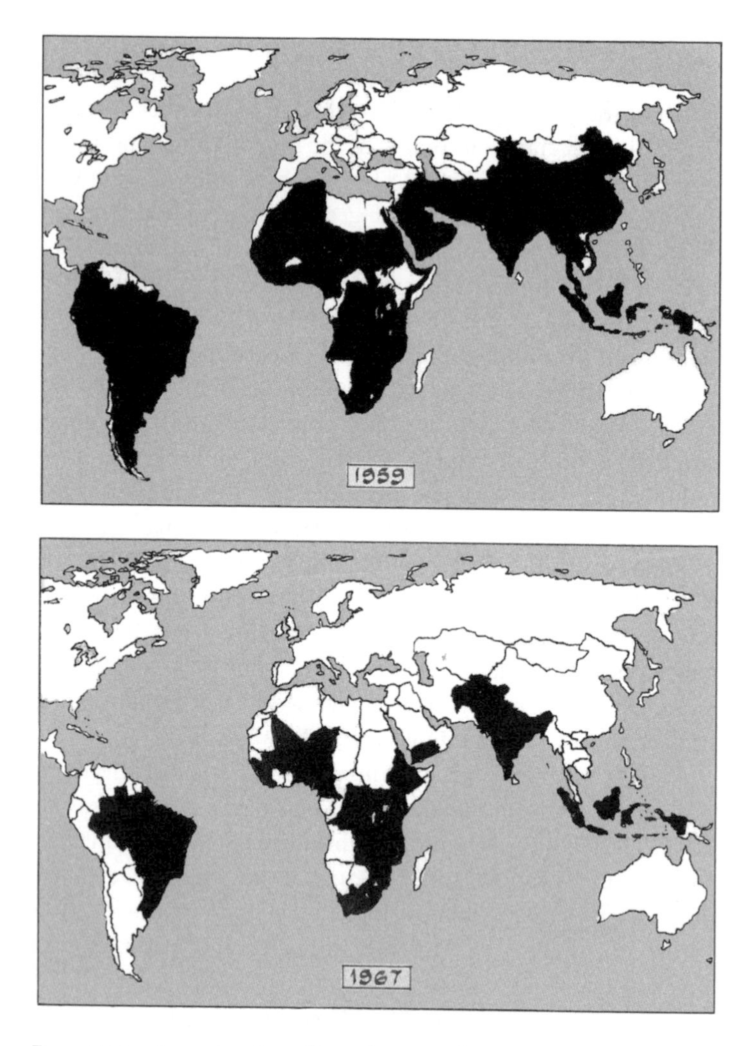

Figure 15.2 Countries affected by endemic smallpox, showing the progressive contraction from 1959 to 1967 and to 1973.

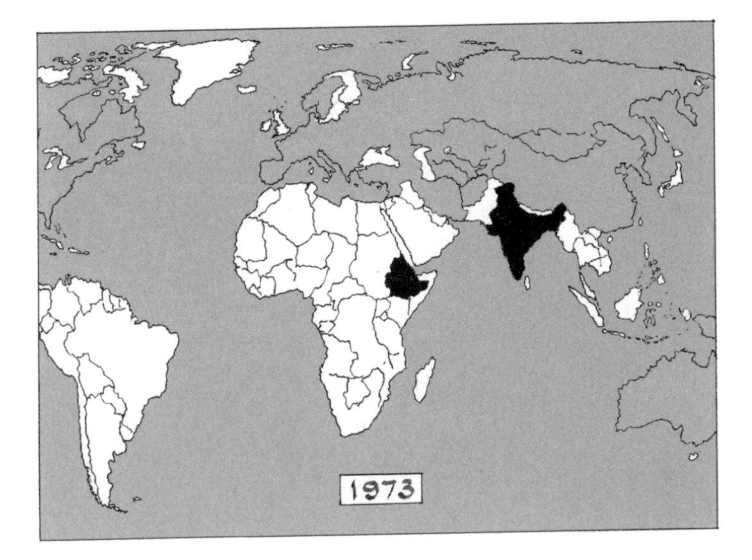

threatened miscarriages. Behind its impressive title, his WHO Smallpox Eradication Unit was tiny – just two other public-health doctors and two secretaries to begin with and barely ten people at its most populous. Operating from a cramped office suite in WHO Headquarters in Geneva, the team's influence spread out across the world like a gigantic pyramid-selling enterprise, channelled through the WHO's Regional Offices to a front-line force of up to 150,000 locally-recruited workers.[21]

The stages of the campaign during the twelve-year countdown to 'Target Zero' are laid out in the monumental WHO publication, *Smallpox and its Eradication* (1980),[22] which is familiarly known as 'The Big Red Book', and by Henderson himself in his excellent book, *Smallpox – Death of a Disease*. The planning was as concerted as it could be when even the short-term future was buffeted by the unforeseen: famine, civil war, mass movements of refugees in and out of infected zones, political wrangling and deliberate undermining by WHO oligarchs. The features that distinguished the campaign and guided it through to its successful conclusion were the infectious evangelism of the top team and the outstanding qualities of their colleagues in the field: resourceful, alert to local sensitivities and determined to learn from setbacks rather than be beaten by them.

This was not the first time that plans had been hatched to wipe out smallpox. Reference to Chapters 9 and 10 will remind us that John Haygarth's paper exercise in 1793 to exterminate smallpox from Britain using isolation and variolation was followed in 1801 by Jenner's bold but strategy-free vision of a world liberated from the disease by vaccination. The first real proof of concept was probably Sacco's claim in 1803 to have swept clear the Italian province of Lombardy, reinforced in 1895 when Sweden declared that it was the first country to have chased out endemic smallpox. Even the idea of a vaccination campaign that encircled the globe was not new, as de Balmis and Salvany had shown in their great Philanthropic Expedition of 1802.

The notion of large-scale international cooperation to rid global regions of smallpox crystallised in 1950, when the Pan-American Health Organisation began a programme to eliminate the disease from Latin America.[23] This aggressive mass-vaccination campaign was led and largely funded by the US. Like the Philanthropic Expedition 150 years earlier, this was not totally altruistic. The US had been badly shaken by the New York outbreak of 1947 which had originated from Mexico, and was determined to eliminate reservoirs of smallpox that could threaten its borders. Progress was slow but steady, especially against *Variola major* which held sway in all South American countries except Brazil. Here, *Variola minor* was endemic and clung on stubbornly.

The WHO was the obvious vehicle to wipe out smallpox across the planet. In 1953, the WHO's first Director-General, Brock Chisholm, proposed that the Organisation should set out to eradicate smallpox from the world.[24] His suggestion, rejected as over-ambitious, unrealistic and too expensive, was the first of many casualties of political and financial wrangling within the WHO. The Organisation's attention shifted to the extermination of malaria and yellow fever – two essentially ineradicable diseases that still grip the world today.

A new proposal was brought to the WHO in 1958 by Professor Viktor Zhdanov, the Russian deputy minister of health.[25] This initiative marked the return of the Russians to the international table after a Cold War-related absence of 9 years and so was warmly received, at least in spirit. The WHO decided that each country with endemic smallpox should vaccinate 80 per cent of its population, the threshold thought to confer general 'herd' immunity, using its own resources. This was a major challenge for poor countries also grappling with famine, malaria, measles and other diseases. The WHO held up the umbrella of moral support but committed only trivial funding ($100,000 per year), barely enough to provide basic technical advice.[26]

Meanwhile, essential ingredients for a successful recipe were coming together: growing evidence from Latin America that large-scale eradication was feasible; the development by Leslie Collier at the Lister Institute in England of freeze-dried vaccine[27] which survived well in tropical heat; and D. A. Henderson. As a relatively fresh medical graduate, Henderson was catapulted into heading the newly-formed Epidemic Intelligence Service (nicknamed the 'disease detectives') within the American Centers for Disease Control (CDC). Rapidly gaining experience and a reputation as a hands-on problem-solver, Henderson became fixated on the global threat still posed by smallpox, and on what had to be done to stamp it out forever. He managed to graft a smallpox eradication programme on to a well-funded but floundering US government campaign to eliminate measles from 20 West African countries, where it was a major killer of children.[28] Henderson's intervention in 1965 saved the programme and brought him to the attention of Marcelino Candau, who had recently succeeded Chisholm as Director-General of the WHO. Candau was a fervent supporter of malaria eradication but believed that plans to exterminate smallpox were doomed to failure and would only steal funding from the struggling malaria programme.

Figure 15.3 D.A. Henderson, photographed in 1982 outside the Temple of Vaccinia in the grounds of the Chantry (now the Edward Jenner Museum), Berkeley.

In 1966, despite Candau's efforts to stall the proposal, the WHO approved a bold initiative: a ten-year programme to exterminate smallpox, in which at least 200 million people, initially across 50 countries, had to be vaccinated each year at a total cost of $180 million.[29] Candau would only commit the WHO to cover less than one-seventh of the costs, leaving some of the world's poorest countries to pick up their own bills. Luckily, other agencies and some countries chipped in; Russia in particular donated 400 million doses of vaccine to India. Candau insisted that Henderson be appointed director of the new Smallpox Eradication Unit. Perhaps he hoped that the 38-year-old, with barely ten years' training in public health, would be crushed by the project's financial and logistical millstones – and so fulfil his own Cassandra-like prophesies that smallpox was here to stay. In fact, Henderson was an inspired choice and his team rose to all the challenges that the programme and its opponents could throw at them.

The campaign proper began in late 1966 at opposite ends of the remaining swathe of endemic smallpox, in Brazil and Indonesia.[30] Brazil, with its population of 96 million, was important psychologically as well as operationally. This was Director-General Candau's home country, where his personal experience of trying to pin down populations that melted away into the Amazonian jungle had persuaded him that smallpox could never be eliminated. Indonesia posed geographical problems of its own, with 120 million inhabitants scattered across 3,000 islands. Some months later, operations began in Africa (346 million people throughout the west, south and east of the continent) and then in India, Nepal and Pakistan, which together contained half of the 1 billion population at risk. The early months of the programme sprang many uncomfortable surprises but also threw in valuable leads which helped to push forward the programme as a whole.

Henderson and his colleagues soon realised that the basic currency of the programme – monitoring numbers of smallpox cases and vaccinating people at risk – had become so debased that it could no longer be trusted. To judge by national statistics, smallpox was dying on its feet, with just 118,000 cases reported worldwide in 1967. Once Henderson's regional teams had dug in, they unearthed a much darker truth. For each smallpox case notified, there could be 20, or 100 or even a thousand that had gone unreported; overall, barely 1 per cent of all cases were being notified. The true size of the remaining threat of smallpox leaped into stark relief: 10–15 million cases worldwide with perhaps 2 million deaths annually.[31] Pre-existing mass vaccination campaigns in Brazil, West Africa and India had lost momentum and the numbers treated were far below target – especially when fraudulent records (which in some cases claimed to have vaccinated over 120 per cent of the population) were put straight.[32] Even vaccination itself could not be relied upon. The failure rate (as judged by absent or trivial residual scars) was 30–40 per cent in some regions, due both to poor technique and to low-quality or inactivated vaccine.

Why was the climate so unfavourable to the eradication programme? Many people sided with Candau's view that this was a lost cause, devouring money that could be better used elsewhere. Candau had even told his regional officers that they could spend the smallpox eradication budget on more deserving projects.[33] There were also powerful vested interests. Brazil's home-produced vaccine was unreliable, especially in the heat, but national pride blocked all attempts to bring in the more stable freeze-dried preparation.[34] Declaring large numbers of smallpox cases was seen as an admission of defeat which some countries could not face, even if they had allowed their national vaccination programme to wither away. Some officials in Bangladesh who reported true numbers were sacked because the sudden increase in cases indicated that they must have failed in their mission.[35] These ingrained behaviours were difficult to turn around. The first year of the campaign in Brazil, 1967, saw just 6 million people recorded as vaccinated out of the target of 30 million, and even this number was inflated by fraudulent returns. After proper records and validation were brought in, it still took another three years for the programme to reach full speed.

Valuable lessons were also learned that streamlined the campaign and allowed its sparse resources to be focused effectively. Freeze-dried vaccine was widely introduced and this helped to cut the failure rate. This preparation remained viable in tropical heat and was reconstituted in water immediately before use.

The act of vaccination itself was revolutionised by a remarkable innovation that was also a triumph of simplicity – the bifurcated needle, invented in 1965 by Dr. Benjamin Rubin at Wyeth Laboratories in the US.[36] This was 5 cm long and had a flattened end cut into two short prongs. Thanks to surface tension, the gap between the prongs held an effective dose when the needle was dipped into liquid vaccine. Holding the needle at right angles to the skin, 15 short scratches were made over a 1–2 cm square. The procedure was only slightly uncomfortable, took a few seconds and was nearly always successful. The stainless steel needles were cheap to produce, easy to sterilise and ideal for large-scale vaccination programmes because unskilled people could be taught the technique in 15 minutes. The bifurcated needle also saved vaccine, eking out a standard 25-dose vial into 100 vaccinations. The needle was introduced in 1968 and rapidly superseded other manual methods such as scratching with a

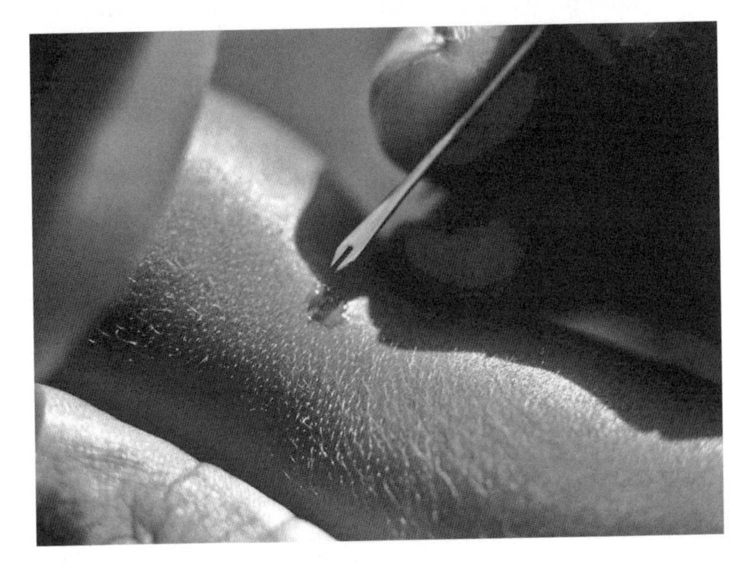

Figure 15.4 The bifurcated vaccination needle, as used in the global eradication campaign. Reproduced by kind permission of the World Health Organisation.

lancet or the painful rotary vaccinator which gouged circular cuts into the skin (for some reason, this remained popular in India until its use was barred and the instruments confiscated in 1972). Another mass-vaccination instrument was the jet injector, a military invention which fired a fine spray of vaccine through the intact skin. Widely used until the bifurcated needle rendered it obsolete, the jet injector could vaccinate up to 1,000 people per hour but wasted vaccine and was prone to break down.[37]

Two important practical advances came out of the early campaign. In Indonesia, a local worker realised that he could find more cases of smallpox by showing photographs of the typical rash to groups of locals, rather than by making the house-to-house visits he was paid to do. Schools proved especially productive, with children aged seven to twelve being particularly good at spotting and reporting cases. His visual aids evolved into the emblematic WHO Smallpox Recognition Cards, which became a key resource throughout the world (see Plate 10).[38] The other breakthrough was the discovery in Brazil that smallpox could be stamped out effectively in a locality by identifying cases early (hence the crucial importance of effective surveillance), then isolating them and vaccinating close contacts and everyone resident within a radius of 50 or 100 metres, rather than the entire population.[39] This 'surveillance-containment' technique was effective, economical and socially acceptable – which makes it strange that it was blocked and resisted by so many WHO officials, still indoctrinated with the dogma of population-wide mass vaccination.

The basic elements of the eradication campaign – vaccination kit, recognition cards and surveillance-containment – were supplemented by determination and imagination. Opponents were quietly but effectively neutralised. Politicians were charmed or otherwise nobbled, while obstructive WHO officials suddenly found themselves doing something else after Candau was succeeded as Director-General in 1973 by the astute Halfdan Mahler. Resistance to vaccination was overcome in numerous ingenious ways. Night-time raids by vaccinators caught up with villagers who had run away by day, while children were enticed out of hiding to see "the tallest man in the world" (actually the six-foot six-inch Dr. William Foege).[40] In Botswana, a Christian sect finally bent its principles and accepted vaccination; the alternative, by Presidential decree, was deportation.[41]

The operation adapted itself to meet the challenges of each area. The standard jeeps, motorbikes and cycles were replaced as required by mules in the mountains of Afghanistan, elephants in India and boats in the Amazon jungle and the floods of Bangladesh. There were many calls on individual enterprise. The pilot of a WHO helicopter captured by Ethiopian rebels managed to vaccinate his kidnappers while awaiting rescue.[42] In Bihar, Henderson had to buy the corpse of a smallpox victim to prevent it from being thrown into the Ganges.[43]

With proper collection and attention to detail, reliable data began to flow into Geneva from across the smallpox belt. Soon after the initial shock of discovering the true scale of residual smallpox, it became clear that the tide was turning against the Angel of Death in Brazil, Indonesia and Western Africa. Experience and confidence built up as smallpox retreated and as the number of outbreaks declined, resources could be concentrated more rapidly and effectively to snuff out any new incursion. The surveillance-containment stratgey was refined and tested by Foege, a CDC smallpox expert turned missionary whom Henderson recruited to head operations in Eastern Nigeria. Containment vaccination of family and neighbours was enough to stop outbreaks, if cases were identified quickly. By following up 754 cases and vaccinating 75,000 people, Foege's team succeeded in eliminating smallpox from the entire region by May 1967.[44] The timing was doubly fortunate. Supplies of vaccine were running out and could never have stretched to the whole population of 12 million, and within days Eastern Nigeria had become Biafra and war had broken out. The conflict and the population shifts which it caused were a severe test of Foege's strategy; the country remained free from smallpox.

West Africa, where 60 million people had been vaccinated in 18 months, declared its last case of smallpox in May 1970, followed by Brazil in 1971 – a fitting landmark for the half-way point of the ten-year programme. Indonesia and most remaining African countries were confirmed smallpox-free by 1972. By mid-1973, the belt of endemic smallpox had shrunk into just two areas – a block comprising India, Pakistan, Bangladesh and Nepal, and Ethiopia and Somalia in the Horn of Africa.[45]

Smallpox had effectively been chased back to its roots. *Variola minor*, which probably sprang up in Africa, was the prevalent form in Ethiopia, while *major* gripped the Indian enclave where it had first emerged.

Showdown at the Zeropox Guest House

The campaign in India had started against resistance from government and the national WHO office, which had shared Candau's prejudice that this was a futile waste of money. The Indian government even considered pulling out of the programme but was persuaded to continue by a flurry of outbreaks in 1967, some perilously close to Delhi. Even then, support was grudging, with stubborn refusals to accept wisdom proven elsewhere; even the bifurcated needle was rejected.[46]

The Indian subcontinent presented some of the knottiest problems of the programme, complicated by famine, poverty, war and tradition. In March 1971, the civil war over Bangladesh's fight for independence sent a flood of 10 million refugees across the border with India and into horrifically overcrowded camps. One camp near Calcutta, holding about 250,000 people, was not checked for illness until Foege, home for Christmas and watching a television report on the crisis, spotted obvious smallpox cases among the refugees. Emergency vaccination began almost immediately but too late to reach the 50,000 refugees who returned to newly independent Bangladesh, where they unleashed an epidemic that tore through their homeland.[47]

The programme also ran up against powerful traditions. Shitala Mata still held sway in Hindu areas, with roadside shrines to remind her devotees that smallpox was "the Goddess's kiss" and that vaccination was therefore sacrilege.[48] The origins of vaccine were similarly anathema to devout Hindus and Jains, who revered the cow; the WHO's own photographs of vaccine being harvested from calves strapped down on to a tilting table would not have helped to sway opinion. A living fossil of medical practice was encountered in Afghanistan. Traditional variolators still practised and, as throughout history, left trails of secondary smallpox in their wake; variolation was estimated to cause one-third of all smallpox cases in the country.[49] With some difficulty, the variolators were persuaded to stop and some were converted to the cause of vaccination. An accommodation was also reached with Shitala, who outlived her disease and remains in evidence today, including her presence on various websites.[50]

Figure 15.5 Calves used to generate vaccine in India during the eradication programme. Reproduced by kind permission of the World Health Organisation.

All successful campaigns depend on the foot-soldiers as much as the generals. Some insights into life in the front line come from accounts by two British doctors, drafted in to coordinate local operations in Bangladesh and India. Christopher Burns-Cox, a consultant physician from Bristol with experience in tropical diseases, spent three months in Bangladesh in summer 1974.[51] Smallpox was just one of many preoccupations. For most Bangladeshis, famine was the greatest threat and many of the huts that Burns-Cox visited contained at least one person dying of starvation. For others, the devaluation of the currency was a disaster. The smallest coins, traditionally thrown to beggars, were now worth nothing, forcing the Beggars' Association to go on strike. For the British High Commissioner, listening to the Test Match while Burns-Cox pleaded for emergency aid, the main worry was whether flooding would damage the library in Dhaka.

Following two days' induction in the 'Yellow Submarine' (an inflatable office in Dhaka), Burns-Cox was out in the field running operations across a region badly hit by floods. His tour of duty was enlivened by a flying visit from Henderson, but otherwise was an unremitting grind of chasing cases and vaccinating contacts. Children caught hiding up trees were vaccinated in situ on the sole of the foot, and there was no hiding-place for the migrants packed on to trains: Burns-Cox vaccinated the captive population on the roof, while the rest of the team dealt with those inside. Like starvation, smallpox was everywhere. While doing his rounds by bicycle, Burns-Cox once came across two badly-affected children whom he described as "oozing". All he could do was lay them over the cross-bar of his bicycle and carry them back to base to be tended by the local nurses as best they could. There was no time to be moved by these or other individual tragedies: "If I'd burst into tears, it wouldn't have helped."

Dr. A.D. Macrae, a virologist from Nottingham, visited Bangladesh and India in spring 1975, by which time the campaign was closing in relentlessly on the last strongholds of smallpox in

Bihar State.[52] There, Macrae was part of a multinational team staying at the Zeropox Guest House in Patna, decorated with a mural showing "the demon smallpox being pranged by a dusky St. George armed with a large two-pronged vaccination needle". Many problems remained. Bodies of Hindu victims were still being consigned to rivers, because Shitala would be insulted if the recipients of her kiss were cremated in the traditional way. Villagers concealed smallpox cases, including babies under blankets, and sent children to hide in the fields when the vaccinators approached. Their main fear was loss of livelihood – people locked away in quarantine (guarded by sentries paid by the WHO) could not work, and neither could children with a sore arm following vaccination. WHO policy prevailed over the rights of the individual: dissenters were held down, but only for the few seconds needed to wield the bifurcated needle.

Macrae's account shows an almost surreal humbling of the Angel of Death. Macrae had worked as a biologist for 25 years but had only ever seen one case before coming to Bihar, in the isolation hospital that had replaced the hospital boats moored in the Thames at Long Reach. Now, with the death of smallpox imminent, doctors were flooding in from around the world to see the last cases – to the point where the Bangladeshi government was "irritated by the numbers of foreign doctors trying to enter the country". Ten years earlier, smallpox had been a scourge that killed 2 million people every year; now, doctors were jostling to see this clinical curiosity before it finally disappeared. Even then, with textbook chapters on the diagnosis of smallpox about to be closed for the last time, the assembled experts could still get the diagnosis wrong, sometimes confusing the rash with that of chickenpox. There were still flashes of the old Angel's power in its dying spasms. A twelve-year-old girl hidden away to recover from mild smallpox infected six other people. Four of them, all fathers and breadwinners, died. The outbreak was quickly contained by vaccination, but four families were wrecked.

Soon after Macrae returned home, Bihar and the rest of India was declared smallpox-free in time to celebrate Independence Day on 15 August 1975.[53] *Variola major* was now cornered in Bangladesh, where the outbreaks steadily fell away. The last case was notified in December 1975: a three-year-old girl, Rahima Banu, found hiding under a sack in a village on the densely-populated island of Bhola in the Mouth of the Ganges.[54] Rahima turned out to be the tip of an iceberg, as over 140 other cases had been infected but had been hushed up by the local medical officer. Fortunately, she recovered and the rest of the iceberg had safely melted away. A last intensive burst of containment vaccination swung into action on Bhola, and no more cases were detected.

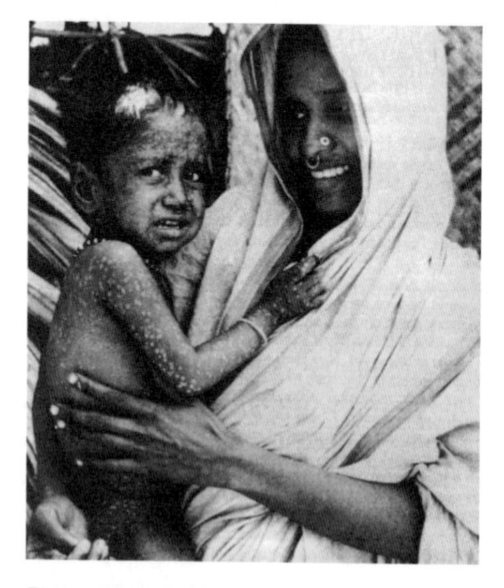

Figure 15.6 Rahima Banu, the last victim of naturally-occurring *Variola major*, 1975. Reproduced by kind permission of the World Health Organisation.

This small-scale success marked a double landmark of global importance. Incredibly, the whole of Asia had been swept clear of smallpox and *Variola major*, the incarnation of the Angel of Death, had been expunged from the planet.

Last stand

This just left *Variola minor* in Ethiopia and neighbouring Somalia, where the government had carefully covered up smallpox outbreaks and blocked the WHO's attempts to investigate them. Other challenges included the vast scrublands of the Ogaden Desert which swallowed up entire tribes of nomads with ease, competition from native variolators, and a nail-biting countdown to the annual pilgrimage to Mecca which could have injected smallpox (albeit *minor*) into a vast social network that penetrated to all parts of the world. At top level, resources were poured in by the WHO, CDC (where Foege was now Director) and foreign governments. On the ground, cases were isolated behind thornbush barriers and guards were stationed at water-holes to wait for the return of wandering nomads.[55]

The world's last victim of natural smallpox (see Plate 13) was infected in the port of Merca, Somalia in late October 1977. *Variola minor* put in a final flourish, while mankind showed that we remained unenlightened to the very end. Ali Maow Maalin was 23 years old, had been a part-time vaccinator and now worked as a cook in the town's hospital. He had never been vaccinated, even though this was mandatory for both his jobs. On 12 October, a jeep arrived carrying two sick nomad children with suspected smallpox, bound for the adjacent isolation camp. Maalin hopped in to show them the way; he was on board for only a few minutes. When he was admitted to his own hospital with a fever two weeks later, nobody thought of smallpox, even though one of the children had just died of it. Maalin was diagnosed initially with malaria and then, when the rash broke out, chickenpox. He was not isolated in hospital or at home, until a nurse friend finally turned him in as a probable case of smallpox. Maalin, his 91 contacts and the eradication campaign in general were lucky and there were no further cases.[56] However, the world's last mini-outbreak of *Variola minor* carried a mortality rate of one in three, more typical of *major*.

Target Zero was achieved just seven months after the programme's intended ten-year lifetime. A decent period had to elapse to make sure that no more cases would spring up.[57] After two tense years and 9,000 false alarms, eradication was judged complete and the declaration that opens this book was made on 9 December 1979.

The eradication programme was an improbably triumphant finale to the story of smallpox, and one of the greatest coups in the history of medicine. Henderson deserves special praise, together with his right-hand man Isao Arita, Foege and other leading players, but this was a team effort on a global scale – in the words of Robert Preston, "One of the noblest and best things we have done, as a species."

When the programme finished, Henderson had lapel badges made from a bifurcated needle bent into a circle to symbolise the long-awaited Target Zero. He sent a badge and a personalised citation certifying membership of the 'Order of the Bifurcated Needle' to all the international staff who had fought in the campaign.[58] The one awarded to Dr. Christopher Burns-Cox is now in the Edward Jenner Museum in Berkeley.

Henderson also sent a tyre to the WHO mandarin who had sworn he would eat one if smallpox has ever wiped out. Like the insignia of the Order of the Bifurcated Needle, the tyre remained a perfect and unbroken Zero.[59]

16 Legacy of an Angel

"Infectious diseases spring up in obscurity, and extend indefinitely; but if opposed with judgment, they might, like empires, be controlled; and would decline and fall."

James Moore *The History of Small Pox*, 1815

The stainless-steel zero of the Order of the Bifurcated Needle would have been a fitting symbol to round off the story of smallpox. In fact, the final chapter was complicated by at least one twist. A brief addendum had to be inserted before the WHO finally committed its declaration to parchment in May 1980, and there may or may not have been a sub-plot, which we can only hope will not turn into a sequel.

The addendum was set in Birmingham, England, and began during the Bank Holiday weekend in late August 1978, just ten months after Ali Maow Maali climbed into a jeep to direct two sick children to the isolation camp at Merca.[1] Dr. Alasdair Geddes, consultant in infectious diseases, was called in to see a 40-year-old woman with fever and a rash who had been admitted to a general ward at the East Birmingham Hospital. Geddes had served in the front line of the eradication campaign in Bangladesh in 1973 and was the region's designated smallpox expert – a role that he had come to regard as nominal until he saw the patient, Janet Parker. The working diagnosis was influenza complicated by an antibiotic allergy; in fact, she was covered with a classical smallpox rash which was already becoming confluent. Geddes knew of Mrs. Parker, who worked as a medical photographer in the Medical School at Birmingham University. Her office was on the floor above a poxvirus research laboratory run by Dr. Henry Bedson, and Geddes immediately guessed that this was the source of the infection. He rang Bedson at home and arranged to meet him in the laboratory, where Bedson would perform urgent electron microscopy on samples of blister fluid to confirm the diagnosis. Bedson was obviously shaken and as the image came up on the electron microscope, he lowered his head over the screen. Geddes had to nudge him to one side and peer over his shoulder to see the result: in his words, "like a building site, with bricks lying everywhere".

Mrs. Parker had not been vaccinated for twelve years, even though all those working on the floor below had been revaccinated routinely every three years. She was immediately transferred to the city's isolation hospital which for years had stood unused but in constant readiness. Despite intensive supportive therapy, she developed severe haemorrhagic smallpox complicated by renal failure and died on 11 September.

By then, the outbreak had already claimed another victim. On 5 September, Bedson had identified the specific strain of *Variola major* that had infected Mrs. Parker. Named 'Abid' after the three-year-old Pakistani boy from whom it had been isolated, it was the subject of research in his laboratory. The following day, Bedson was found at home with a suicide note and knife-cuts to his throat. He was rushed into the emergency unit and resuscitated, but too late to prevent irreversible brain damage; the painful decision was taken the next day to switch off his life support.

Over 340 contacts of Mrs. Parker were traced and vaccinated. Fortunately, only one – her mother – developed smallpox and this was aborted by prophylactic vaccination and methisazone.

This was not the first time that smallpox had broken out of a supposedly secure research laboratory. In 1973, safety standards at the virology laboratory in the London School of Tropical Medicine had slumped to the point where smallpox virus was routinely handled in the general laboratory rather than behind the highest-category isolation barriers. A junior researcher,

Ann Algeo, was hospitalised on 16 March with a fever and a peculiar rash. She had been revaccinated against smallpox the previous year and this presumably explains why nobody made the connection – even though she had watched samples of live *Variola major* being injected into eggs on an open bench a couple of weeks earlier. While being nursed as a case of glandular fever on a general ward, Ann Algeo passed smallpox on to a couple who visited the patient in the next bed; she recovered but they both died in the Long Reach isolation hospital on the banks of the Thames Estuary.[2]

Following this catastrophic lapse, the WHO ordered inspections of all laboratories across the world that were known to hold stocks of variola virus.[3] There turned out to be several hundred. Many were judged dangerous because of poor facilities or shoddy procedures and were instructed to close or wind down. Bedson's laboratory in Birmingham, which had done valuable work on poxviruses and had identified *Variola intermedia*, was scheduled for closure because it failed to meet standards of safe practice. In summer 1978, Bedson and his team were working flat out to finish the last few months of their research programme. The specific safety breach that led to Mrs. Parker's death was never identified. The official report by Professor Reginald Shooter blamed a fault in the duct which carried wiring and pipes from the smallpox handling area past her desk.[4] However, other experts doubted that Bedson's experiments could have spilled enough variola virus into the air. According to their calculations, only huge volumes of viral medium – over 11,000 gallons – could have generated enough airborne viral particles to infect Mrs. Parker.[5]

This accident hastened the demise of many smallpox research programmes. In the late 1950s, over 600 laboratories worldwide had worked with variola but the number fell away rapidly during the 1970s and especially after the escape in London. At the time of the Birmingham incident, all but 14 had destroyed their stocks or passed them to more secure laboratories, and just six remained in 1980.[6] From 1983, all documented stocks were stored in only two centres: the world's main reference smallpox laboratories at the Centers for Disease Control and Prevention (CDC) in Atlanta, Georgia, and the Research Institute for Viral Preparation in Moscow. The Russians later moved their stock to the State Centre for Research on Virology and Biotechnology (known by the James Bond-style acronym, VECTOR) at Koltsovo in Novosibirsk Region.[7]

Viruses will keep forever if frozen to –70 °C in liquid nitrogen. *Variola major* is now a genie in suspended animation, locked away in a bottle that cannot realistically be unstoppered or broken. History is also frozen inside the several hundred screw-topped plastic vials that contain the last mortal remains of the virus. Many of the strains are named after their sources and remind us of particular chapters in the history of smallpox. They include 'Harvey' (which caused the outbreak at the London School of Tropical Medicine, 1973), 'Rahima' (Bhola Island, 1975), 'Abid' (Birmingham, 1978) and 'India', isolated in 1959 and apparently of special interest to the Russians.

We do not know all the precautions in place to prevent the virus from escaping but can assume that terrorist attacks and earthquakes are covered as well as breakdown of the freezers. However, Benjamin Franklin's adage that death and taxes are the only certainties reminds us that, as long as the virus exists, the risk of it escaping can never be completely eliminated.

So why not carry mankind's centuries-long battle against smallpox to its logical conclusion and simply destroy these last scraps? This is the argument pushed forcefully by D. A. Henderson and others. Henderson is clearly frustrated by serial stalling by the WHO, which first decreed in 1985 that the stocks should be killed off but has since repeatedly deferred execution. The decision will next be reviewed in 2011.[8] Those arguing against destruction have included senior representatives of the US, Russia, Britain and France. Paranoia about national security may be an important driver but various reasons have been cited in public, some of which must seem bizarre to those who fought in the front line against smallpox.

One counter-argument is that the variola virus is a resource that could still teach us valuable lessons and so should not be thrown away. There may be some substance to this. We do not know how smallpox kicked the immune system so hard that the immunity it induced usually lasted a lifetime; understanding this quirk could be exploited to make other vaccines more durable. Similarly, discovering how variola disabled components of the body's virus-killing machinery could also in theory be turned to our advantage. However, Henderson maintains that, thanks to advances in molecular biology, intact virus is no longer needed to learn any remaining lessons of variola. By 1994, Craig Venter and other workers had identified every last character of the DNA sequence from several key strains of *Variola major* – a considerable undertaking, as the genome comprises more than 180 genes, over six times the complement of the polio virus which had earlier been sequenced.[9] Specific DNA-cleaving enzymes have been used to chop the variola virus genome into a 'library' of fragments in which particular genes are embedded. Using molecular biological techniques that have now become routine, a gene of interest can be copied countless times and enough of the protein which it encodes can be run off in the test tube to study in detail its structure, what it does and how it works.[10] Novel compounds can be designed to target the active sites of proteins of interest, such as those which drive immune responses or cripple the body's antiviral defences, and the activity of these 'candidate' drugs can be tested in the laboratory without the need for viable virus.

However, other types of experiment may still need intact viruses. Tiny samples of variola virus in the CDC have been thawed out to evaluate the effects of new drugs such as cidofovir, which inhibits DNA replication and therefore the multiplication of poxviruses.[11] Cidofovir was developed after smallpox had been eradicated and could not be tested under proper clinical trial conditions. It was found to be beneficial in rats and mice experimentally infected with their own native poxviruses, but rodents cannot be given smallpox (which only attacks man and higher primates). Cidofovir also prevented variola virus from killing human cells grown in culture, but we do not know whether it could protect contacts from developing full-blown smallpox or, best of all, could fill that critical gap in the therapeutic cupboard and rescue victims who are already in the grip of the infection. There might seem to be no logic in reawakening an enemy from a state that is effectively death to see how easy it can be killed – unless there is a chance that the enemy will somehow reappear to do battle once more. In that eventuality, discussed below, this academic exercise suddenly becomes an essential plank of forward planning.

Philosophical arguments have also been advanced to preserve the variola virus from final extinction. Against the horrors of smallpox, some of these views might seem ethereal to the point of absurdity. It has been stated that humans, being the planet's dominant life-form, must respect biodiversity and do not have the right to play God by deliberately snuffing out other species.[12] The premeditated murder of another life-form is evidently different from the unintended extinction of other species – estimated at 27,000 species *per year* – which occurs as a regrettable but inevitable consequence of civilisation. Stranger still, the decision to annihilate variola could fall foul of international environmental law,[13] even though the species in question has killed billions of people and is not alive by any conventional definition. Moreover, 'biodiversity' implies some dynamic interaction with other forms of life. Imprisoned in its mausoleum of liquid nitrogen and with no prospect of release back into the biosphere, the variola virus is already as good as extinct.

For now, the virus remains locked in suspended not-quite animation and the WHO in indecision. It is now a quarter of a century since the WHO executive first voted to destroy the last stocks, but mankind's 'Greatest Killer' may remain with us for some time to come.

Return of the angel?

"Fate keeps on happening."
Anita Loos, *Gentlemen Prefer Blondes*, 1925

Fallen angels do not necessarily disappear for ever. Their archetype managed a strikingly effective comeback and has already turned up in this book – as the first inoculator who cursed Job with boils and later as the short black man, about the height of a walking stick, who stalked Cotton Mather's flock in New England. More recently, if some accounts from the post-eradication era are accurate, Satan may also have helped to direct operations in certain biotechnology facilities.

Could smallpox really return? The complete absence of any cases for over 30 years can only mean that it has been annihilated from free existence on the planet, and it is vanishingly unlikely (but not beyond the bounds of possibility) that it could escape from the high-security vaults in the CDC or Koltsovo. The hypothetical spectre has been raised that smallpox could break out from the thawing of infected corpses which had lain deep-frozen in the permafrost of the tundra.[14] Such relics could well exist, but it seems unlikely that the variola virus could survive prolonged entombment in ordinary ice which, even in the high Arctic, is much warmer than the –70 °C of liquid nitrogen that is needed to abolish all possibility of degradation. Even if it were to happen, an outbreak triggered in this way could be quickly crushed by isolation and vaccination, as practised during the eradication programme.

In theory, a virus similar to smallpox could arise by mutation from an existing poxvirus that currently infects another species. This might not have to recapitulate the two-step process thought to have brought *Variola major* into the world from its taterapox grandparent. Monkeypox is already most of the way there. It can infect humans, causing a severe smallpox-like illness with a high mortality in children.[15] However, its infectivity is relatively low (it is more commonly caught from rodents, in which it is also endemic, than monkeys). Also, it has not acquired the knack of jumping easily from one human to another, which would be essential for it to match the killing power of smallpox. Poxviruses are notoriously fluid – as evidenced by the spontaneous disappearance of horsepox and camelpox and the enigmatic arrival of vaccinia – but it remains to be seen whether the magic wand of mutation could conjure up these essential credentials. As with an outbreak triggered by the thawing of a smallpox-riddled corpse, a freak mutation of monkeypox or some other poxvirus would produce a single focus of infection that could be readily contained by isolation and vaccination.

There may exist another reservoir of smallpox infection. We have to hope that it does not, because the threat it holds could easily turn into a nightmare: smallpox rampaging across the world once again, feeding off the largest susceptible population in the history of our species, and thriving on mankind's collective inability to deal with this forgotten scourge.

Like the contents of the vaults in Atlanta and Koltosovo, this reservoir would be of human making. Its origins lie in the devious gifts of smallpox-contaminated blankets which English traders presented to North American Indians in the 1760s.[16] The tradition was perpetuated during the Second World War by the Pingfan Institute in Japan, which dispensed with all niceties and sprayed Chinese prisoners of war with a mist of variola virus.[17] It is depressing to realise that mankind, which was united in the 'noble cause' of eradicating a common enemy, is also capable of exploiting that same enemy to kill other people. This is smallpox, transformed into a weapon of biological warfare.

Throughout history, *Variola major* was always an agent of mass destruction. However, it is not ideally suited for use in warfare and certain refinements need to be added to make it worthy of the title of 'weapons grade'. Some strains are more lethal than others and so more attractive for military applications. 'India', isolated from a citizen of that country who set off a serious outbreak in Moscow in 1959, was admired because of its virulence.[18] The infectivity of smallpox, although terrifying, is too uncertain to neutralise a dispersed enemy population within a military timeframe. Possible solutions to this thorny problem range from portable devices that can rapidly spread an invisible aerosol of the virus, to the scattering of 'weaponised' variola virus over a large area from showers of miniature bomblets released by a ballistic missile.[19] To ensure that there could be no escape from Doomsday, Russian researchers may even have tried to develop a radiation-resistant strain of *Variola major* which would thrive in a radioactive wasteland and pick off survivors of a nuclear strike.[20]

The extent to which smallpox has been exploited as a biological weapon is uncertain. The US, Britain, France and other countries all had active germ-warfare programmes in the past but have denied working on smallpox. So did Russia, but two prominent defectors – Vladimir Pasechnik and Kanatjan Alibekov, who escaped to the US in 1989 and 1992 – both claimed that they had been involved in a highly active programme that focused on 'improving' methods of delivering *Variola major*.[21] Pride of place went to the bomblets, which could be carried in the warhead of an SS-18 intercontinental ballistic missile. The warhead had to be specially modified to carry a refrigeration unit to keep its cargo cool and viable throughout the flight.[22]

There was some supporting evidence, but none of it conclusive. In 1971, a young woman studying plankton samples on a Russian fisheries ship fell ill with smallpox and sparked off an outbreak that caused several deaths. According to the official Russian report, the source must have been someone travelling from Afghanistan, the nearest endemic zone at the time.[23] The ship docked in various potentially infected ports but the young woman rarely disembarked. An alternative source, not mentioned in the report, was an island in the Aral Sea used for secret biological weapons experiments and where the dispersal of a half-kilo charge of *Variola major* was allegedly being tested.[24] The island was surrounded by a 40-kilometre exclusion zone; the fisheries ship had breached regulations and sailed to within 15 kilometres of the island. Later, while the Americans were monitoring test flights of the SS-18, they noted an unusual spin of some warheads, which could have been explained by their modification to carry the smallpox cooling unit.[25]

Then, in 1991, President Mikhail Gorbachev yielded to pressure from the US and Britain and allowed a team of their biological weapons experts to visit VECTOR in Koltsovo. There, the observers saw a state-of-the-art germ warfare establishment.[26] The facilities included a test chamber with a radiating 'octopus' arrangement of tubes which they thought could be used to test the dispersal of infectious aerosols, as released by an exploding bomblet. Most suggestive, but also most peculiar, was a technician who told the visitors, via a British interpreter, that he was testing smallpox. This news came as a bombshell, as the Russians had assured the world that they had long since given up their smallpox programme. Both the technician and interpreter insisted that 'smallpox' was the correct word. Shortly after, the visit was abruptly ended by an official who angrily denied that they were working with variola.

The truth about the Russians' military interest in smallpox is currently beyond our reach and may well remain so. It is unfortunate that the Russian smallpox holding was removed to VECTOR, given the facility's history as a germ warfare research centre. In another unhappy turn of events, Viktor Zhdanov, the energetic Health Minister who had brought the notion of global smallpox eradication to the WHO in 1958, was later appointed to head up Russia's network of germ-warfare facilities, known as Biopreparat.[27] However, there is no hard evidence – at least, not in the public domain – that the Russians were developing smallpox-based biological weapons and some of the verbal claims may be dubious. It seems incredible, for example, that personnel in a top-secret establishment, working on a project that was illegal under international law, would not be briefed against making unguarded comments to foreign inspectors. Furthermore, the most damning and explicit accounts come from two individuals who have made new lives and good money from their defection, including the best-seller, *Biohazard*, by the restyled 'Ken Alibeck'.[28]

Figure 16.1 Professor Viktor Zhdanov, WHO visionary and later chief of the Russian germ warfare establishment. © RIA Novosti, Moscow.

Yet many respected figures, notably the redoutable D.A. Henderson, are convinced that the threat is real and that the world must plan for the possibility that smallpox will be used as an offensive weapon.[29] The most likely aggressor would be a small nation or a terrorist group, rather than a superpower. If the defectors Pasechnik and Alibekov have painted an accurate picture, then there will have been plenty of opportunities for interested parties to buy up know-how and smuggled samples of 'battle' strains of *Variola major*, from disaffected employees who were shed when the Biopreparat network was closed down in the early 1990s. According to Alibekov,[30] Biopreparat dealt in an obscenely extravagant currency – literally tonnes of microbiological ordinance, which puts into terrifying perspective the notion that a half-kilo of *Variola major* could infect people over an effective range of 15 kilometres across the Aral Sea. Much less sophisticated facilities, readily adaptable from the type of research laboratories which the WHO closed down in their hundreds during the 1970s, could produce a kilo or more of variola virus. This amount would still contain enough theoretical doses to infect all 6.8 billion of the world's inhabitants hundreds of times over. Under real-life conditions, the yield would obviously be much lower but, if used effectively and with imagination, could still wreak havoc on a global scale.

The horrors of a terrorist attack using smallpox, and the practicalities of trying to deal with its aftermath, have been tackled in scenarios played out in public (and no doubt in others, behind closed doors). The 'Dark Winter' simulation planned and executed by Henderson and colleagues in 2001 lived up to its sombre name.[31] This began with several outbreaks of smallpox in the Midwest of America following the deliberate release of aerosolised *Variola major* in several shopping malls, and ended two weeks later with 16,000 people infected, vaccine running out and medical services paralysed. 'Dark Winter' excited the interest of policy-makers as well as the media and was given extra impact by its timing – just a few weeks before the 9/11 terrorist attacks of 2001, which were followed a month later by a cluster of anthrax-contaminated letters, sent through the US Mail, which killed several people.

With these sharp reminders of the horrific potential biological warfare, vaccination production was restarted in the US and no doubt in other countries as well. This was a useful precaution, as the WHO had cut its operating costs in 1990 by sacrificing 9.5 million doses of smallpox vaccine (leaving just 5 million doses), in order to save freezer maintenance costs which came to about $20,000 per year.[32]

Homework (optional)

"The potential consequences of a competently executed smallpox attack have not been adequately considered by policy makers."

Martin M. Weiss, Peter D. Weiss, Glenn Mathisen and Phyllis Guze
Rethinking Smallpox, 2004

Interested readers may wish to play out their own variola warfare scenario. If so, imagine smallpox appearing out of the blue where you live and plot its course across the world that you know. You will need to think big and remember all those unlearned lessons which so often crippled mankind's ability to deal with smallpox.

An outbreak triggered by a terrorist attack using smallpox could be frighteningly large. We note with approval that Thomas Jefferson's words to Edward Jenner eventually came true: "Future generations will know by history only that the loathsome smallpox has existed." The corollary is that, with no routine vaccinations for 30 years, the susceptible population – comprising an entire never-vaccinated generation and older people whose vaccinations will have worn off – is now bigger than ever before. In the absence of effective countermeasures, the model of spread would be closer to the rampage that annihilated the Aztecs and Incas in the sixteenth century than to any outbreaks within living memory.

Rapid identification of cases was the cornerstone of the surveillance-containment strategy that underpinned the eradication programme. Once an outbreak is established and publicised, most diagnoses would be straightforward. However, if smallpox dropped in unannounced – which is the most likely plan for terrorists wishing to maximise damage and panic – vital days could be wasted before the disease was recognised. Think of all the cases during the 1970s that were misdiagnosed as chickenpox, malaria, glandular fever, and so on – and these in a time when smallpox was still around. Now, virtually no practising doctors have ever seen a case and the disease has also been eradicated from present-day medical training. You, having read this book, would be more likely to think of the diagnosis than the medical and nursing staff who would first meet new cases. The initial delay in diagnosis could be catastrophic. The man from Mexico City who visited New York in 1947 and the case of Ann Algeo in London, 1973 will remind us how easily missed cases could spread smallpox at home and at work, and particularly in hospital. With a vast non-immune population, dissemination could be far wider, perhaps approaching the epidemic that was triggered by one infected soldier returning home to Kamchatka, Siberia, in 1778 and which ultimately affected over 20,000 people.[33]

How effectively could we fight back? There may or may not be enough vaccine to go around in each country (probably not, in many cases). How would extra vaccine be raised, when large-scale production facilities have long since been dismantled? How well are individual countries prepared for a large-scale emergency vaccination programme? Topical examples such as the various national responses to swine flu are not encouraging and are puny in scale beside what would be needed to contain smallpox. We would have to factor in the effects of mass panic, which would probably be worse than that which drove 6 million New Yorkers (far more than was necessary) to demand vaccination in the 1947 outbreak – which affected only a dozen people.[34] Even if faced by the threat of smallpox, would everyone at risk consent to be vaccinated? Not unless human behaviour has changed dramatically. Would vaccination be made compulsory, and how would this be enforced? Would conscientious objection be allowed under these conditions? Alternative forms of protection are largely untested and drugs such as methisazone and cidofovir are in short supply and expensive. What could be offered to those who refused vaccination or could not be vaccinated for medical reasons?

Surely modern medicine could rescue smallpox victims? No: specific treatment has not advanced during the 30 years since the disease was last managed, when the outcomes were no better than 50 years or even a century earlier. There is still no known drug capable of stopping the established disease in its tracks. The untried cidofovir may have some use as a prophylactic but is very unlikely to prove a miracle cure for those already in the grip of full-blown smallpox. Where would severely ill victims be nursed? Intensive care units would initially be flooded with serious cases, especially before the diagnosis became clear. These people would have little hope of recovery and would block the facilities for patients with other, potentially treatable diseases. A pragmatic but draconian policy would probably have to be implemented, barring all smallpox cases from intensive care and high-dependency units.

Where would cases be isolated until they were no longer infectious? Remember the imaginative solutions of the past – Klondike in Middlesbrough, the forest of army tents in the grounds of the isolation hospital in Copenhagen – then wind forward to today, and think of where hundreds, possibly thousands, of smallpox victims could be accommodated in cities such as London, New York or Tokyo. Perhaps Central Park, or Wembley Stadium?

Don't forget the knock-on effects – transport, schools, the wealth as well as the health of nations, and travel in a world where the advent of swine flu has recently reminded us that we take free passage for granted. How long would it be before schools reopened after an outbreak? If the underground railways in London, New York or Tokyo were to be contaminated, could other modes of transport compensate? How long would it be before an infected underground system could be declared microbiologically safe, and how long after that before people would be confident enough to use it? How far would the value of the Stock Exchange or Wall Street fall if a large outbreak hit London or New York, and how long would these take to recover?

And finally, don't forget to look to the future. If terrorists could do this once, they could probably do it again. The original perpetrators might be hunted down but by then there would be plenty of material for others who might wish to follow their example. How could we raise and maintain effective defences against the threat of smallpox recurring at any moment?

At that point, the eradication of smallpox would slip back to being a landmark of purely historical interest, and a world free from smallpox would be remembered as a brief period of luxury that would be impossible to relive. And as for so much of our history, we would have to share our planet with the Angel of Death.

Smallpox: a retrospective

"Soften your tread.
Methinks the earth's surface is but bodies of the dead.
Walk slowly in the air,
So you do not trample on the remains of God's servants."

Abù al'Ala' al-Ma'arri (975–1057) Syrian humanist and poet,
blinded by smallpox at the age of four 'Elegy on the Death of a Relative'

The millennia of our coexistence with smallpox have been a mixed experience. The dominant memory must be of a hideous disease that killed thousands of millions of people and mutilated countless more, and that thoroughly merited its title of the Angel of Death.

Smallpox was notable not only for the viciousness of the virus's attack but also for its ability to bring out man's inhumanity to man. Noble responses to the victims of smallpox, such as compassion and the determination to alleviate their suffering, have surfaced repeatedly throughout this story, but so also have other traits that show mankind at its most base. We have seen people unlucky enough to catch smallpox being dragged away to die outside city walls; Conquistadores and priests thanking God for calling down smallpox to annihilate their enemies; and scientists working to improve the virus's natural potential as a Doomsday weapon. In its defence, smallpox is by no means the only infection to have inspired such excesses of human cruelty. People with plague or cholera were also hounded and killed – and victims of cholera were beaten to death on the streets of Paris as recently as 1832[35] – while other germs conscripted into the arms race have included plague and anthrax. Throughout its lifetime, variola virus was passive and constant, unthinking and doing only what it was programmed to do; the key ingredient that determined whether smallpox would trigger philanthropy or savagery was human nature, and nothing else.

Smallpox brought some blessings as well as curses. The quirky relationships between particular poxviruses dangled an opportunity that could be seized by the observant, and also provided a test-bed on which to prove the concept of vaccination. Jenner's view of vaccination was strictly monochromatic; he had no idea that this would become one of the most valuable transferable technologies in the whole of medicine. Jenner would have been gratified by the success of the global eradication campaign which translated his far-fetched dream into a reality so complete that we have now largely forgotten about it; he would probably have been dumbfounded to learn how many viral and bacterial infections can now be prevented by vaccination. Our battle against smallpox also helped to spin off the science of immunology, which now permeates every medical discipline and holds the promise of novel treatments for diseases as diverse as arthritis, diabetes, multiple sclerosis and certain cancers.

We squandered many opportunities to learn from smallpox. Right up to our last encounters with the disease, we continued to make the same basic errors of neglecting and hopelessly underestimating the threat. We also wasted huge amounts of time and energy in squabbling among ourselves when our efforts should have been channelled into fighting the common enemy. Both sides of the vaccination debate showed how effectively beliefs and emotions – even if obviously illogical – can triumph over facts and common sense. The echoes of that confrontation still reverberate today, in the campaigns against polio, measles-mumps-rubella (MMR) and other vaccines, and in the often ill-judged attempts to counter these attacks.

Looking back, the story of smallpox is so tightly entwined with our own that we must not forget it simply because the Angel of Death no longer stalks the world. This was one of the grandest epics in the history of medicine, if not in the history of our species. A story on a scale this monumental deserves an unforgettable ending. We can only hope that this has already been written.

Appendix
Case History: Erysipelas following vaccination

As reported to the Sixth Report of the Royal Commission on Vaccination, 1894, by Dr. Thomas Skinner of Liverpool.

Q: Will you give the Commission the particulars of the case?

A: A young lady, fifteen years of age, living at Grove Park, Liverpool, was revaccinated by me at her father's request, during an outbreak of small-pox in Liverpool in 1865. [I used lymph from] a young girl, the picture of health, and whose vaccine vesicle was matured, and as perfect in appearance as it is possible to conceive. On the eighth day I took off the lymph in a capillary glass tube, almost filling the tube with clear, transparent lymph. Next day, 7th March, 1865, I revaccinated the young lady from this same tube, and from the same tube and at the same time I revaccinated her mother and the cook. Before opening the tube I remember holding it up to the light and requesting the mother to observe how perfectly clear and homogeneous, like water, the lymph was, neither pus nor blood corpuscles were visible to the naked eye. All three operations were successful, and on the eighth day all three vesicles were matured "like a pearl upon a rose petal," as Jenner described a perfect specimen. On that day, the eighth day after the operation, I visited my patient, and to all appearance she was in the soundest health and spirits, with her usual bright eyes and ruddy cheeks …

Between the tenth and the eleventh day after the revaccination—that is, about three days after the vesicle had matured and begun to scab over—I was called in haste to my patient the young lady, whom I found, in one of the most severe rigors I ever witnessed, such as generally precedes or ushers in surgical, puerperal, and other forms of fever. This would be on the 18th March, 1865. Eight days from the time of this rigor my patient was dead, and she died of the most frightful form of blood poisoning that I ever witnessed, and I have been forty-five years in the active practice of my profession. After the rigor, a low form of acute peritonitis set in, with incessant vomiting and pain, which defied all means to allay. At last stercoraceous [i.e. like faeces] vomiting, and cold, clammy, deadly sweats of a sickly odour set in, with pulselessness, collapse, and death, which closed the terrible scene on the morning of the 26th March, 1865.

Within twenty minutes of death rapid decomposition set in, and within two hours so great was the bloated and discoloured condition of the whole body, more especially of the head and face, that there was not a feature of this once lovely girl recognisable.

Q: To what do you attribute the death there ?

A: I can attribute the death there to nothing but vaccination.

Notes

CHAPTER 1

1. Robertson RG. Rotting face: smallpox and the American Indian. Caldwell, Idaho: Caxton, 2001.
2. Macaulay, Lord Thomas B. The history of England from the Accession of James II, ed. CH Firth. London: Macmillan, 1914.
3. Behbehani AM. The smallpox story: life and death of an old disease. Microbiol Rev 1983; 455–509, at p. 456.
4. Lloyd Davies M, Lloyd Davies TA. The Bible: medicine and myth. Cambridge: Silent Books, 1991.
5. Inhorn MC, Brown PJ. The anthropology of infectious disease. Ann Rev Anthropol 1990; 19:89–117.
6. Morens DM, Littman RJ. 'Thucydides syndrome' reconsidered: new thoughts on the 'Plague of Athens.' Am J Epidemiol 1994; 140:621–8.
7. Behbehani, The smallpox story, pp. 456–7; Porter R. The greatest benefit to mankind. A medical history of humanity from antiquity to the present. London: Fontana Press, 1997, pp. 96–7.
8. Hopkins DR. The greatest killer. Smallpox in history. Chicago: University of Chicago Press, 2002, pp. 16–18.
9. Ibid., p. 28.
10. Porter, The greatest benefit to mankind, pp. 55–62.
11. Hopkins, The greatest killer, pp. 9–12.
12. Ibid., p. 33; Glynn I, Glynn J. The life and death of smallpox. Cambridge: Cambridge University Press, 2004, pp. 40–1.
13. Fracastoro G. De contagione et contagiosis morbis (On contagion, contagious diseases and their cure), 1546. Also Hopkins, The greatest killer, pp. 29–30; Porter, pp. 428–30.
14. Porter, The greatest benefit to mankind, pp. 436–7.
15. Copeman SM. The bacteriology of vaccinia and variola. Brit Med J 1896; no. 1847:1277–9.
16. Guarnieri G. Ricerche sulla patogenesi ed etiologia dell' infezione vaccinica e vaiolosa. Archivio per le scienze mediche 1892; xvi:403–23.
17. Osler W. The principles and practice of medicine. New York: Appleton, 1892, p. 48.
18. Nagler FPO, Rake G. The use of the electron microscope in diagnosis of variola, vaccinia and varicella. J Bacteriol 1948; 55:45–51.
19. Behbehani, The smallpox story, pp. 482–3.
20. Smith GL. Poxviruses. Chapter 7.10.4 in Warrell DA, Cox TM, Firth JD (eds), Oxford Textbook of Medicine, 4th edition. Oxford: Oxford University Press, 2003, pp. 345–9.
21. Marsden JP. Smallpox. Chapter 5 in Modern practice in infectious fevers. London: Butterworth, 1951, pp. 569–83.
22. Baxby D, Ashton DG, Jones DM, Thomsett LR. An outbreak of cowpox in captive cheetahs: virological and serological studies. J Hyg 1982; 89:365–72; Chantrey J,

Meyer H, Baxby D et al. Cowpox: reservoir hosts and geographic range. Epidemiol Inf 1999; 122:455–60.
23. Baxby D, Bennett M, Getty B. Human cowpox 1969–93; a review based on 54 cases. Brit J Dermatol 1994; 131:598–607.
24. Baxby D. Jenner's smallpox vaccine: the riddle of vaccinia virus and its origin. London: Heinemann Educational Books, 1981.
25. Lourie B, Nakano JH, Kemp GE, Setzer HW. Isolation of poxvirus from an African rodent. J Inf Dis 1975; 132:677–81.
26. Li Y, Carroll DS, Gardner SN et al. On the origins of smallpox: correlating variola phylogenetics with historical smallpox records. Proc Natl Acad Sci 2007; 104:15787–92.
27. Behbehani, The smallpox story, pp. 483–7; Ramsay AM, Emond RTD. Smallpox. Chapter 14 in Infectious Diseases, 2nd edition. London: William Heinemann, 1978, pp. 159–95, at pp. 166–8.
28. Pankhurst R. The history and traditional treatment of smallpox in Ethiopia. Med Hist 1965; 9:343–55.
29. Department of Health. Directive for dealing with outbreaks of smallpox. London: HM Stationery Office, 1962, p. 24.
30. Downie AW, Dumbell KR. Survival of variola virus in dried exudate and crusts from smallpox patients. Lancet 1947:1:550–3.
31. Li, Carroll, Gardner et al., On the origins of smallpox, pp. 15787–92.
32. Behbehani, The smallpox story, pp. 483–7; Ramsay and Emond, Smallpox, pp. 166–8.
33. Behbehani, The smallpox story, pp. 483–7; Ramsay and Emond, Smallpox, pp. 166–8.
34. Ramsay and Emond, Smallpox, p. 167.
35. Bras G. Observations on the formation of smallpox scars. Arch Pathol 1952; 54:149–56.
36. Marsden JP. On the diagnosis of smallpox. Brit J Clin Pract 1958; 12:1–9; Behbehani, The smallpox story, pp. 483–4.
37. Bras, Observations on the formation of smallpox scars, pp. 149–56.
38. Ibid..
39. Anonymous. Smallpox before Jenner. The terrible legacies of the disease. Brit Med J 1896; no. 1847:1263.
40. Behbehani, The smallpox story, p. 483; Ramsay and Emond, Smallpox, pp. 159–60.
41. Hopkins, The greatest killer, p. 131.
42. Ricketts TF. The diagnosis of smallpox. London: Cassell, 1908; Ramsay and Emond, Smallpox, pp. 170–3.
43. Marsden, On the diagnosis of smallpox, pp. 1–9; Behbehani, The smallpox story, pp. 483–4.
44. Marsden JP. Variola minor. A personal analysis of 13,686 cases. Bull Hyg 1948; 23:735–46.
45. Ricketts, The diagnosis of smallpox; Ramsay and Emond, Smallpox, pp. 170–3.
46. Ramsay and Emond, Smallpox, pp. 174–8.
47. Behbehani, The smallpox story, pp. 485–6; Ramsay and Emond, Smallpox, pp. 178–9.

CHAPTER 2

1. Hopkins DR. The greatest killer. Smallpox in history. Chicago: University of Chicago Press, 2002, p. 42.
2. Ibid., pp. 237–8.
3. Baron J. The life of Edward Jenner, MD, with illustrations of his doctrines and selections from his correspondence, vol. 1. London: Henry Colbourn, 1827, p. 261.
4. Porter R. The greatest benefit to mankind. A medical history of humanity from antiquity to the present. London: Fontana Press, 1997, pp. 236–7.
5. Baron, The life of Edward Jenner, p. 261; Bernoulli D. Histoire de l'Académie Royale des Sciences, Part 2, pp. 1–45. Paris, 1766; Glynn I, Glynn J. The life and death of smallpox. Cambridge: Cambridge University Press, 2004, p. 70; Jurin J. A letter to the learned Dr Caleb Cottesworth, containing a comparison between the danger of

the natural smallpox, and of that given by inoculation. Phil Trans Roy Soc 1722; 32:213–17.
6. Morrison GE. An Australian in China. London, 1895.
7. Porter, The greatest benefit to mankind, pp. 311–12.
8. Ibid., p. 24.
9. Ibid., pp. 122–4.
10. Macaulay, TB. The history of England from the Accession of James II, ed. CH Firth. London: Macmillan, 1914.
11. Stewart I. Reminiscences of a smallpox epidemic. Nursing Record, 12 April 1888, p. 30.
12. Marsden JP. Smallpox. Chapter 5 in Modern practice in infectious fevers. London: Butterworth, 1951, pp. 569–83.
13. Hopkins, The greatest killer, p. 131.
14. Roddis LH. Edward Jenner and the discovery of smallpox vaccination. Menasha WI: George Banta Publishing Co., 1930, p. 2.
15. Anonymous. Brit Med J 1896; No. 1847, p. 1261.
16. Hopkins, The greatest killer, p. 33; Taylor PA. Speech to the House of Commons, Hansard 11 June 1880.
17. Porter, The greatest benefit to mankind, p. 230.
18. Hopkins, The greatest killer, pp. 295–8; Trevelyan M. Folklore and folk stories of Wales, 1st edition. 1909.
19. Coxe W. History of the House of Austria, 2nd edition, vol. 4. London: Longman, 1820, p. 117.
20. Porter, The greatest benefit to mankind, pp. 312–13.
21. Boëns H. La variole, la vaccine et les vaccinides en 1884. Bruxelles: A Manceaux, 1884, pp. 116–20.
22. Ibid., pp. 116–20.
23. Porter, The greatest benefit to mankind, pp. 60–1.
24. Coxe, History of the House of Austria, p. 117.
25. Glynn, Glynn, The life and death of smallpox, pp. 25, 41.
26. Ibid.
27. Porter, The greatest benefit to mankind, pp. 266–70.
28. Wesley J. Primitive Physic: or, an easy and natural method of curing most diseases. 1747.
29. Hopkins, The greatest killer, p. 32.
30. Glynn, Glynn, The life and death of smallpox, p. 25; Hopkins, The greatest killer, p. 33; Moore J. The history of the smallpox. London: Longman, Hurst, Rees, Orme and Brown, 1815.
31. Hopkins, The greatest killer, p. 33; Miller G. The adoption of inoculation for smallpox in England and France. Philadelphia: University of Pennsylvania Press, 1957, p. 39.
32. Hopkins, The greatest killer, p. 33; Sandwith FM. Smallpox and its early history. Clin J 1910; 36:29–302.
33. Burkhardt JL. Travels in Nubia, 2nd edition. London: John Murray, 1822, pp. 133, 211–12.
34. Radford E, Radford MA. Encyclopedia of superstition. Kessinger Publishing, 1949.
35. Macaulay, The history of England.
36. Crantz D. History of Greenland, 2nd edition, vol. 2. London: Longman, 1820, pp. 12–16.
37. Edward Jenner Museum archives, BEKJM/708.
38. Stewart, Reminiscences of a smallpox epidemic, pp. 30–1.
39. Edward Jenner Museum archives, BEKJM/609.
40. Edward Jenner Museum archives, BEKJM/647.
41. Hopkins, The greatest killer, pp. 204–21.
42. Foster EA. Motolinia's history of the Indians of New Spain. Berkeley: Cortes Society, 1950, p. 39.
43. Hopkins, The greatest killer, p. 211.
44. Ibid., pp. 212–15.

45. Patterson KB, Runge T. Smallpox and the American Indian. Am J Med Sci 2002; 323:216–22.
46. Heagerty JJ. Four centuries of medical history in Canada and a sketch of the medical history of Newfoundland, vol. 1. Toronto: MacMillan, 1928; Stearn EW, Stearn AE. The effect of smallpox on the destiny of the Amerindian. Boston: Humphries, 1945.
47. Simpson HM. The impact of disease on American history. New Engl J Med 1954; 250:679–87.
48. Woodward SB. The story of smallpox in Massachusetts. New Engl J Med 1932; 206:1181–91.
49. Porter, The greatest benefit to mankind, p. 123.
50. Patterson, Runge, Smallpox and the American Indian; Roland CG. Early history of smallpox in North America. Ann Roy Coll Phys Surg Canada 1992; 25:121–3.
51. Glynn, Glynn, The life and death of smallpox, pp. 38–9.
52. Hopkins, The greatest killer, p. 73.
53. Mahan AT. The major operation of the Navies in the War of American Independence. London: Sampson Low, Marston, 1913, pp. 117–19.

CHAPTER 3

1. Charles BG. The placenames of Pembrokeshire. Aberystwyth: National Library of Wales, 1992, pp. 610–11.
2. Williams P. Part of two letters containing a method of procuring the small pox, used in South Wales. Phil Trans Roy Soc 1723; 32:262–4; Williams P. Part of a letter from the same learned and ingenious gentleman, upon the same subject, to Dr. Jurin, R.S. Sect. Phil Trans Roy Soc 1723; 32:264–6; Wright R. A letter on the same subject from Mr. Richard Wright, Surgeon of Haverford West, to Mr. Sylvanus Bevan, Apothecary in London. Phil Trans Roy Soc 1723; 32:267–9.
3. Charles, The placenames of Pembrokeshire, pp. 610–11.
4. Williams, Part of two letters containing a method of procuring the small pox, pp. 262–4.
5. Williams, Part of a letter from the same learned and ingenious gentleman, pp. 264–6.
6. Wright, A letter on the same subject, pp. 267–9.
7. Temple R. The genius of China: 3,000 years of science, discovery and invention. New York: Simon and Schuster, 1986, p. 136; Needham J. Science and civilisation in China, vol. 6, part 6. Medicine. Cambridge: Cambridge University Press, 1999, p. 154.
8. Janetta A. The vaccinators: smallpox, medical knowledge and the 'opening' of Japan. Stanford: Stanford University Press, 1989, pp. 13–14.
9. Herbert EW. Smallpox inoculation in Africa. J African History 1975; XVI:539–59; Pankhurst R. The history and traditional treatment of smallpox in Ethiopia. Med Hist 1965; 9:343–55; Russell A. An account of inoculation in Arabia, in a letter from Dr. Patrick Russell, Physician at Aleppo, to Alexander Russell, MD FRS. Phil Trans Roy Soc 1768; 58:140–50.
10. Bruce J. Travels to discover the source of the Nile, vol. 2. Edinburgh, 1790, p. 227.
11. Burkhardt JL. Travels in Nubia. London, 1819, p. 229.
12. Glynn I, Glynn J. The life and death of smallpox. Cambridge: Cambridge University Press, 2004, p. 52; Hopkins DR. The greatest killer. Smallpox in history. Chicago: University of Chicago Press, 2002, p. 46; Vollgnad H. Globus vitellinus, misc. curiosa sive ephem. nat. cur. Jena, 1671, pp. 181–2.
13. Janetta, The vaccinators, pp. 13–14.
14. Behbehani AM. The smallpox story: life and death of an old disease. Microbiol Rev 1983; 455–509, at p. 459.
15. Behbehani, The smallpox story, p. 459.
16. Ibid.
17. Timonius E. An account, or history, of procuring the smallpox by incision, or inoculation, as it has for some time been practised at Constantinople. Phil Trans Roy Soc 1714; 29 (339):72–92.
18. Ibid., pp. 72–92.

19. Pylarinus J. Variolas excitandi per transplantationem methodus. Venice, 1715.
20. Pylarini J. Nova et tuta variolas excitandi per transplantationem methodus, nuper inventa & in usum tracta. Phil Trans Roy Soc 1716; 24 (347): 393–9.
21. Behbehani, The smallpox story, p. 459.
22. Ibid.
23. Anonymous. A recipe: or the ingredients of a medicine for the spreading mortal distemper amongst cows: lately sent over from Holland, where a like distemper rages amongst the black cattle. Phil Trans Roy Soc 1714; 29:50.
24. Mather C. An extract of several letters from Cotton Mather, DD to John Woodward, MD and Richard Waller Esq: S.R. Secr. Phil Trans 1714: 29:62–71.
25. Wagstaffe W. A letter to Dr. Freind; showing the danger and uncertainty of inoculating the small pox. London, 1722.

CHAPTER 4

1. Grundy I. Lady Mary Wortley Montagu. Oxford: Oxford University Press, 1999, pp. 5–13.
2. Stuart L. Biographical anecdotes of Lady Mary Wortley Montagu. In R Halsband, I Grundy (eds) Lady Mary Wortley Montagu. Essays and poems. Oxford: Clarendon Press, 1977, p. 9.
3. Grundy, Lady Mary Wortley Montagu, pp. 16–21.
4. Ibid., p. 25.
5. Halsband R. The complete letters of Lady Mary Wortley Montagu. 3 volumes: Vol. 1, 1708–1720; vol. 2, 1721–1751; vol. 3, 1752–1762. Oxford: Oxford University Press, 1965. Vol. 1, p. 141 [Letter to Wortley, 6 August 1712].
6. Grundy, Lady Mary Wortley Montagu, pp. 30–56; Halsband, The complete letters, vol. 1, pp. 146–65 [Letters to Wortley, 7–18 August 1712].
7. Lady Mary Wortley Montagu. The Lover: a ballad. In R Dodsley (ed) A Collection of Poems by Several Hands. London: R. Dodsley, 1748.
8. Halsband, The complete letters, vol. 1, p. 181. [Letter to Wortley, 22 June 1713].
9. Ibid., pp. 181–2 [Letter to Wortley, 25 June 1713].
10. Ibid., pp. 182–4 [Letter to Wortley, 3 July 1713].
11. Ibid., pp. 204, 210–11, 213–15 [Letters to Wortley, January–August 1714].
12. Ibid., pp. 237–9 [Letters to Wortley, 27 November–9 December, 1714].
13. Lady Mary Wortley Montagu. Six Town Eclogues. With some other poems. London: M. Cooper, 1747.
14. Grundy, Lady Mary Wortley Montagu, pp. 99–100.
15. Ibid., p. 101.
16. Lady Mary Wortey Montagu. Saturday, The Small Pox [usually known as 'Flavia', after the subject's name]. In Six Town Eclogues. With some other poems. London: M. Cooper, 1747.
17. MacMichael W. The gold-headed cane. London, 1827. Republished in facsimile edition by the Royal College of Physicians of London, 1968.
18. Grundy, Lady Mary Wortley Montagu, pp. 114–33.
19. Ibid., pp. 152–66.
20. Lady Mary Wortley Montagu. Letters of the Right Honourable Lady M—y W—y M—e: written during her travels in Europe, Asia and Africa, to persons of distinction, men of letters, & C in different parts of Europe: which contain, among other curious relations, accounts of the policy and manners of the Turks: drawn from sources that have been inaccessible to other travellers. London: T Becket, PS De Hondt, 1763.
21. Grundy, Lady Mary Wortley Montagu, pp. 143, 148, 152, 159, 163; Halsband, The complete letters, vol. 1, pp. 312–15 [Letter to Lady — 1 April 1717]; Halsband, The complete letters, vol. 1, pp. 340–4 [Letter to Anne Thistlethwayte 1 April 1717]; pp. 347–52 [Letter to Lady March 18 April 1717]; pp. 380–7 [Letter to Lady March 10 March 1718]; pp. 396–403 [Letter to Lady Bristol 10 April 1718].
22. Halsband, The complete letters, vol. 1, pp. 368–9 [Letter to Lady — 17 June 1717].

23. Ibid., pp. 337–40 [Letter to Sarah Chiswell 1 April 1717].
24. Ibid., pp. 391–2 [Letter to Wortley 23 March 1718].
25. Maitland C. Mr. Maitland's account of inoculating the smallpox. London, J. Downing, 1722, pp. 3–7.
26. Ibid.
27. Halsband, The complete letters, vol. 1, pp. 391–2 [Letter to Wortley 23 March 1718].
28. Ibid., p. 393 [Letter to Wortley 1 April 1718].
29. Grundy, Lady Mary Wortley Montagu, pp. 167–78.
30. Ibid., p. 209.
31. Maitland, Mr. Maitland's account, pp. 9–11.
32. Henderson GD (ed). Mystics of the North-East. Including: 1. Letters of James Keith MD and others to Lord Deskford. Aberdeen: The Third Spalding Club, 1934, p. 58; Maitland, Mr. Maitland's account, p. 11.
33. Grundy, Lady Mary Wortley Montagu, pp. 211–14.
34. Behbehani AM. The smallpox story: life and death of an old disease. Microbiol Rev 1983; 455–509, at p. 462; Miller G. The adoption of inoculation for smallpox in England and France. Philadelphia: University of Pennsylvania Press, 1957, p. 84; Sloane H. An account of inoculation (communicated to the Royal Society 1736). Phil Trans Roy Soc 1756; 49:516–20.
35. Behbehani, The smallpox story, p. 462; Sloane, An account of inoculation.
36. Miller G. Smallpox inoculation in England and America: a reappraisal. The William and Mary Quarterly 1956; series 3, vol. 13:480–1; Miller, The adoption of inoculation, pp. 88–9.
37. Grundy, Lady Mary Wortley Montagu, p. 214; London Gazette, 17–24 March 1722; Miller, Smallpox inoculation, pp. 480–1.
38. Behbehani, The smallpox story, p. 463; Miller, Smallpox inoculation, p. 483; Miller, The adoption of inoculation, pp. 92–4.
39. Stearn RP. Remarks upon the introduction of inoculation for smallpox in England. Bull Med Hist 1950; 24:103–22.

CHAPTER 5

1. Rudolf R, Musher DM. Inoculation in the Boston smallpox epidemic of 1721. Arch Int Med 1905; 115:692–6.
2. Hopkins DR. The greatest killer. Smallpox in history. Chicago: University of Chicago Press, 2002, pp. 237–8.
3. Winslow D. A destroying angel: the conquest of smallpox in Colonial Boston. Boston: Houghton Mifflin, 1974.
4. Beall OT, Shryock RH. Cotton Mather, first significant figure in American medicine. Baltimore: Johns Hopkins University Press, 1954.
5. Mather C. The Angel of Bethesda. New London: Timothy Green, 1722; Mather C. Christi Magnalia Americana, ed K Murdock, 2 vols. Cambridge MA: Harvard University Press, 1977.
6. Mather C. The Negro Christianised. An essay to excite and assist that good work, the instruction of Negro-servants in Christianity. Boston: B Green, 1706.
7. Simpson HM. The impact of disease on American history. New Engl J Med 1954; 250:679–87.
8. Godbeer R. The Devil's dominion. Magic and religion in early New England. Cambridge: Cambridge University Press, 1994.
9. Mather C. Remarkable providences relating to witchcrafts and possessions. A faithful account of many wonderful and surprising things. Boston, 1689.
10. Boyer P, Nissenbaum S. Salem possessed. The social origins of witchcraft. Cambridge, Mass: Harvard University Press, 1974.
11. Craker WD. Spectral evidence, non-spectral acts of witchcraft and confessions at Salem in 1692. Hist J 1997; 40:331–58.

12. La Plante E. Salem Witch Judge: the life and repentance of Samuel Sewall. New York: HarperOne, 2007, pp. 304–11.
13. Calef R. More wonders of the invisible world, display'd in five parts. London: Nath. Hillar, 1700.
14. Mather C. The wonders of the invisible world. Observations as well historical as theological, upon the nature, the number, and the operations of the Devils. Boston, 1693.
15. Calef, More wonders of the invisible world.
16. Thacher J. American medical biography; or, memoirs of eminent physicians who have flourished in America. 2 vols, 1828. Republished, New York: Milford House, 1967, vol. 1, pp. 255–7.
17. Mather C. An extract of several letters from Cotton Mather, DD to John Woodward, MD and Richard Waller, Esq; S R Secr. Phil Trans Roy Soc 1714; 29 (no. 339):62–71.
18. Ibid, p. 64 [Letter to Woodward 19 November 1712].
19. Ibid, pp. 70–1 [Letter to Waller 28 November 1712].
20. Ibid, pp. 61–2 [Letter to Woodward 17 November 1712].
21. Ibid.
22. Ibid, p. 66.
23. Kittredge GL. Further notes on Cotton Mather and the Royal Society. Publ Colonial Soc Massachusetts 1913; 14:281–92.
24. Behbehani AM. The smallpox story: life and death of an old disease. Microbiol Rev 1983; 455–509, at p. 465; Mather C. The diary of Cotton Mather. Ed WC Ford, 2 vols. New York: Frederick Ungar (n.d.), entry 16 December 1706.
25. Hopkins, The greatest killer, pp. 248–9.
26. Mather, An extract of several letters.
27. Hopkins, The greatest killer, pp. 248–9.
28. Carrell JL. The specked monster. A historical tale of battling smallpox. New York: Dutton, 2003, sources cited for Chapter 3, pp. 120–41; Hopkins, The greatest killer, pp. 247–8; Thacher, American medical biography, vol. 1, pp. 20–2.
29. Mather, The diary of Cotton Mather, entry 2 June 1721.
30. Thacher, American medical biography, vol. 1, pp. 21–2; Van de Wettering M. A reconsideration of the inoculation controversy. New Engl Quarterly 1985; 58:46–67, at pp. 49–50.
31. Mather, The diary of Cotton Mather, vol. 2, pp. 620–1, 624.
32. Thacher, American medical biography, vol. 1, pp. 21–2.
33. Ibid., pp. 185–92.
34. Behbehani, The smallpox story, p. 464; Thacher, American medical biography, vol. 1, pp. 21, 187.
35. Blake JB. The inoculation controversy in Boston: 1721–2. New Engl Quarterly 1952; XXV:493; Miller G. Smallpox inoculation in England and America: a reappraisal. The William and Mary Quarterly 1956; series 3, vol. 13:480–1, at pp. 477–9; Van de Wettering, A reconsideration of the inoculation controversy, p. 46.
36. Blake JB. Public health in the town of Boston, 1630–1822. Cambridge, Mass.: Harvard University Press, 1959, p. 57; Miller, Smallpox inoculation, p. 478; The Flying Post; or, Post-Master. London, 30 Nov–2 Dec 1721.
37. Blake, Public health in the town of Boston, p. 55.
38. Ibid., p. 58.
39. Ibid., p. 57.
40. Glynn I, Glynn J. The life and death of smallpox. Cambridge: Cambridge University Press, 2004, p. 60.
41. Douglass A [Anonymous]. Abuses and scandals of some late pamphlets in favour of inoculation of the small pox, modestly obviated, and inoculation further consider'd in a letter to A— S— MD and FRS in London. Boston: J Franklin, 1722, pp. 17–18; Van de Wettering, A reconsideration of the inoculation controversy, p. 54.
42. Williams J. Several arguments, proving that inoculating the small pox is not contained in the Laws of Physick, natural or divine, and is therefore unlawful. Boston: J Franklin, 1721.

43. Boylston Z. An historical account of the small-pox inoculated in New England, upon all sorts of persons, Whites, Blacks, and of all ages and constitutions. London: S Chandler, 1726.
44. Douglass W. Letter to Dr. Alexander Stuart, RS, 25 September 1721. Roy Soc Journal Book XII, p. 163.
45. Mather C. The way of proceeding in the small pox inoculated in New England. Phil Trans Roy Soc 1722; 32 (no. 370):33–5.
46. Mather C. An account of the method and success of inoculating the small-pox in Boston in New England. London: Jeremiah Dummer, 1724.
47. Williams, Several arguments, p. 12.
48. Ibid.
49. Grainger S [Anonymous]. The imposition of inoculation as a duty religiously considered in a letter to a gentleman in the country. Boston: New England Courant, 28 August 1721.
50. Williams, Several arguments.
51. Blake, Public health in the town of Boston, p. 60; Woodward SB. The story of smallpox in Massachusetts. New Engl J Med 1932; 206:1181–91.
52. Archbald F. A letter from one in the Country to his friend in the City. Boston, 1722.
53. Blake, Public health in the town of Boston, p. 57; Colman B. A narrative of the method and success of inoculating the small pox in New England. With a reply to the objections made against it from principles of conscience. In a letter from a Minister at Boston, ed. D Neal. London 1722, pp. 16–17.
54. Douglass W, cited in Greenwood I [Academicus]. A friendly debate, or, a dialogue between Academicus, and Sawny & Mundungus, two eminent physicians, about some of their recent performances. Boston, 1722, p. 23.
55. Blake, Public health in the town of Boston, p. 70; Mather C., cited in Douglass, Abuses and scandals, p. 16; Mather I. Several reasons proving that inoculating or transplanting the small pox is a lawful practice, and that it has been blessed by God for the saving of many a life, ed. GL Kittredge. Cleveland, 1921.
56. Thacher, American medical biography, vol. 1, pp. 21,187.
57. Mather, The diary of Cotton Mather, vol. 2, pp. 657–8.
58. Ibid.
59. Hopkins, The greatest killer, pp. 250–1.
60. Greenwood, A friendly debate.
61. Blake, Public health in the town of Boston, pp. 72–3.
62. Ibid., p. 67.
63. Thacher, American medical biography, vol. 1, pp. 21, 187.

CHAPTER 6

1. Behbehani AM. The smallpox story: life and death of an old disease. Microbiol Rev 1983; 455–509, at p. 462; Maitland C. Mr. Maitland's account of inoculating the small-pox. London: J Downing, 1722, pp. 26–30.
2. Maitland, Mr. Maitland's account, p. 2; Miller G. Smallpox inoculation in England and America: a reappraisal. The William and Mary Quarterly 1956; series 3, vol 13:480–1, at p. 483.
3. Miller, Smallpox inoculation, p. 483; Post-Boy, London, 22–24 May 1722.
4. Mead R. A discourse on the small-pox and measles, 1747. In Mead R. The medical works of Richard Mead, MD. Edinburgh, 1775, reprinted New York: AMS Press, 1978, pp. 223–95.
5. Wagstaffe W. Letter to Dr. Freind, showing the danger and uncertainty of inoculating the small pox. London, 1722.
6. Miller, Smallpox inoculation, p. 485; Wagstaffe W. Lettre au Docteur Freind; montrant le danger et l'incertitude d'insérer la petite vérole. London, 1722. Also in Journal des Savants, February 1723, pp. 133–41.

7. Nettleton T. An account of the success of inoculating the small-pox. Phil Trans Roy Soc 1722; 32:35–48.
8. Jurin J. Letter to the learned Caleb Cotesworth, containing a comparison between the mortality of the natural small pox, and that given by inoculation. London, 1723.
9. Massey, E. A sermon against the dangerous and sinful practice of inoculation. Preach'd at St. Andrew's Holborn, on Sunday, July the 8th, 1722.
10. Roddis LH. Edward Jenner and the discovery of smallpox vaccination. Menasha, WI: George Banta Publishing Company, 1930, pp. 15–16.
11. Anonymous. Self-murther and duelling: the effects of cowardice and atheism. London, 1728.
12. Applebee's Original Weekly Journal, London, 5 August 1721, p. 2129 and 3 February 1722, p. 2285; Miller, Smallpox inoculation, pp. 479, 484.
13. Miller, Smallpox inoculation, p. 479; Post-Boy, London, 7–9 September 1721.
14. Montagu, MW [Anonymous]. An account of the inoculating the small pox at Constantinople. The Flying-Post, London, 11–13 September 1722.
15. Jurin, Letter to the learned Caleb Cotesworth, p. 5; Miller, Smallpox inoculation, pp. 485–6.
16. Mather C. The diary of Cotton Mather. Ed WC Ford, 2 vols. New York: Frederick Ungar (n.d.); vol. 2, pp. 620–1, 624.
17. Boylston Z. An historical account of the small-pox inoculated in New England, upon all sorts of persons, Whites, Blacks, and of all ages and constitutions. London: S Chandler, 1726, pp. 42–7.
18. Newman H. The way of proceeding in the small pox inoculated in New England. Communicated by Henry Newman, Esq.: of the Middle Temple. Phil Trans Roy Soc 1722; 32:33–5.
19. Nettleton, An account of the success of inoculating the small-pox.
20. Behbehani, The smallpox story, p. 462; Maitland, Mr. Maitland's account, p. 20.
21. May M. Inoculating the urban poor in the late eighteenth century. Brit J Hist Sci 1997; 30:291–305, at pp. 294–6.
22. Dimsdale T. The present method of inoculating for the small pox. London, 1767.
23. Royal Vaccination Commission. Final report of the Royal Commission appointed to inquire into the subject of vaccination. London: Her Majesty's Stationery Office, 1896, p. 150; Sutton R. A new method of inoculating for the small pox, Ipswich Journal, 1 May 1762; van Zwanenberg D. The Suttons and the business of inoculation. Med Hist 1978; 22:71–82, at p. 78.
24. Van Zwanenberg, The Suttons, pp. 71–6.
25. Ibid., pp. 73–6.
26. Pead PJ. Vaccination rediscovered. Chichester: Timefile Books, 2006, p. 40.
27. Houtton R. Indisputable facts relative to the Suttonian art of inoculation, with observations on its discovery, progress, encouragement, opposition, etc. Dublin: WG Jones, 1768.
28. Glynn I, Glynn J. The life and death of smallpox. Cambridge: Cambridge University Press, 2004, p. 76.
29. Behbehani, The smallpox story, p. 466; Miller, Smallpox inoculation, p. 490.
30. Sutton D. The inoculator or Suttonian system of inoculation, fully set forth in a plain and familiar manner. London, 1796.
31. Glynn, Glynn, The life and death of smallpox, p. 91; van Zwanenberg, The Suttons, pp. 79–80.
32. Jones H. Beauty's triumph: a poem in two cantos. Cited in Deutsch H, Nussbaum F. Defects. Michigan: University of Michigan Press, 2000, p. 318.
33. May, Inoculating the urban poor, pp. 297–8; Prince T. Gentleman's Magazine, London 1753; XXIII, p. 414.
34. May, Inoculating the urban poor, p. 297.
35. Ibid., p. 296.
36. Extract from the Register of Mickleover, 1788–1789, p. 127. Edward Jenner Museum MS BEKJM/1783.
37. May, Inoculating the urban poor, p. 298.

38. Miller, Smallpox inoculation, p. 486; Thacher J. American medical biography; or, memoirs of eminent physicians who have flourished in America. 2 vols, 1828. Republished, New York: Milford House, 1967. Vol. 1, pp. 189–90.
39. Boylston, An historical account of the small-pox.
40. Blake JB. Public health in the town of Boston, 1630–1822. Cambridge, Mass.: Harvard University Press, 1959, pp. 75–7; Hopkins DR. The greatest killer. Smallpox in history. Chicago: University of Chicago Press, 2002, pp. 254–5.
41. Blake, Public health in the town of Boston, pp. 78–82; Hopkins, The greatest killer, p. 256.
42. Chambers E. Cyclopaedia: or, an universal dictionary of arts and sciences, 2nd edition. London, 1738.
43. Roddis, Edward Jenner, p. 31.
44. Franklin B, Heberden W. Some account of the success of inoculation for the smallpox in England and America. London, 1754.
45. Behbehani, The smallpox story, p. 464.
46. Ibid., p. 464.
47. Bishop WJ. Thomas Dimsdale, MD, FRS (1712–1800) and the inoculation of Catherine the Great of Russia. Ann Med Hist 1932; New Series, 4:321–88; Clendenning PH. Dr. Thomas Dimsdale and smallpox inoculation in Russia. J Hist Med 1973; 28:109–25; Dimsdale T. Tracts on inoculation, written and published at St. Petersburgh in the year 1768, with additional observations, by command of Her Imperial Majesty the Empress of All the Russias. London: J Phillips, 1781.
48. Behbehani, The smallpox story, p. 466; Glynn,Glynn, The life and death of smallpox, p. 70.
49. Behbehani, The smallpox story, p. 466; Glynn, Glynn, The life and death of smallpox, pp. 82–3.
50. Behbehani, The smallpox story, p. 459.
51. De la Condamine CM. Mémoire sur l'inoculation de la petite vérole. Mémoires de l'Académie Royale des Sciences 1754; 623–4.
52. Glynn, Glynn, The life and death of smallpox, p. 69.
53. Daniels MH. French engraved portraits. Metropolitan Museum of Art Bull 1925; 20:73.
54. Bardell D. Smallpox during the American War of Independence. ASM News 1976; 42:526–30.
55. Stark RB. Immunization saves Washington's Army. Surg Gynecol Obstetr 1977; 144:425–31.
56. Janetta A. The vaccinators: smallpox, medical knowledge and the 'opening' of Japan. Stanford: Stanford University Press, 2007.
57. Blake, Public health in the town of Boston, pp. 75–80.
58. May, Inoculating the urban poor, p. 296; Maddox I. A sermon presented before His Grace John Duke of Marlborough, President, the Vice-presidents and Governors of the Hospital for the Small-Pox, and for Inoculation, at the Parish Church of St. Andrew Holborn, on Thursday, March 5th, 1752. London, 1752.
59. Woodville W. Medical botany, containing systematic and general descriptions, with plates, of all the medicinal plants, indigenous and exotic, comprehended in the Catalogues of the Materia Medica, as published by the Royal Colleges of Physicians of London and Edinburgh. London: J Phillips, 1790–93.
60. Woodville W. The history of the inoculation of the small-pox in Great Britain, vol. 1. London: J Phillips, 1796.
61. Haygarth J. An inquiry how to prevent the smallpox. London: J Johnson, 1784.
62. Booth C. John Haygarth, FRS (1740–1827): a physician of the Enlightenment. Philadelphia: American Philosophical Society, 2005, pp. 86–8.
63. Haygarth J. A sketch of a plan to exterminate the casual small-pox from Great Britain; and to introduce general inoculation. London: J Johnson, 1793.

CHAPTER 7

1. Dod, P. Several cases in physick: and one in particular, giving an account of a person who was inoculated for the small-pox, and had the small-pox upon the inoculation, and yet had it again. To which is added, a letter to Dr. Lee, giving him an account of a letter of Dr. Freind's. Together with the said letter. London: C Davis, 1746.
2. Kirkpatrick J, Barrowby W, Schomberg I. A letter to the real and genuine Pierce Dod, MD, exposing the low absurdity of a late spurious pamphlet falsely ascrib'd to that learned physician: with a full answer to the mistaken case of a natural small-pox, after taking it by inoculation. By Dod Pierce, MS. London: M Cooper, 1746.
3. Royal Vaccination Commission. Final report of the Royal Commission appointed to inquire into the subject of vaccination. London: Her Majesty's Stationery Office, 1896, p. 158.
4. Sutton D. The inoculator or Suttonian system of inoculation. London, 1796; Brit Med J 1896; 1847:1265.
5. Ibid.
6. Newman H. The way of proceeding in the small pox inoculated in New England. Phil Trans Roy Soc 1722; 32:34.
7. Brit Med J 1896; No. 1847:1265.
8. Steer FW. A relic of an 18th-century isolation hospital. Lancet 1956; 270:200–1.
9. May M. Inoculating the urban poor in the late eighteenth century. Brit J Hist Sci 1997; 30:291–305.
10. Kumar, Clark, Clinical medicine, pp. 208–13.
11. Ibid., pp. 204–5.
12. Ibid., pp. 201–2.
13. Ibid., pp. 162–3.
14. Baron J. Life of Edward Jenner, MD with illustrations from his doctrines, and selections from his correspondence, 2 vols. London: H Colbourn, 1827; vol. 1, p. 231.
15. Hopkins DR. The greatest killer. Smallpox in history. Chicago: University of Chicago Press, 2002, pp. 63, 254.
16. Roberts J. The madness of Queen Maria. The remarkable life of Maria I of Portugal. Langley Burrell: Templeton Press, 2009, pp. 101–2.
17. Thacher J. American medical biography; or, memoirs of eminent physicians who have flourished in America, 2 vols, 1828. Republished New York: Milford House, 1967; vol. 1, pp. 191–2.
18. Blake JB. Public health in the town of Boston, 1630–1822. Cambridge, Mass.: Harvard University Press, 1959, p. 75; Douglass W. Letter to Cadwallader Codden, 1 May 1722. Mass Hist Soc Collections 1854, 4th series, 32 (2):170; Thacher, American medical biography, pp. 256–7.
19. Silverman K. The life and times of Cotton Mather. New York: Harper & Row, 1984.
20. Gates HL, Higginbotham EB (eds). African American National Biography. Oxford: Oxford University Press, 2008.
21. Forbes E. Paul Revere and the world he lived in. Houghton Mifflin, 1942. Reprinted 1999, p. 75.
22. Sloane H, Birch T. An account of inoculation by Sir Hans Sloane, Bart. Given to Mr. Ranby, to be published. Anno 1736. Phil Trans Roy Soc 1755–56; 49:516–20.
23. Halsband R. The complete letters of Lady Mary Wortley Montagu, 3 vols. Oxford: Oxford University Press, 1965. Vol. 2, p. 26 [Letter to Lady Mar July 1723]; pp. 66–7 [Letter to Lady Mar June 1726]; pp. 211–12 [Letter to Wortley 23 November 1740]; Vol. 3, p. 294 [Lady Mary's will 23 June 1762].
24. Grundy I. Lady Mary Wortley Montagu. Oxford: Oxford University Press, 1999. See Chapter 29 and p. 566.
25. Halsband, The complete letters of Lady Mary Wortley Montagu, vol. 2, p. 31 [Letter to Lady Mar 31 October 1723].

26. Boringdon, Lord. An Act for preventing the spread of the small pox (1813). With two drafts (1808–11) bearing corrections by Edward Jenner. Wellcome Library Collection, MS 3025/1, 3.
27. Baron, Life of Edward Jenner. Vol. 2, pp. 59–60.
28. Act 3 and 4 Vict., c. 29, sec VIII, passed July 23 1840; Brit Med J 1896; No. 1847:1265.
29. Herbert EN. Smallpox inoculation in Africa. J African Hist 1975; 16: 544.
30. Fosbroke TD. Berkeley manuscripts. Abstracts and extracts of Smyth's Lives of the Berkeleys ... and biographical anecdotes of Dr. Jenner. London: John Nichols, 1821, pp. 221–2.

CHAPTER 8

1. Fewster J. Cowpox and its ability to prevent smallpox. Unpublished paper read to the Medical Society of London, 1765.
2. Baron J. Life of Edward Jenner, MD with illustrations from his doctrines, and selections from his correspondence, 2 vols. London: H Colbourn, 1827 (vol. 1), 1838 (vol. 2). Vol. 2, p. 49; Fewster J. Extracts of a letter to Mr. Rolph. In Pearson G. An inquiry concerning the history of the cowpox. London: J Johnson, 1798, p. 116.
3. Peachey GC. John Fewster, an unpublished chapter in the history of vaccination. Ann Hist Med (NS) 1929; 1:229–40.
4. Adams J. Observations on morbid poisons, phagedena and cancer. Chapter 5. London: Johnson, 1795; Darmon P. Vaccins et vaccinations avant Jenner: une querelle d'antériorité. Histoire, économie et société 1984; 3:583–92, at p. 588.
5. Barquet N, Domingo P. Smallpox: the triumph over the most terrible of the Ministers of Death. Ann Int Med 1997; 127:635–42, at p. 638.
6. Glynn I, Glynn J. The life and death of smallpox. Cambridge: Cambridge University Press, 2004, p. 100.
7. Pead PJ. Vaccination rediscovered. New light in the dawn of man's quest for immunity. Chichester: Timefile Books, 2006, pp. 43–6; Wallace EM. The first vaccinator – Benjamin Jesty of Yetminster and Worth Matravers and his family. Wareham: Anglebury-Bartlett, 1981, pp. 1–6.
8. Pead, Vaccination rediscovered, pp. 48–55; Wallace, The first vaccinator, pp. 7–8.
9. Pead, Vaccination rediscovered, pp. 61–2; Southey R, Southey CC. The life of the Rev. Andrew Bell. 3 vols. London: J Murray, 1844; Wallace, The first vaccinator, p. 14.
10. Pead, Vaccination rediscovered, pp. 61–2.
11. Ibid., p. 106; Wallace, The first vaccinator, p. 15.
12. Darmon, Vaccins et vaccinations avant Jenner, p. 587; Fourier E. L'antiquité de la vaccine. Chronique Médicale, Paris, 1896, p. 328.
13. Böse J [Anonymous]. Von der Seuche unter den Kindern; über Stellen aus dem Livio. Allgemeine Unterhaltungen Göttingen bei Rosenbusch 1769; 39:305–12; Stern BJ. Society and medical progress. Princeton: Princeton University Press, 1941. Reprinted McGrath Publishing, 1970, p. 57.
14. Darmon, Vaccins et vaccinations avant Jenner, p. 587.
15. Bruce W. Cowpox in Persia – similar disease in milch sheep. In Gay-Lussac J et al (eds), Annales de chimie et de physique 1819; X:330–1.
16. Stern, Society and medical progress, p. 58; Wallace, The first vaccinator, p. 19.
17. Pearson G. An examination of the report of the Committee of the House of Commons on the claims of remuneration for the vaccine pock inoculation. London: Johnson, 1802; Stern, Society and medical progress, p. 58.
18. Barquet, Domingo, Smallpox: the triumph, p. 638; Plett PC. Peter Plett und die übrigen Entdecker der Kuhpockenimpfung vor Edward Jenner. Sudhoffs Archiv, Zeitschrift für Wissenschaftgeschichte 2006; 90 (2); S219–32; Reich F. Peter Plett, a forerunner of Jenner. Contribution to the history of vaccination. Medizinische 1953; 21 (15):520–1; Schmidt JS. Wo sind die ersten Kuhblatten inoculiert worden? Schleswig-Holsteinischen Provinzialberichte 1815; 1:77–8.
19. Darmon, Vaccins et vaccinations avant Jenner, p. 588.

20. Wujastuk, D. 'A pious fraud': the Indian claim for pre-Jennerian smallpox vaccination. Chapter 9 in Meulenberg GJ, Wujastuk D (eds), Studies on Indian medical history. Delhi: Motilal Banrsidass, 2001, pp. 121–54.
21. Darmon, Vaccins et vaccinations avant Jenne, p. 587; Fourier, L'antiquité de la vaccine, p. 328.
22. Baron, Life of Edward Jenner, vol. 1, pp. 360–7; Coley NG. George Pearson MD, FRS (1751–1828). The greatest chemist in England? Notes and Records of the Royal Society of London 2003; 57(2):161–75, at p. 169.
23. Pead, Vaccination rediscovered, pp. 62–4; Wallace, The first vaccinator, p. 14.
24. Pead, Vaccination rediscovered, pp. 67–72; Pead PJ. The first vaccinator's 'lost' portrait is found. Wellcome History 2006; Issue 30: 7.
25. Pearson G, Nihell L, Nelson T et al. Report of the Original Vaccine Pock Institution. Edin Med Surg J 1805; 1:513–14.
26. Baron, Life of Edward Jenner, vol. 2, p. 383 [Letter to James Moore, Esq, 18 November 1812.
27. Baron, Life of Edward Jenner, vol. 1, pp. 47–8.
28. Ibid., p. 48.
29. Peachey, John Fewster.
30. Baron, Life of Edward Jenner, vol. 1, pp. 49–50.
31. Pead, Vaccination rediscovered, pp. 90–100.
32. Pead, Vaccination rediscovered, pp. 72–3; Pead PJ. Benjamin Jesty: new light in the dawn of vaccination. Lancet 2003: 362:2104–9.

CHAPTER 9

1. Baron J. Life of Edward Jenner, MD with illustrations from his doctrines, and selections from his correspondence, 2 vols. London: H Colbourn, 1827 (vol. 1), 1838 (vol. 2); vol. 2, p. 381; Fisher RB. Edward Jenner 1749–1823. London: André Deutsch, 1991, pp. 5–12.
2. Fisher, Edward Jenner, p. 13.
3. Ibid., p. 14; Fosbroke TD. Berkeley Manuscripts. London: John Nichols & Son, 1821, p. 222.
4. Fisher, Edward Jenner, pp. 19–20.
5. Bailey I. Edward Jenner (1749–1823): naturalist, scientist, country doctor, benefactor to mankind. J Med Biogr 1996; 4:63–70.
6. Baron, Life of Edward Jenner, vol. 1, p. 121.
7. Fisher, Edward Jenner, pp. 21–6; Moore W. The Knife Man. London: Bantam Press, 2005, pp. 77–85.
8. Porter R. The greatest benefit to mankind. A medical history of humanity from antiquity to the present. London: Fontana Press, 1999, p. 277. Riches E. The history of lithotomy and lithotrity. Ann R Coll Surg Engl 1968, 43: 185–99.
9. Baron, Life of Edward Jenner, vol. 1, p. 105; Fisher, Edward Jenner, p. 59.
10. Moore, The Knife Man, pp. 40–65.
11. Ibid., pp. 130–42.
12. Ibid., pp. 14–17, 156–62, 210–19.
13. Ibid., p. 179.
14. Ibid., pp. 174–5, 266–73.
15. Baron, Life of Edward Jenner, vol. 1, pp. 3–5; Fisher, Edward Jenner, pp. 25–6.
16. Moore, The Knife Man, pp. 170–3, 179, 235–6.
17. Ibid., pp. 180–1, 192–6. Murley R. Also Carter R. Letters to the Editor (John Hunter). World J Surg 1994; 18:290.
18. Baron, Life of Edward Jenner, vol. 1, pp. 27–64; Fisher, Edward Jenner, p. 50; Hunter J. Letters from the past from John Hunter to Edward Jenner. Eds EH Cornelius, AJ Harding Rains. London: Royal College of Surgeons of England, 1976, p. 30; Le Fanu

W. A bibliography of Edward Jenner, 2nd edition. Winchester: St Paul's Bibliographies, 1985, p. 19.
19. Hunter, Letters from the past, pp. 9–31; Le Fanu, A bibliography of Edward Jenner, pp. 8–13.
20. Hunter, Letters from the past, pp. 17–19.
21. Ibid., p. 9.
22. Ibid.
23. Hunter, Letters from the past, p. 22.
24. Ibid., p. 23.
25. Ibid., p. 39.
26. Baron, Life of Edward Jenner, vol. 1, pp. 38–41.
27. Hunter, Letters from the past, p. 9.
28. Jenner E. Notebook, 1787. MS 372, Royal College of Physicians of London.
29. Jenner E [Anonymous]. Cursory observations on emetic tartar. Wotton-under-Edge: J Bence, 1780; Jenner E. A process for preparing pure emetic tartar by re-crystallization. Transactions of a Society for the Improvement of Medical and Chirurgical Knowledge 1793; 1:30–3.
30. Fisher, Edward Jenner, pp. 54–5; Jenner, Notebook, pp. 24–8.
31. Fisher, Edward Jenner, pp. 46–50.
32. Baron, Life of Edward Jenner, vol. 1, pp. 77–85; Jenner E. Observations on the natural history of the cuckoo. By Mr. Edward Jenner, in a letter to John Hunter, Esq: FRS. Phil Trans Roy Soc 1788; 78:219–37; Le Fanu, A bibliography of Edward Jenner, pp. 10–16.
33. Baron, Life of Edward Jenner, vol. 1, pp. 27–64; Fisher, Edward Jenner, p. 50; Hunter, Letters from the past, p. 30; Le Fanu, A bibliography of Edward Jenner, p. 19.
34. Regulations and Transactions of the Glostershire Medical Society, initiated May 1788. MS 736, Royal College of Physicians of London.
35. Baron, Life of Edward Jenner, vol. 1, pp. 50–1; Fisher, Edward Jenner, pp. 51–2.
36. Fisher, Edward Jenner, pp. 21–6.
37. Jenner E. Collection of poems and songs, 1794. MS 3017 (vol. 1) and 3016 (vol. 2), Wellcome Library Archives.
38. Baron, Life of Edward Jenner, vol. 1, pp. 15–16.
39. Ibid., p. 71; Gloucestershire Journal, 6 September 1784, p. 3.
40. Fisher, Edward Jenner, p. 33.
41. Baron, Life of Edward Jenner, vol. 1, p. 103.
42. Saunders P. Edward Jenner. The Cheltenham years, 1795–1823. Hanover, NH: University Press of New England, 1982.
43. Moore, The Knife Man, pp. 392–3.
44. Ibid., p. 394.
45. Fisher, Edward Jenner, p. 53; Moore, The Knife Man, pp. 394–5.
46. Fosbroke, Berkeley Manuscripts, p. 64.
47. Jenner E. Diary of visits to patients, fees received, expenses, etc. 1794. MS 3018, Wellcome Library Archives.
48. Ibid.
49. Regulations and Transactions of the Glostershire Medical Society.
50. Fisher, Edward Jenner, p. 53.
51. Parry CH. An inquiry into the symptoms and causes of the syncope anginosa, commonly called angina pectoris. Bath: Cruttwell, 1799.
52. Fisher, Edward Jenner, p. 58.
53. Baxby D. Edward Jenner's unpublished cowpox Inquiry and the Royal Society: Everard Home's report to Sir Joseph Banks. Med Hist 1999; 43: 108–10; Jenner E. An inquiry into the natural history of a disease known in Glostershire by the name of the cox-pox. MS in the Royal College of Surgeons of England, London; n.d. (1796).
54. Fisher, Edward Jenner, pp. 66–7.
55. Ibid., p. 67; Jenner E. Letters to John Baron and others, Royal College of Surgeons of England, London.
56. Baxby, Edward Jenner's unpublished cowpox Inquiry.
57. Baxby D. The genesis of Edward Jenner's Inquiry of 1798; a comparison of the two unpublished manuscripts and the published version. Med Hist 1985; 29:193–5.

58. Baxby D. Edward Jenner's Inquiry; a bicentenary analysis. Vaccine 1999; 17:301–7, at pp. 302–4.
59. Jenner E. An Inquiry into the causes and effects of the Variolae Vaccinae: a disease discovered in some of the Western Counties of England, particularly Gloucestershire, and known by the name of the Cow Pox. London: Sampson Low, 1798.
60. Baxby, Edward Jenner's Inquiry, p. 303.
61. Ibid., p. 305; see also Baxby D, Bennett M, Getty B. Human cowpox 1969–93: a review based on 54 cases. Brit J Dermatol 1994; 131:598–607.
62. Jenner E. An Inquiry into the causes and effects of the Variolae Vaccinae: a disease discovered in some of the Western Counties of England, particularly Gloucestershire, and known by the name of the Cow Pox, 2nd edition. London: Sampson Low, 1799.
63. Jenner E. Further observations on the variolae vaccinae or cow-pox. London: Sampson Low, 1799.
64. Jenner E. A continuation of facts and observations relative to the variolae vaccinae. London: Sampson Low, 1800.
65. Brunton, D. Woodville, William (1752–1805). In Oxford Dictionary of National Biography. Oxford University Press, 2004 (http://www.oxforddnb.com/view/article/29940, accessed 6 December 2009).
66. Woodville W. Medical botany, containing systematic and general descriptions, with plates, of all the medicinal plants, indigenous and exotic, comprehended in the Catalogues of the Materia Medica, as published by the Royal Colleges of Physicians of London and Edinburgh. London: James Phillips, 1790–93.
67. Woodville W. The history of the inoculation of the small-pox in Great Britain, vol. 1. London: James Phillips, 1796.
68. Woodville W. Report of a series of inoculations for the variolae vaccinae, or cow-pox. London: James Phillips, 1799.
69. Baxby, Edward Jenner's Inquiry, p. 306; McVail JC. Cow-pox and small-pox: Jenner, Woodville and Pearson. Brit Med J 1896; 1:1271–6; Jenner, Woodville and Pearson. Brit Med J Jenner Centenary Issue, no. 1847 (23 May 1896), 1271–6.
70. Royal Vaccination Commission. Final report of the Royal Commission appointed to inquire into the subject of vaccination. London: Her Majesty's Stationery Office, 1896. Dissent by W Collins and A Picton, p. 44.
71. Baron, Life of Edward Jenner, vol. 1, p. 134.
72. Beale N, Beale E. Evidence-based medicine in the eighteenth century: the Ingen Housz–Jenner correspondence revisited. Med Hist 2005; 49:80–1.
73. Baron, Life of Edward Jenner, vol. 1, pp. 289–96; Beale, Beale, Evidence-based medicine, pp. 83–98.
74. Baron, Life of Edward Jenner, vol. 2, p. 62.
75. Abraham JJ. Lettsom. His life, times, friends and descendants. London: William Heinemann, 1933, pp. 329–32; Coley NG. George Pearson MD, FRS (1751–1828). The greatest chemist in England? Notes and Records of the Royal Society of London 2003; 57(2):161–75; Pearson GC. An inquiry concerning the history of the cow pox. Principally with a view to supersede and extinguish the small pox. London: J Johnson, 1798; Pearson GC. On the progress of the variolae vaccinae. Med Phys J 1799; III:97–8, 399.
76. Abraham, Lettsom. His life, p. 339; Baron, Life of Edward Jenner, vol. 1, p. 362; Fisher, Edward Jenner, pp. 88, 95–6.
77. Abraham, Lettsom. His life, p. 339; Baron, Life of Edward Jenner, vol. 1, p. 362; Fisher, Edward Jenner, pp. 88, 95–6.
78. Baron, Life of Edward Jenner, vol. 2, p. 62.
79. Squirrel R. Observations addressed to the public in general on the cow-pox, showing it to originate in the scrophula. London: W Smith & Co., 1805.
80. Abraham, Lettsom. His life, p. 250; Fisher, Edward Jenner, p. 93; Moseley B. A treatise on Lues Bovilla; or cow pox, 2nd edition. London: J Nichols & Son, 1805.
81. Rowley W. Cox-pox inoculation no security against small-pox infection. London: J Harris, 1805.
82. Paré A. Deux livres de chirurgie, de la génération de l'homme, & manière d'extraire les enfans hors du ventre de la mère, ensemble ce qu'il faut faire pour la faire mieux,

& plus tost accoucher, avec la cure de plusieurs maladies qui luy peuvent survenir. Paris: A Wechel, 1573.
83. Fisk D. Dr. Jenner of Berkeley. London: Heinemann, 1959, p. 180.
84. Abraham, Lettsom. His life, p. 351.
85. Fisher, Edward Jenner, p. 59.
86. Abraham, Lettsom. His life, pp. 329–55.
87. Abraham, Lettsom. His life, p. 336.
88. Ring J. An answer to Mr. Goldson, proving that vaccination is a permanent security against the small-pox. London: J Murray, 1804.
89. Dunning R. Some observations on vaccination or the inoculated cow-pox. London, 1800.
90. Jenner E. The origin of the vaccine inoculation. London: DN Shury, 1801.

CHAPTER 10

1. Brit Med J (BMJ). Jenner Centenary number (No. 1847). 23 May 1896, pp. 1265–9.
2. Baron J. Life of Edward Jenner, MD with illustrations from his doctrines, and selections from his correspondence, 2 vols. London: H Colbourn, 1827 (vol. 1), 1838 (vol. 2); vol. 1, p. 389; Glynn I, Glynn J. The life and death of smallpox. Cambridge: Cambridge University Press, 2004, p. 125.
3. Abraham JJ. Lettsom. His life, times, friends and descendants. London: William Heinemann, 1933, pp. 333–4; Baron, Life of Edward Jenner, vol. 1, p. 386; Hopkins DR. The greatest killer. Smallpox in history. Chicago: University of Chicago Press, 2002, p. 263.
4. Brit Med J, Jenner Centenary number, p. 1269; Hopkins, The greatest killer, pp. 263–4; Waterhouse B. A prospect of exterminating the small pox; being the history of the variolae vaccinae, or kine pox, commonly called the cow pox, as it has appeared in England, with an account of a series of inoculations performed for the kine pox in Massachusetts. Cambridge: Cambridge Press, W Hillard, 1800 (part 1), 1802 (part 2).
5. Hopkins, The greatest killer, p. 264; Blake JB. Benjamin Waterhouse and the introduction of vaccination. Philadelphia: University of Philadelphia Press, 1957.
6. Fisher RB. Edward Jenner 1749–1823. London: André Deutsch, 1991, p. 111; Halsey R. How the President Thomas Jefferson, and Doctor Benjamin Waterhouse established vaccination as a public health procedure. New York: New York Academy of Medicine, 1936; Hopkins, The greatest killer, p. 265.
7. Hopkins, The greatest killer, pp. 265–6.
8. Baron, Life of Edward Jenner, vol. 1, pp. 410–19.
9. Ibid., pp. 420–4; Hopkins, The greatest killer, pp. 144–6, 149–53.
10. Baron, Life of Edward Jenner, vol. 1, pp. 424–6; 525–9.
11. Ibid., p. 426; Brit Med J, Jenner Centenary number, p. 1267; Hopkins, The greatest killer, pp. 146–9.
12. Baron, Life of Edward Jenner, vol. 1, pp. 390–3; Brit Med J, Jenner Centenary number, p. 1268.
13. Baron, Life of Edward Jenner, vol. 1, pp. 324, 395–403.
14. Ibid., p. 401.
15. Brit Med J, Jenner Centenary number, p. 1268; Bowers JZ. The odyssey of smallpox vaccination. Bull Med Hist 1981; 55:17–33.
16. Baron, Life of Edward Jenner, vol. 1, pp. 459–65; Zielonka JS. Inoculation: pro and con. J Hist Med Allied Sci 1972; 27:447.
17. Hopkins, The greatest killer, pp. 127–8, 316.
18. Brit Med J, Jenner Centenary number, pp. 1267–8.
19. Hopkins, The greatest killer, pp. 187–8.
20. Baron, Life of Edward Jenner, vol. 1, pp. 605–6; Brit Med J, Jenner Centenary number, p. 1269; Hopkins, The greatest killer, pp. 224–6.

21. Franco-Paredes C, Lammoglia L, Santos-Preciado JI. The Spanish Royal Philanthropic Expedition to bring smallpox vaccination to the New World and Asia in the 19th century. Clin Inf Dis 2005; 41:1285–9.
22. Ibid.
23. Brit Med J, Jenner Centenary number, p. 1269.
24. Behbehani AM. The smallpox story: life and death of an old disease. Microbiol Rev 1983; 455–509, at pp. 478–9; Smith MM. The 'Real Expedición Marítima de la Vacuna' in New Spain and Guatemala. Trans Amer Phil Soc 1974; N.S. 64, part 1:1–74.
25. Royal Vaccination Commission. Final report of the Royal Commission appointed to inquire into the subject of vaccination. London: Her Majesty's Stationery Office, 1896, p. 145.
26. Baron, Life of Edward Jenner, vol. 1, pp. 150–4.
27. Glynn, Glynn, The life and death of smallpox, pp. 172–3.
28. Brit Med J, Jenner Centenary number, p. 1282; Glynn, Glynn, The life and death of smallpox, pp. 169–70; Hopkins, The greatest killer, p. 88.
29. Hime TH. Animal vaccination. In Brit Med J, Jenner Centenary Number, 23 May 1896, pp. 1279–89.
30. Glynn, Glynn, The life and death of smallpox, pp. 170–5.
31. Baron, Life of Edward Jenner, vol. 2, pp. 99–100; Glynn, Glynn, The life and death of smallpox, p. 112; Fisher, Edward Jenner, pp. 115–16. Hill, R. Cow-pock inoculation vindicated and recommended from matters of fact. London: JB Nichols, 1806.
32. Royal Vaccination Commission, Final report, pp. 292–6.
33. Baron, Life of Edward Jenner, vol. 2, pp. 14–15; BMJ, Jenner Centenary number, p. 1270; Fisher, Edward Jenner, pp. 203–4; Jenner E. The origin of the vaccine inoculation. London: DN Shury, 1801.
34. Brit Med J, Jenner Centenary number, pp. 1265–7.
35. Ibid.
36. Fisher, Edward Jenner, p. 91; Glynn, Glynn, The life and death of smallpox, p. 109.
37. Baron, Life of Edward Jenner, vol. 2, pp. 6–7.
38. Baron, Life of Edward Jenner, vol. 1, pp. 479–96; Brit Med J, Jenner Centenary number, pp. 1251–2; Fisher, Edward Jenner, pp. 113–31.
39. Fisher, Edward Jenner, p. 124; Glynn, Glynn, The life and death of smallpox, p. 112.
40. Fisher, Edward Jenner, p. 124; Wheeler A. Vaccination. Opposed to science and a disgrace to English Law. London: EW Allen, 1879.
41. Baron, Life of Edward Jenner, vol. 1, pp. 497–504; Fisher, Edward Jenner, pp. 127–9.
42. Birch J. Serious reasons for uniformly objecting to the practice of vaccination. London: J Smeeton, 1806; Glynn, Glynn, The life and death of smallpox, p. 111.
43. Baron, Life of Edward Jenner, vol. 1, pp. 490–1.
44. Fisher, Edward Jenner, p. 126.
45. Ibid.
46. Baron, Life of Edward Jenner, vol. 1, p. 510; Brit Med J, Jenner Centenary number, p. 1251; Fisher, Edward Jenner, pp. 130–1.
47. Baron, Life of Edward Jenner, vol. 1, pp. 50–10; Brit Med J, Jenner Centenary number, p. 1251.
48. Brit Med J, Jenner Centenary number, pp. 1265–7.
49. Baron, Life of Edward Jenner, vol. 2, pp. 55–68; Brit Med J, Jenner Centenary number, p. 1252.
50. Baron, Life of Edward Jenner, vol. 2, p. 9.
51. Brit Med J, Jenner Centenary number, p. 1252.
52. Baron, Life of Edward Jenner, vol. 1, pp. 565–77.
53. Ibid., pp. 567–82.
54. Ibid., p. 582; Fisher, Edward Jenner, pp. 186–212.
55. Baron, Life of Edward Jenner, vol. 1, pp. 449–57, 459–62; vol. 2, pp. 33–5, 101–5, 166–8; Brit Med J, Jenner Centenary number, pp. 1260–1; Fisher, Edward Jenner, p. 111.
56. Baron, Life of Edward Jenner, vol. 1, p. 526.
57. Baron, Life of Edward Jenner, vol. 2, pp. 37–8, 116–17; Fisher, Edward Jenner, p. 249.

58. Baron, Life of Edward Jenner, vol. 2, pp. 113–16; Brit Med J, Jenner Centenary number, p. 1250.
59. Brit Med J, Jenner Centenary number, p. 1260.
60. Baron, Life of Edward Jenner, vol. 2, pp. 94–5.
61. Fisher, Edward Jenner, p. 147.
62. Ibid., p. 182; Royal College of Physicians. Report of the Royal College of Physicians of London on Vaccination. Med Phys J 1807; 18:97–102.
63. Glynn, Glynn, The life and death of smallpox, p. 131.
64. Baron, Life of Edward Jenner, vol. 2, p. 163.
65. Fisher, Edward Jenner, pp. 213–54.
66. Baron, Life of Edward Jenner, vol. 2, pp. 297–8; Brit Med J, Jenner Centenary number, p. 1254; Fisk D. Dr. Jenner of Berkeley. London: Heinemann, 1959, p. 209.
67. Fisher, Edward Jenner, p. 147; Jenner E. Letter to Dr. John de Carro, 30 March 1803; and Letter to the Rev. Robert Ferryman, 17 February 1805. MSS in Royal College of Physicians of London.
68. Baron, Life of Edward Jenner, vol. 2, p. 15.
69. Ibid.; Fisk, Dr. Jenner of Berkeley, p. 218.
70. Baron, Life of Edward Jenner, vol. 2, pp. 72–6; Fisher, Edward Jenner, pp. 213–15.
71. Baron, Life of Edward Jenner, vol. 2, pp. 140–2; Brit Med J, Jenner Centenary number, p. 1246.
72. Baron, Life of Edward Jenner, vol. 2, pp. 219–20.
73. Baron, Life of Edward Jenner, vol. 1, p. 467; vol. 2, p. 277; Jenner E. Some observations on the migration of birds. By the late Edward Jenner, MD FRS; with an introductory letter to Sir Humphry Davy, Bart. Pres RS. By the Rev. GC Jenner. Phil Trans Roy Soc 1824; 114:11–44.
74. Letter about Jenner's plesiosaur fossil, Edward Jenner Museum, MS BEKJM/1618.
75. Baron, Life of Edward Jenner, vol. 2, pp. 222–3.
76. Ibid., pp. 314, 317–22; Baron J. Edward Jenner. Obituary in Gloucester Journal, 3 February 1823; Fisher, Edward Jenner, p. 292.
77. Baron, Life of Edward Jenner, vol. 2, pp. 314, 317–22; Baron, Edward Jenner, Obituary; Fisher, Edward Jenner, p. 292.
78. Brit Med J, Jenner Centenary number, p. 1253.

CHAPTER 11

1. Brit Med J. Jenner Centenary number (No. 1847). 23 May 1896, pp. 1270–2.
2. Wolfe RM, Sharp LK. Anti-vaccinationists past and present. Brit Med J 2002; 325:430–2.
3. Tebb W. Compulsory vaccination in England: with incidental references to foreign states. London: EW Allen, 1884, pp. 3–4, 13.
4. Taylor PA. The vaccination question. Speech of Mr. P.A. Taylor on Dr. Cameron's resolution respecting animal vaccine. Delivered in the House of Commons, 11 June 1880. From Hansard's Parliamentary Debates, Vol. CCLII. London: Waterlow, p. 8.
5. Taylor PA, Hopwood CH. Speeches of Mr. P.A. Taylor and Mr. C.H. Hopwood on vaccination in the House of Commons, 19 June 1883. London: EW Allen, 1883, p. 4.
6. Ibid.
7. Tebb, Compulsory vaccination in England, pp. 47–50.
8. Ibid., p. 8; citing Daily Chronicle, 26 August 1882.
9. Royal Vaccination Commission. Final report of the Royal Commission appointed to inquire into the subject of vaccination. London: Her Majesty's Stationery Office, 1896, pp. 9–10, 266–70.
10. Ibid., pp. 130–2.
11. Tebb, Compulsory vaccination in England, p. 7.
12. Ibid., pp. 14, 28.
13. Royal Vaccination Commission, Final report, p. 12.

14. Tebb, Compulsory vaccination in England, p. 28.
15. Glynn I, Glynn J. The life and death of smallpox. Cambridge: Cambridge University Press, 2004, p. 158.
16. Wheeler A. Vaccination in the light of history. London: EW Allen, 1878; Wheeler A. Vaccination. Opposed to science and a disgrace to English law. An address delivered at Darlington, 10 November 1879. London: EW Allen, 1879. Wheeler A. Vaccination. London: EW Allen, 1883.
17. Taylor, Hopwood, Speeches of Mr. P.A. Taylor and Mr. C.H. Hopwood, pp. 6–7.
18. Ibid., pp. 19–20; British Medical Association (BMA). Debate on compulsory vaccination in the House of Commons on Friday May 13th, 1893. London: British Medical Assocation, p. 6.
19. Baldwin P. Contagion and the State in Europe, 1830–1930. Chapter 4, Smallpox faces the lancet. Cambridge: Cambridge University Press, 1999; BMA, Debate on compulsory vaccination, p. 6.
20. Creighton C. Vaccination. In Encyclopaedia Britannica, 9th edition. 1888; Wallace AR. Vaccination a delusion. Its penal enforcement a crime. Chapter 2, Much of the evidence adduced for vaccination is worthless. London: Swan Sonnenschein, 1898.
21. Taylor, Hopwood, Speeches of Mr. P.A. Taylor and Mr. C.H. Hopwood, p. 19.
22. Disraeli B. Attributed by Twain M in Autobiography. New York: Harper & Brothers, 1924.
23. BMA, Debate on compulsory vaccination, pp. 6–10, 13–14.
24. Royal Vaccination Commission, Final report, p. 13.
25. Guy WA. 250 years of smallpox in London. Paper read before the Statistical Society, 20 June 1882. Stat Soc J, September 1882. London: E Stanford; Wheeler, Vaccination, pp. 8–10.
26. Taylor, The vaccination question, p. 9.
27. BMA, Debate on compulsory vaccination, pp. 6–10; Hopkins DR. The greatest killer. Smallpox in history. Chicago: University of Chicago Press, 2002, p. 42 and Fig. 3.
28. BMA, Debate on compulsory vaccination, pp. 6–10.
29. Royal Vaccination Commission, Final report, pp. 91–2.
30. Taylor, Hopwood, Speeches of Mr. P.A. Taylor and Mr. C.H. Hopwood.
31. Ibid., p. 16.
32. Crookshank E. History and pathology of vaccination, 2 vols. London: HK Lewis, 1888.
33. Royal Vaccination Commission, Final report, p. 295.
34. Boëns H. La variole, la vaccine et les vaccinides en 1884. Reprinted from the Bulletin de l'Académie Royale Médicale de Belgique, vol. 18, no. 1. Brussels: A Manceaux, 1884, p. 14.
35. Ibid., pp. 14–15; Brit Med J 23 July 1881, pp. 138–9.
36. Boëns, La variole, p. 15.
37. Ibid.
38. Ibid., p. 109.
39. Ibid.
40. Appel L. A non-vaccinator's comment on the medical aspect of the Imperial Vaccination League's 'Ten answers to questions on the subject of vaccination.' London: Personal Rights Association, 1902.
41. Glynn, Glynn, The life and death of smallpox, p. 158.
42. Wallace AR. 45 years of registration statistics proving vaccination to be both useless and dangerous. London: EW Allen, 1885.
43. Circular letter from Miss L Loat, Secretary of the National Anti-vaccination League, London, 9 February 1911. St. Deiniol's Library MS A/31/1–26.
44. Altschuler G. From religion to ethics: Andrew D White and the dilemma of a Christian rationalist. Church History 1978; 47:308–24.
45. Knaggs HV. Smallpox – a healing crisis and the truth about vaccination. London, 1924.
46. Taylor, The vaccination question, p. 22.
47. McCormack E. Is vaccination a disastrous illusion? London: The National Anti-Vaccination League, 1909, p. 16.

48. Pamphlets in the collection of St. Deiniol's Museum, MS A/31/1–26.
49. 'Dr. Sly'. Germ culture: or a scientific method of rearing and floating microscopic canards at will. A translation from the French. London: W Young, 1882.
50. Gümpel G. Smallpox. An inquiry into its real nature and its possible prevention showing the extent and duration of the protection afforded by vaccination. An attempt to solve this vexed question without statistics. London: Swan Sonnenschein, 1902.
51. Kitson J. Sanitary lessons to working women in Leeds, during the winters of 1872 and 1873. Leeds: Ladies' Council of the Yorkshire Board of Education, 1873.
52. Buckton CM. Two winters' experience in giving lectures to my fellow townswomen of the working classes, on physiology and hygiene. Leeds: W Wood, 1873.
53. Parker WW. The ancient and modern physician – St. Luke and Jenner. Reprinted from the Transactions of the Medical Society of Virginia, 1891.

CHAPTER 12

1. Little LC. Crimes of the cowpox ring. Some moving pictures thrown on the dead wall of official silence. Minneapolis: The Liberator Publishing Co., 1906.
2. Johnston RD. The radical middle class. Populist democracy and the question of capitalism in progressive era Portland, Oregon. Princeton: Princeton University Press, 2003, pp. 199–201.
3. Little, Crimes of the cowpox ring.
4. Johnston, The radical middle class, pp. 201–3.
5. Glynn I, Glynn J. The life and death of smallpox. Cambridge: Cambridge University Press, 2004, pp. 158–9.
6. McBean E. The poisoned needle. Suppressed facts about vaccination. Mokelumne Hill, CA: Health Research, 1957. Reprinted by Health Research Books, 1993, p. 44.
7. Johnston, The radical middle class, pp. 197–8, 203–4.
8. Glasgow Herald, 4 March 1878, cited in Tebb W. Compulsory vaccination in England: with incidental references to foreign states. London: EW Allen, 1884.
9. Creighton C. Vaccination. In Encyclopaedia Britannica, 9th edition, 1888.
10. Boëns H. La variole, la vaccine et les vaccinides en 1884. Reprinted from the Bulletin de l'Académie Royale Médicale de Belgique, vol. 18, no. 1. Brussels: A Manceaux, 1884, pp. 107–8.
11. Tebb, Compulsory vaccination in England, p. 9.
12. Porter R. The greatest benefit to mankind. A medical history of humanity from antiquity to the present. London: Fontana Press, 1997, pp. 166–8.
13. Moseley B. A treatise on Lues Bovilla; or cow pox, 2nd edition. London: J Nichols & Son, 1805.
14. Creighton C. The natural history of cowpox and vaccinal syphilis. London, 1887; McCormack E. Is vaccination a disastrous illusion? London: The National Anti-Vaccination League, 1909, pp. 27–8.
15. Hodge JW. The vaccination superstition: prophylaxis to be realized through the attainment of health, not by the propagation of disease; can vaccination produce syphilis? Niagara Falls, 1901, p. 41.
16. Glynn, Glynn, The life and death of smallpox, pp. 153–5; Hopkins DR. The greatest killer. Smallpox in history. Chicago: University of Chicago Press, 2002, pp. 285–7.
17. Taylor PA. The vaccination question. Speech of Mr. P.A. Taylor on Dr. Cameron's resolution respecting animal vaccine. Delivered in the House of Commons, June 11th, 1880. From Hansard's Parliamentary Debates, Vol. CCLII. London: Waterlow, p. 16.
18. Pachiotti X. Sifilide trasmessa per mezzo della vaccinazione in Rivalta presso Acqui. Turin, 1862. Cited in Després A. Traité théorique et pratique de la syphilis. Paris: Germer Baillière, 1873, p. 215.
19. Ballard E. On vaccination: its value and alleged dangers. London: Longman & Green, 1868.

20. Mortimer P. Robert Cory and the vaccine syphilis controversy: a forgotten hero? Lancet 2006; 367:1112–15; Taylor PA, Hopwood CH. Speeches of Mr. P.A. Taylor and Mr. C.H. Hopwood on vaccination in the House of Commons, June 19th, 1883. London: EW Allen, 1883, p. 23.
21. Taylor, The vaccination question, p. 20.
22. Tebb, Compulsory vaccination in England, pp. 24–5.
23. British Medical Association. Debate on compulsory vaccination in the House of Commons on Friday May 13th, 1893. London: British Medical Association, pp. 10–11; Taylor, The vaccination question, pp. 23–4.
24. McBean, The poisoned needle, pp. 53–8; Taylor, Hopwood, Speeches of Mr. P.A. Taylor and Mr. C.H. Hopwood, p. 15.
25. The Vaccination Act, 1898. See Brit Med J 1898; 2 (1963), p. 431.
26. Kant I. Fragments littéraires. Paris: MV Cousin, 1843; McCormack, Is vaccination a disastrous illusion?, p. 17.
27. McCormack, Is vaccination a disastrous illusion?, p. 12.
28. Farr WA. Letter to the Registrar General on causes of mortality, in 35th Annual Report of the Registrar General, 1867, pp. 213–24.
29. Tebb, Compulsory vaccination in England, p. 9.
30. Porter, The greatest benefit to mankind, pp. 369–70.
31. Greene J. Good vaccine lymph. An inquiry as to what extent it is desirable to employ heifer vaccination, with details of that method. Birmingham: C Edmonds, 1871.
32. McCormack, Is vaccination a disastrous illusion?, p. 15, citing Lancet, 7 June 1902.
33. Little, Crimes of the cowpox ring, p. 6.
34. Glynn, Glynn, The life and death of smallpox, pp. 169–72; Loat L. The truth about vaccination and immunization. London, 1951; McCormack, Is vaccination a disastrous illusion?, p. 37.
35. McCormack, Is vaccination a disastrous illusion?, p. 15, citing Indian Lancet, 1 March 1897.
36. Collins WJ. A review of the Norwich Vaccination Inquiry. Read to the London Society for the Abolition of Compulsory Vaccination, December 18, 1882. London: EW Allen, 1883; Tebb, Compulsory vaccination in England, pp. 4–6.
37. Glynn, Glynn, The life and death of smallpox, pp. 172–3.
38. Osler W. The principles and practice of medicine. New York: D Appleton & Co., 1802, pp. 65–7.
39. Royal Commission on Vaccination. Final report. London: HM Stationery Office, 1896.
40. Tebb, Compulsory vaccination in England, p. 26.
41. Ibid., p. 46.
42. Creighton, Vaccination.
43. Royal Commission on Vaccination. Final report, p. 128.
44. Creighton, Vaccination; Little, Crimes of the cowpox ring, pp. 12–14.
45. Collins, A review of the Norwich Vaccination Inquiry; Tebb, Compulsory vaccination in England, pp. 4–6, 7.
46. Cobb FP. The medical profession and its morality. Modern Review, July 1881, p. 35.
47. Behbehani AM. The smallpox story: life and death of an old disease. Microbiol Rev 1983; 455–509, at pp. 489–90; Creighton, Vaccination; Glynn, Glynn, The life and death of smallpox, pp. 152–3; Goldstein JA, Neff JM, Lane JM, Koplan JP. Smallpox vaccination reactions, prophylaxis and therapy of complications. Pediatrics 1975; 55:342–7.
48. Tebb, Compulsory vaccination in England, p. 9.
49. Porter, The greatest benefit to mankind, p. 427.
50. McBean, The poisoned needle, pp. 53–8; Taylor, Hopwood, Speeches of Mr. P.A. Taylor and Mr. C.H. Hopwood, p. 15.
51. McCormack, Is vaccination a disastrous illusion?, p. 37.
52. Tebb, Compulsory vaccination in England, p. 9.
53. Royal Commission on Vaccination. Final report, p. 194.
54. Anti-Vaccination League, London. Pamphlets in St. Deiniol's Library, MS A/31/1–26.
55. Johnston, The radical middle class, pp. 201–6.

56. Ibid., pp. 209–11.
57. Ibid., p. 212.
58. Little, Crimes of the cowpox ring, p. 7.
59. Johnston, The radical middle class, pp. 197–8, 203–4.
60. Porter, The greatest benefit to mankind, pp. 438–9.

CHAPTER 13

1. Taylor PA, Hopwood CH. Speeches of Mr. P.A. Taylor and Mr. C.H. Hopwood on vaccination in the House of Commons, June 19th, 1883. London: EW Allen, 1883, p. 19.
2. McBean E. The poisoned needle. Suppressed facts about vaccination. Mokelumne Hill, CA: Health Research, 1957. Reprinted by Health Research Books, 1993, p. 22.
3. Wheeler A. Vaccination. Opposed to science and a disgrace to English law. An address delivered at Darlington, 10th November, 1879. London: EW Allen, 1879, pp. 2–6; Wheeler A. Vaccination. 1883. London: EW Allen, 1883, p. 14.
4. Cited in McCormack E. Is vaccination a disastrous illusion? London: The National Anti-Vaccination League, 1909, p. 24; also in Milnes A. Statistics of small-pox and vaccination, with special reference to age-incidence, sex-incidence, and sanitation. J Roy Stat Soc 1897; 60:557.
5. Royal Vaccination Commission. Final report of the Royal Commission appointed to inquire into the subject of vaccination. London: Her Majesty's Stationery Office, 1896, pp. iii–iv. Commission by Queen Victoria dated 29 May 1889.
6. Johnston RD. The radical middle class. Populist democracy and the question of capitalism in progressive era Portland, Oregon. Princeton: Princeton University Press, 2003, p. 208.
7. Royal Vaccination Commission. Final report, p. 1. Preface by Lord Herschel, Chair of the Commission.
8. McCormack, Is vaccination a disastrous illusion?, p. 30.
9. Royal Vaccination Commission. Final report, p. 194.
10. Ibid., p. 142.
11. Ibid., p. 134.
12. Ibid., pp. 141, 137–40.
13. Glynn I, Glynn J. The life and death of smallpox. Cambridge: Cambridge University Press, 2004, pp. 163–4; see also the Vaccination Act 1898, in Brit Med J 1898; 2(1974):1351–4.
14. Johnston, The radical middle class, pp. 182–5.
15. Ibid., p. 182.
16. Meade T. 'Civilising Rio de Janeiro': the public health campaign and the riot of 1904. J Soc Hist 1986; 20:301–22.
17. Hopkins DR. The greatest killer. Smallpox in history. Chicago: University of Chicago Press, 2002, pp. 285–7.
18. Osler W. The principles and practice of medicine. New York: D Appleton & Co., 1802, p. 240.
19. Hopkins, The greatest killer, pp. 285–7.
20. Glynn, Glynn, The life and death of smallpox, pp. 158–60.
21. Boëns H. La variole, la vaccine et les vaccinides en 1884. Reprinted from the Bulletin de l'Académie Royale Médicale de Belgique, vol. 18, no. 1. Brussels: A Manceaux, 1884, pp. 19–20.
22. Ibid., p. 20; Glynn, Glynn, The life and death of smallpox, pp. 158–60.
23. Glynn, Glynn, The life and death of smallpox, pp. 159–60; see also Millard's obituary, Brit Med J 1952; 1 (4759):660–2.
24. Glynn, Glynn, The life and death of smallpox, pp. 158–60; Boëns, La variole, p. 20.
25. Royal Vaccination Commission. Final report, p. 261.

26. Glynn, Glynn, The life and death of smallpox, p. 153; Kidd BE. Hadwen of Gloucester. Man, medico, martyr. London: John Murray, 1933.
27. Glynn, Glynn, The life and death of smallpox, p. 161.
28. Ibid.
29. Dramatic scenes at Gloucester Assizes, Gloucester Chronicle, 8 November 1924; Kidd, Hadwen of Gloucester, pp. 308–9.
30. Brit Med J. Jenner Centenary number. 23 May 1896, p. 1270; Goldson W. Cases of smallpox subsequent to vaccination. Portsea, 1804; Royal Vaccination Commission. Final report, p. 171.
31. Baron J. The life of Edward Jenner, MD, with illustrations of his doctrines and selections from his correspondence. 2 vols. London: H Colbourn, 1827 (vol. 1), 1838 (vol. 2); vol. 2, p. 29 [Letter to Mr. Dunning 5 July 1804]; see also pp. 24–5, 348–9, 352.
32. McCormack, Is vaccination a disastrous illusion?, p. 24.
33. Pickering J. Anti-vaccination: the statistics of the medical officers of the Leeds Small-pox Hospital exposed and refuted, in a letter to the Leeds Board of Guardians. London: F Pitman, 1876.
34. BMJ, Jenner Centenary number, p. 1270; Royal Vaccination Commission. Final report, p. 140.
35. Taylor PA. The vaccination question. Speech of Mr. P.A. Taylor on Dr. Cameron's resolution respecting animal vaccine. Delivered in the House of Commons, June 11th, 1880. From Hansard's Parliamentary Debates, Vol. CCLII. London: Waterlow.
36. Palmer DD, Palmer BJ. The science of chiropractic: its principles and adjustments. Palmer School of Chiropractic, 1906.
37. Ibid., pp. 350–4.
38. Ibid., p. 379.
39. McBean, The poisoned needle, pp. 44, 161–212.
40. Ibid.
41. British Medical Association. Debate on compulsory vaccination in the House of Commons on Friday May 13th, 1893. London: British Medical Association, pp. 13–15.
42. Hopkins, The greatest killer, p. 303.
43. Royal Vaccination Commission. Final report, pp. 159–65.
44. Ibid., p. 20.
45. Ibid., p. 90.
46. Ballard E. On vaccination: its value and alleged dangers. London: Longman & Green, 1868.
47. Brit Med J, Jenner Centenary number, pp. 1268–9
48. Conlon CP, Warrell DA. Travel and expedition medicine. In Warrell DA, Cox TM, Firth JD, eds. Oxford Textbook of Medicine, 4th edition. Oxford: Oxford University Press, 2003, p. 313.
49. Baxby D. Inoculation and vaccination: smallpox, cowpox and vaccinia. Med Hist 1965; 9:383–5; Downie AW, Dumbell KR. Poxviruses. Ann Rev Microbiol 1956; 10:237–52.
50. Baxby D. Jenner's smallpox vaccine: the riddle of vaccinia virus and its origin. Chapter 13. London: Heinemann Educational Books, 1981.
51. Creighton C. Jenner and vaccination. A strange chapter of medical history. London: Swan Sonnenschein, 1889, pp. 279–80 Creighton C. A history of epidemics in Britain. Vol. 1 (1891), Vol. 2 (1894). Republished London: Frank Cass, 1965; Fisher RB. Edward Jenner 1749–1823. London: André Deutsch, 1991, p. 80; Glynn, Glynn, The life and death of smallpox, pp. 166–7.
52. Glynn, Glynn, The life and death of smallpox, p. 188.
53. Royal Vaccination Commission. Final report, p. 170. See also McCormack, Is vaccination a disastrous illusion?, p. 25.
54. Fisher, Edward Jenner, pp. 81–5; Laurance B. Cowpox in man: and its relationship to milkers' nodules. Lancet 1955; 265:764–6.
55. Baxby, Jenner's smallpox vaccine.
56. Baron, The life of Edward Jenner, vol. 2, p. 47; Wheeler, Vaccination. Opposed to science, 1879, pp. 11–12.
57. Wheeler A. Vaccination. Opposed to science, 1879, p. 10.

58. Isaac Newton, quoted in More LT. Isaac Newton: a biography. New York: Charles Scribner's Sons, 1934, p. 664.
59. Brit Med J, Jenner Centenary number, p. 1258.
60. Pasteur L. Address at St. James' Hall, London, 8 August 1881. Trans Int Med Congress 1881; i:85.
61. Wilkinson JJG. Pasteur and Jenner. An example and a warning. London, 1881, p. 3.

CHAPTER 14

1. Behbehani AM. The smallpox story: life and death of an old disease. Microbiol Rev 1983; 455–509, at p. 483.
2. Dingle CV. A short account of the Middlesbrough small-pox epidemic – 1897–8. Lancet 1898; 1:1104–6, at p. 1105; Dingle CV. The story of the Middlesbrough small-pox outbreak and some of its lessons. Reprinted from Public Health, 1898. London: Rebman Publishing Company, 1898, p. 1.
3. 'The story of the smallpox epidemic in Middlesbrough.' Portraits of the workers. Views of the wards. Supplement to the Northern Weekly Gazette. Middlesbrough, 1898, p. 1.
4. Dingle, Public Health, pp. 1–2.
5. Dingle, Lancet 1898, pp. 1104–5; Dingle, Public Health, pp. 3–4.
6. 'The story of the smallpox epidemic in Middlesbrough', pp. 6–8.
7. Ibid., pp. 9–11.
8. Ibid., p. 19; Hardy C. Images of Teesside. Derby: Breedon Books, 1992. Caption to Fig. 14a, p. 127.
9. Dingle, Public Health, pp. 9–10; 'The story of the smallpox epidemic in Middlesbrough', pp. 12–13.
10. 'The story of the smallpox epidemic in Middlesbrough', p. 13.
11. Ibid., pp. 13, 26.
12. Ibid., p. 28.
13. Ibid.
14. Ibid.
15. Ibid., p. 26.
16. Ibid.
17. Ibid., p. 11.
18. Dingle, Lancet 1898, pp. 1105–6.
19. 'The story of the smallpox epidemic in Middlesbrough', pp. 25–6.
20. Ibid., p.1.
21. Lillie W. The history of Middlesbrough. An illustration of the evolution of English industry. Middlesbrough: County Borough of Middlesbrough, 1968, pp. 182–8.
22. Personal account by Mrs. Kitty Hutchinson, Edward Jenner Museum MS BEKJM/1082/1–2.
23. Tovey D. The Bradford smallpox outbreak in 1962: a personal account. J Roy Soc Med 2004; 97:244–7. Also his personal recollections in the Jenner Museum MS BEKJM/484.
24. Ikeda K. The blood in purpuric smallpox. J Amer Med Assoc 1925; 84:1807.
25. Logan JS. A memoir of Dr. Norman Ainley (1924–1962), and a last look at smallpox and vaccination. Ulster Med J 1993; 63:145–52.
26. The account is taken from a letter to Professor R.A. Shooter from Dr. Viggo Faber, retired professor of infectious diseases at the Rigshospitalet, Copenhagen, dated 24 August 1992. Jenner Museum MS BEKJM/518/1.
27. Henderson DA. Importations of smallpox into Europe 1961–1973. WHO Bulletin, Geneva: WHO/SE/74.62.
28. Behbehani, The smallpox story, pp. 487–8; Henderson, Importations of smallpox.
29. Copeman SM. Antitoxins and other organic remedies. Lecture at the Westminster Hospital Medical School. Brit Med J 1895; 2:835–6; Dingle CV. Vaccination. Paper read on June 12th 1900, to Middlesbrough and District Medical Society, by Dr.

Charles V. Dingle MD, Medical Officer of Health, Middlesbrough. Jenner Museum MS BEKJM/412/36.

30. Hopkins DA. The greatest killer. Smallpox in history. Chicago: University of Chicago Press, 2002, pp. 298–300; Ricketts TH, Byles JB. The red light treatment of smallpox. Lancet 1904; 2:287–90.
31. Letters to the Medical Officer of Health in Bradford following the 1962 outbreak. Jenner Museum MS BEKJM/513/3.
32. Marsden JP. Case of malignant smallpox treated with compound 33T57. Brit Med J 1962; 2:524–5.
33. Anonymous. Today's drugs. Methisazone. Brit Med J 1964; 2:621; Bauer DJ, St. Vincent L, Kempe CH, Young PA, Downie AW. Prophylaxis of smallpox with methisazone. Amer J Epid 1969, 90:130–45.
34. De Clercq E. Cidofovir in the treatment of poxvirus infections. Antiviral Res 2007; 55:1–13.
35. Franklin B. Letter to Jean-Baptiste Le Roy, 13 November 1789.
36. Letter from Dr. Viggo Faber to Professor Shooter, 24 August 1992. Jenner Museum MS BEKJM/518/1.

CHAPTER 15

1. Henderson DA. Smallpox – the death of a disease. Amherst: Prometheus Books, 2009, p. 119.
2. Chapin CV, Smith J. Permanency of the mild type of smallpox. J Prev Med 1932; 1:1–29; Fenner F, Henderson DA, Arita I, Jezek Z, Ladynyi ID. Smallpox and its eradication. Geneva: WHO, 1988, pp. 329–30; Glynn, p. 188.
3. Behbehani AM. The smallpox story: life and death of an old disease. Microbiol Rev 1983; 455–509, at p. 483; Bedson HS, Dumbell KR, Thomas WRG. Variola in Tanganyika. Lancet 1965; 2:1085–8.
4. Jenner E. An Inquiry into the causes and effects of the Variolae Vaccinae: a disease discovered in some of the Western Counties of England, particularly Gloucestershire, and known by the name of the Cow Pox. London: Sampson Low, 1798; p. 60.
5. Chapin CV. Variation in the type of infectious disease as shown by the history of smallpox in the United States, 1895–1912. J Infect Dis 1913; 13:171–96; Hopkins DR. The greatest killer. Smallpox in history. Chicago: University of Chicago Press, 2002, pp. 287–90.
6. Ibid.
7. Probst CO. Smallpox in Ohio. J Amer Med Ass 1899; 33:1589–90; Hopkins, The greatest killer, p. 289; Bishop ER. The epidemic of smallpox in Western New York. Buffalo Med J 1898–9; 38:420–38.
8. Koplan JP, Foster SO. Smallpox: clinical types, causes of death and treatment. J Inf Dis 1979; 140:440–1.
9. Chapin, Variation in the type of infectious disease; Hopkins, The greatest killer, pp. 287–90.
10. Hopkins, The greatest killer, p. 291.
11. Williams LR. The smallpox epidemic at Niagara Falls. Am J Publ Hlth 1915; 5:423–37.
12. Behbehani, The smallpox story, pp. 486–7; Hopkins, The greatest killer, p. 294.
13. Weinstein I. An outbreak of smallpox in New York City. Am J Publ Hlth 1947; 37:1376–84.
14. Hopkins, The greatest killer, p. 97.
15. Glynn I, Glynn J. The life and death of smallpox. Cambridge: Cambridge University Press, 2004, p. 142; Hopkins, The greatest killer, p. 97.
16. Marsden JP. Variola minor. A personal analysis of 13,686 cases. Bull Hygiene 1948; 23:735–46.
17. Gordon CW, Donnely JD, Fothergill R et al. Variola minor. A preliminary report from the Birmingham Hospital Region. Lancet 1966; 1:1311–13.

18. Glynn, Glynn, The life and death of smallpox, p. 192.
19. Hopkins, The greatest killer, p. 134.
20. Fenner et al., Smallpox and its eradication, p. 517; Henderson, Smallpox, pp. 79–105.
21. Fenner et al., Smallpox and its eradication, pp. 431–7; Henderson, pp. 79–87.
22. Fenner et al., Smallpox and its eradication.
23. Henderson, Smallpox, pp. 57–60.
24. Fenner et al., Smallpox and its eradication, p. 366; Henderson, Smallpox, pp. 60–1.
25. Fenner et al., Smallpox and its eradication, pp. 366–7; Henderson, Smallpox, pp. 61–2.
26. Fenner et al., Smallpox and its eradication, pp. 368–70.
27. Collier LH. The development of a stable smallpox vaccine. J Hygiene 1955; 53:76–101; Henderson, Smallpox, p. 53.
28. Henderson, Smallpox, pp. 25–6, 69–73.
29. Ibid., pp. 62–6, 74–8.
30. Ibid., pp. 107–27.
31. Ibid., pp. 87–8, 91.
32. Glynn, Glynn, The life and death of smallpox, p. 203; Henderson, Smallpox, pp. 64–5.
33. Henderson, Smallpox, pp. 86–7.
34. Ibid., pp. 93–4, 116–17.
35. Ibid., pp. 209–10.
36. Rubin BA. A note on the development of the bifurcated needle for smallpox vaccination. WHO Chron 1980; 34: 180–1.
37. Fenner et al., Smallpox and its eradication, pp. 472–4; Henderson, Smallpox, pp. 68–9, 95.
38. Henderson, Smallpox, pp. 122–3.
39. Ibid., pp. 114–15, 117.
40. Dr. D. A. Henderson, personal communication, October 2009.
41. Henderson, Smallpox, p. 56.
42. Fenner et al., Smallpox and its eradication, pp. 1023–4; Henderson, Smallpox, p. 224.
43. Henderson, Smallpox, p. 182.
44. Fenner et al., Smallpox and its eradication, p. 495; Foege WH, Millar JD, Henderson DA. Smallpox eradication in West and Central Africa. Bull World Hlth Org 1975; 52:209–22.
45. Fenner et al., Smallpox and its eradication, pp. 530–2.
46. Henderson, Smallpox, pp. 159–62.
47. Fenner et al., Smallpox and its eradication, pp. 1820–1; Henderson, Smallpox, pp. 164–5.
48. Fenner et al., Smallpox and its eradication, pp. 714–16; Hopkins, The greatest killer, pp. 157–63.
49. Henderson, Smallpox, pp. 189–95.
50. For an example, see http://www.hindu-blog.com/2008/08/goddess-sheetala-shitala-mata.html.
51. Dr. Christopher Burns-Cox, interview October 2009.
52. Dr. A.D. Macrae, eye witness account. Edward Jenner Museum, MS BEKJM/483/1.
53. Henderson, Smallpox, pp. 210–1.
54. Fenner et al., Smallpox and its eradication, pp. 842–7.
55. Ibid., pp. 1004–31; Henderson, Smallpox, pp. 213–39.
56. Behbehani, The smallpox story, p. 506; Henderson, Smallpox, pp. 244–6.
57. Henderson, Smallpox, pp. 26–7.
58. Ibid.
59. Ibid., p. 119.

CHAPTER 16

1. Professor Alasdair Geddes, interview October 2009; Behbehani AM. The smallpox story: life and death of an old disease. Microbiol Rev 1983; 455–509, at pp. 501–3.

2. Behbehani, The smallpox story, p. 488.
3. Ibid., p. 502.
4. Shooter RA. Report of the investigation into the cause of the 1978 Birmingham smallpox occurrence. London: Her Majesty's Stationery Office, 1980.
5. Behbehani, The smallpox story, p. 503.
6. Henderson DA. Smallpox – the death of a disease. Amherst: Prometheus Books, 2009, pp. 256–7.
7. Glynn I, Glynn J. The life and death of smallpox. Cambridge: Cambridge University Press, 2004, p. 234; Henderson, Smallpox, pp. 256–7, 265.
8. Henderson, Smallpox, pp. 258–67.
9. Mahy BWJ, Esposito JJ, Venter JC. Sequencing the smallpox virus genome: prelude to destruction of a virus species. ASM News 1991; 57:577–80; Cello J, Paul AV, Wimmer E. Chemical synthesis of poliovirus cDNA: generation of infectious virus in the absence of natural template. Science 2002; 297:1016–18.
10. Micklos D, Freyer GA, Crotty DA. DNA science: a first course in recombinant DNA technology. Cold Spring Harbor: Cold Spring Harbor Laboratory Press, 2003.
11. De Clercq E. Cidofovir in the treatment of poxvirus infections. Antiviral Res 2007; 55:1–13.
12. Koplow DA. Smallpox. The fight to eradicate a global scourge. Berkeley: University California Press, 2003, pp. 106–11; 158–78.
13. Koplow, Smallpox, pp. 112–36.
14. Henderson, Smallpox, pp. 259–61.
15. Behbehani, The smallpox story, pp. 503–5; Henderson, pp. 252–4.
16. Stearn EW, Stearn AE. The effect of smallpox on the destiny of the Amerindian. Boston: Bruce Humphries, 1945.
17. Harris S. In E Geissler, JE van C Moon, eds. Biological and toxin weapons. Oxford: Oxford Univ Press, 1999; pp. 127–52.
18. Alibek K. Biohazard. New York: Random House, 1999; p. 112.
19. Glynn, Glynn, The life and death of smallpox, p. 234; Henderson, Smallpox, p. 272; Preston R. The demon in the freezer. The New Yorker, 12 July 1999; pp. 44–61, at pp. 45, 58.
20. Tucker JB. Scourge: the once and future threat of smallpox. New York: Atlantic Monthly Press, 2001; pp. 148, 158.
21. Glynn, Glynn, The life and death of smallpox, pp. 234–5; Henderson, Smallpox, pp. 269–70, 274–5.
22. Preston, The demon in the freezer, p. 56.
23. Henderson, Smallpox, p. 272.
24. Tucker JB, Zilinskas RA. The 1971 smallpox epidemic in Aralsk, Kazakhstan, and the Soviet biological warfare program. Occasional paper no. 9. Monterey, CA: Monterey Institute and Center for Nonproliferation Studies, 2002.
25. Preston, The demon in the freezer, p. 56.
26. Henderson, Smallpox, pp. 273–4; Preston, The demon in the freezer, pp. 56–7.
27. Glynn, Glynn, The life and death of smallpox, p. 232.
28. Alibek K. Biohazard. New York: Random House, 1999.
29. Henderson DA, Inglesby TV, Bartlett JG et al., for the Working Group on Civil Biodefense. Smallpox as a biological weapon. In Henderson DA, Inglesby TV, O'Toole T (eds). Bioterrorism. Guidelines for medical and public health management. Chicago: AMA Press, 2002; pp. 99–120.
30. Alibek, Biohazard.
31. Henderson, Smallpox, pp. 284–6.
32. Ibid., p. 255.
33. Thornton J. Second letter to Mr. Tilloch on the cow-pock. Philosoph Magazine 1805; 20:145–6.
34. Behbehani, The smallpox story, pp. 486–7.
35. Porter R. The greatest benefit to mankind. A medical history of humanity from antiquity to the present. London: Fontana Press, 1997, p. 403.

Bibliography

Abraham JJ. Lettsom. His life, times, friends and descendants. London: William Heinemann, 1933.
Adams J. Observations on morbid poisons, phagedena and cancer. Chapter 5. London: Johnson, 1795.
Alibek K. Biohazard. New York: Random House, 1999.
Altschuler G. From religion to ethics: Andrew D White and the dilemma of a Christian rationalist. Church History 1978; 47:308–24.
Anonymous. A recipe: or the ingredients of a medicine for the spreading mortal distemper amongst cows: lately sent over from Holland, where a like distemper rages amongst the black cattle. Phil Trans Roy Soc 1714; 29:50.
Anonymous. Self-murther and duelling: the effects of cowardice and atheism. London, 1728.
Anonymous. Brit Med J No. 1847, 23 May 1896.
Anonymous. Today's drugs. Methisazone. Brit Med J 1964; 2:621.
Anonymous. Smallpox before Jenner. The terrible legacies of the disease. Brit Med J 1896; no. 1847:1263.
Appel L. A non-vaccinator's comment on the medical aspect of the Imperial Vaccination League's 'Ten answers to questions on the subject of vaccination.' London: Personal Rights Association, 1902.
Archbald F. A letter from one in the Country to his friend in the City. Boston, 1722.
Bailey I. Edward Jenner (1749–1823): naturalist, scientist, country doctor, benefactor to mankind. J Med Biogr 1996; 4:63–70.
Baldwin P. Contagion and the State in Europe, 1830–1930. Chapter 4, Smallpox faces the lancet. Cambridge: Cambridge University Press, 1999.
Ballard E. On vaccination: its value and alleged dangers. London: Longman & Green, 1868.
Bardell D. Smallpox during the American War of Independence. ASM News 1976; 42:526–30.
Baron J. Edward Jenner. Obituary in Gloucester Journal, 3 February 1823.
Baron J. The life of Edward Jenner, MD, with illustrations of his doctrines and selections from his correspondence. 2 vols. London: H Colbourn, 1827 (vol. 1), 1838 (vol. 2).
Barquet N, Domingo P. Smallpox: the triumph over the most terrible of the Ministers of Death. Ann Int Med 1997; 127:635–42.
Bauer DJ, St. Vincent L, Kempe CH, Young PA, Downie AW. Prophylaxis of smallpox with methisazone. Amer J Epid 1969, 90:130–45.
Baxby D. Inoculation and vaccination: smallpox, cowpox and vaccinia. Med Hist 1965; 9:383–5.
Baxby D. Jenner's smallpox vaccine: the riddle of vaccinia virus and its origin. London: Heinemann Educational Books, 1981.
Baxby D, Ashton DG, Jones DM, Thomsett LR. An outbreak of cowpox in captive cheetahs: virological and serological studies. J Hyg 1982; 89:365–72.
Baxby D. The genesis of Edward Jenner's Inquiry of 1798; a comparison of the two unpublished manuscripts and the published version. Med Hist 1985; 29:193–5.

Baxby D, Bennett M, Getty B. Human cowpox 1969–93; a review based on 54 cases. Brit J Dermatol 1994; 131:598–607.
Baxby D. Edward Jenner's unpublished cowpox Inquiry and the Royal Society: Everard Home's report to Sir Joseph Banks. Med Hist 1999; 43: 108–10.
Baxby D. Edward Jenner's Inquiry; a bicentenary analysis. Vaccine 1999; 17:301–7.
Baxby D, Bennett M, Getty B. Human cowpox 1969–93: a review based on 54 cases. Brit J Dermatol 1994; 131:598–607.
Beale N, Beale E. Evidence-based medicine in the eighteenth century: the Ingen Housz–Jenner correspondence revisited. Med Hist 2005; 49:80–1.
Beall OT, Shryock RH. Cotton Mather, first significant figure in American medicine. Baltimore: Johns Hopkins University Press, 1954.
Bedson HS, Dumbell KR, Thomas WRG. Variola in Tanganyika. Lancet 1965; 2:1085–8.
Behbehani AM. The smallpox story: life and death of an old disease. Microbiol Rev 1983; 455–509.
Bernoulli D. Histoire de l'Académie Royale des Sciences, Part 2, pp. 1–45. Paris, 1766.
Birch J. Serious reasons for uniformly objecting to the practice of vaccination. London: J Smeeton, 1806.
Bishop ER. The epidemic of smallpox in Western New York. Buffalo Med J 1898–9; 38:420–3.
Bishop WJ. Thomas Dimsdale, MD, FRS (1712–1800) and the inoculation of Catherine the Great of Russia. Ann Med Hist 1932; New Series, 4:321–88.
Blake JB. The inoculation controversy in Boston: 1721–2. New Engl Quarterly 1952; XXV:493.
Blake JB. Benjamin Waterhouse and the introduction of vaccination. Philadelphia: University of Philadelphia Press, 1957.
Blake JB. Public health in the town of Boston, 1630–1822. Cambridge, Mass.: Harvard University Press, 1959.
Boëns H. La variole, la vaccine et les vaccinides en 1884. Reprinted from the Bulletin de l'Académie Royale Médicale de Belgique, vol. 18, no. 1. Brussels: A Manceaux, 1884.
Boringdon, Lord. An Act for preventing the spread of the small pox (1813). With two drafts (1808–11) bearing corrections by Edward Jenner. Wellcome Library Collection, MS 3025/1, 3.
Böse J [Anonymous]. Von der Seuche unter den Kindern; über Stellen aus dem Livio. Allgemeine Unterhaltungen Göttingen bei Rosenbusch 1769; 39:305–12.
Bowers JZ. The odyssey of smallpox vaccination. Bull Med Hist 1981; 55:17–33.
Boyer P, Nissenbaum S. Salem possessed. The social origins of witchcraft. Cambridge, Mass: Harvard University Press, 1974.
Booth C. John Haygarth, FRS (1740–1827): a physician of the Enlightenment. Philadelphia: American Philosophical Society, 2005.
Boylston Z. An historical account of the small-pox inoculated in New England, upon all sorts of persons, Whites, Blacks, and of all ages and constitutions. London: S Chandler, 1726.
Bras G. Observations on the formation of smallpox scars. Arch Pathol 1952; 54:149–56.
British Medical Association (BMA). Debate on compulsory vaccination in the House of Commons on Friday May 13th, 1893. London: British Medical Assocation.
British Medical Journal (BMJ). Jenner Centenary number (No. 1847). 23 May 1896.
Bruce J. Travels to discover the source of the Nile, vol. 2. Edinburgh, 1790.
Bruce W. Cowpox in Persia – similar disease in milch sheep. In Gay-Lussac J et al (eds), Annales de chimie et de physique 1819; X:330–1.
Buckton CM. Two winters' experience in giving lectures to my fellow townswomen of the working classes, on physiology and hygiene. Leeds: W Wood, 1873.
Burkhardt JL. Travels in Nubia. London, 1819; 2nd edition. London: John Murray, 1822.
Calef R. More wonders of the invisible world, display'd in five parts. London: Nath. Hillar, 1700.
Carrell JL. The specked monster. A historical tale of battling smallpox. New York: Dutton.
Cello J, Paul AV, Wimmer E. Chemical synthesis of poliovirus cDNA: generation of infectious virus in the absence of natural template. Science 2002; 297:1016–18.

Chambers E. Cyclopaedia: or, an universal dictionary of arts and sciences, 2nd edition. London, 1738.
Chantrey J, Meyer H, Baxby D et al. Cowpox: reservoir hosts and geographic range. Epidemiol Inf 1999; 122:455–60.
Chapin CV. Variation in the type of infectious disease as shown by the history of smallpox in the United States, 1895–1912. J Infect Dis 1913; 13:171–96.
Chapin CV, Smith J. Permanency of the mild type of smallpox. J Prev Med 1932; 1:1–29.
Charles BG. The placenames of Pembrokeshire. Aberystwyth: National Library of Wales, 1992.
Clendenning PH. Dr. Thomas Dimsdale and smallpox inoculation in Russia. J Hist Med 1973; 28:109–25.
Cobb FP. The medical profession and its morality. Modern Review, July 1881.
Coley NG. George Pearson MD, FRS (1751–1828): The greatest chemist in England? Notes and Records of the Royal Society of London 2003; 57(2):161–75.
Collier LH. The development of a stable smallpox vaccine. J Hygiene 1955; 53:76–101.
Collins WJ. A review of the Norwich Vaccination Inquiry. Read to the London Society for the Abolition of Compulsory Vaccination, December 18, 1882. London: EW Allen, 1883.
Colman B. A narrative of the method and success of inoculating the small pox in New England. With a reply to the objections made against it from principles of conscience. In a letter from a Minister at Boston, ed. D Neal. London 1722.
Conlon CP, Warrell DA. Travel and expedition medicine. In Warrell DA, Cox TM, Firth JD, eds. Oxford Textbook of Medicine, 4th edition. Oxford: Oxford University Press, 2003.
Copeman SM. Antitoxins and other organic remedies. Lecture at the Westminster Hospital Medical School. Brit Med J 1895; 2:835–6.
Copeman SM. The bacteriology of vaccinia and variola. Brit Med J 1896; no. 1847:1277–9.
Coxe W. History of the House of Austria, 2nd edition, vol. 4. London: Longman, 1820.
Craker WD. Spectral evidence, non-spectral acts of witchcraft and confessions at Salem in 1692. Hist J 1997; 40:331–58.
Crantz D. History of Greenland, 2nd edition, vol. 2. London: Longman, 1820.
Creighton C. The natural history of cowpox and vaccinal syphilis. London, 1887.
Creighton C. Vaccination. In Encyclopaedia Britannica, 9th edition. 1888.
Creighton C. Jenner and vaccination. A strange chapter of medical history. London: Swan Sonnenschein, 1889.
Creighton C. A history of epidemics in Britain. Vol. 1 (1891), Vol. 2 (1894). Republished London: Frank Cass, 1965.
Crookshank E. History and pathology of vaccination, 2 vols. London: HK Lewis, 1888.
Daniels MH. French engraved portraits. Metropolitan Museum of Art Bull 1925; 20:73.
Darmon P. Vaccins et vaccinations avant Jenner: une querelle d'antériorité. Histoire, économie et société 1984; 3:583–92.
De Clercq E. Cidofovir in the treatment of poxvirus infections. Antiviral Res 2007; 55:1–13.
De la Condamine CM. Mémoire sur l'inoculation de la petite vérole. Mémoires de l'Académie Royale des Sciences 1754; 623–4.
Department of Health. Directive for dealing with outbreaks of smallpox. London: HM Stationery Office, 1962.
Després A. Traité théorique et pratique de la syphilis. Paris: Germer Baillière, 1873.
Dimsdale T. The present method of inoculating for the small pox. London, 1767.
Dimsdale T. Tracts on inoculation, written and published at St. Petersburgh in the year 1768, with additional observations, by command of Her Imperial Majesty the Empress of All the Russias. London: J Phillips, 1781.
Dingle CV. A short account of the Middlesbrough small-pox epidemic – 1897–8. Lancet 1898; 1:1104–6.
Dingle CV. The story of the Middlesbrough small-pox outbreak and some of its lessons. Reprinted from Public Health, 1898. London: Rebman Publishing Company, 1898.
Dingle CV. Vaccination. Paper read on June 12th 1900, to Middlesbrough and District Medical Society, by Dr. Charles V. Dingle MD, Medical Officer of Health, Middlesbrough. Jenner Museum MS BEKJM/412/36.
Disraeli B. Attributed by Twain M in Autobiography. New York: Harper & Brothers, 1924.

Dod, P. Several cases in physick: and one in particular, giving an account of a person who was inoculated for the small-pox, and had the small-pox upon the inoculation, and yet had it again. To which is added, a letter to Dr. Lee, giving him an account of a letter of Dr. Freind's. Together with the said letter. London: C Davis, 1746.
Douglass A [Anonymous]. Abuses and scandals of some late pamphlets in favour of inoculation of the small pox, modestly obviated, and inoculation further consider'd in a letter to A— S— MD and FRS in London. Boston: J Franklin, 1722.
Douglass W. Letter to Dr. Alexander Stuart, RS, 25 September 1721. Roy Soc Journal Book XII.
Douglass W. Letter to Cadwallader Codden, 1 May 1722. Mass Hist Soc Collections 1854, 4th series, 32 (2):170.
Downie AW, Dumbell KR. Survival of variola virus in dried exudate and crusts from smallpox patients. Lancet 1947:1:550–3.
Downie AW, Dumbell KR. Poxviruses. Ann Rev Microbiol 1956; 10:237–52.
'Dr. Sly'. Germ culture: or a scientific method of rearing and floating microscopic canards at will. A translation from the French. London: W Young, 1882.
Edward Jenner Museum archives, various MS.
Farr WA. Letter to the Registrar General on causes of mortality, in 35th Annual Report of the Registrar General, 1867.
Fenner F, Henderson DA, Arita I, Jezek Z, Ladynyi ID. Smallpox and its eradication. Geneva: WHO, 1988.
Fewster J. Cowpox and its ability to prevent smallpox. Unpublished paper read to the Medical Society of London, 1765.
Fewster J. Extracts of a letter to Mr. Rolph. In Pearson G. An inquiry concerning the history of the cowpox. London: J Johnson, 1798.
Fisher RB. Edward Jenner 1749–1823. London: André Deutsch, 1991.
Fisk D. Dr. Jenner of Berkeley. London: Heinemann, 1959.
Foege WH, Millar JD, Henderson DA. Smallpox eradication in West and Central Africa. Bull World Hlth Org 1975; 52:209–22.
Forbes E. Paul Revere and the world he lived in. Houghton Mifflin, 1942. Reprinted 1999.
Fosbroke TD. Berkeley manuscripts. Abstracts and extracts of Smyth's Lives of the Berkeleys ... and biographical anecdotes of Dr. Jenner. London: John Nichols, 1821.
Foster EA. Motolinia's history of the Indians of New Spain. Berkeley: Cortes Society, 1950.
Fourier E. L'antiquité de la vaccine. Chronique Médicale, Paris, 1896.
Fracastoro G. De contagione et contagiosis morbis (On contagion, contagious diseases and their cure), 1546.
Franco-Paredes C, Lammoglia L, Santos-Preciado JI. The Spanish Royal Philanthropic Expedition to bring smallpox vaccination to the New World and Asia in the 19th century. Clin Inf Dis 2005; 41:1285–9.
Franklin B. Letter to Jean-Baptiste Le Roy, 13 November 1789.
Franklin B, Heberden W. Some account of the success of inoculation for the smallpox in England and America. London, 1754.
Gates HL, Higginbotham EB (eds). African American National Biography. Oxford: Oxford University Press, 2008.
Glynn I, Glynn J. The life and death of smallpox. Cambridge: Cambridge University Press, 2004.
Godbeer R. The Devil's dominion. Magic and religion in early New England. Cambridge: Cambridge University Press, 1994.
Goldson W. Cases of smallpox subsequent to vaccination. Portsea, 1804.
Goldstein JA, Neff JM, Lane JM, Koplan JP. Smallpox vaccination reactions, prophylaxis and therapy of complications. Pediatrics 1975; 55:342–7.
Gordon CW, Donnely JD, Fothergill R et al. Variola minor. A preliminary report from the Birmingham Hospital Region. Lancet 1966; 1:1311–13.
Grainger S [Anonymous]. The imposition of inoculation as a duty religiously considered in a letter to a gentleman in the country. Boston: New England Courant, 28 August 1721.
Greene J. Good vaccine lymph. An inquiry as to what extent it is desirable to employ heifer vaccination, with details of that method. Birmingham: C Edmonds, 1871.

Greenwood I [Academicus]. A friendly debate, or, a dialogue between Academicus, and Sawny & Mundungus, two eminent physicians, about some of their recent performances. Boston, 1722.

Grundy I. Lady Mary Wortley Montagu. Oxford: Oxford University Press, 1999.

Guarnieri G. Ricerche sulla patogenesi ed etiologia dell' infezione vaccinica e vaiolosa. Archivio per le scienze mediche 1892; xvi:403–23.

Gümpel G. Smallpox. An inquiry into its real nature and its possible prevention showing the extent and duration of the protection afforded by vaccination. An attempt to solve this vexed question without statistics. London: Swan Sonnenschein, 1902.

Guy WA. 250 years of smallpox in London. Paper read before the Statistical Society, 20 June 1882. Stat Soc J, September 1882. London: E Stanford.

Halsband R. The complete letters of Lady Mary Wortley Montagu. 3 vols: Vol. 1, 1708–1720; vol. 2, 1721–1751; vol. 3, 1752–1762. Oxford: Oxford University Press, 1965.

Halsey R. How the President Thomas Jefferson, and Doctor Benjamin Waterhouse established vaccination as a public health procedure. New York: New York Academy of Medicine, 1936.

Hardy C. Images of Teesside. Derby: Breedon Books, 1992

Harris S. In E Geissler, JE van C Moon, eds. Biological and toxin weapons. Oxford: Oxford Univ Press, 1999.

Haygarth J. An inquiry how to prevent the smallpox. London: J Johnson, 1784.

Haygarth J. A sketch of a plan to exterminate the casual small-pox from Great Britain; and to introduce general inoculation. London: J Johnson, 1793.

Heagerty JJ. Four centuries of medical history in Canada and a sketch of the medical history of Newfoundland, vol. 1. Toronto: MacMillan, 1928.

Henderson DA. Importations of smallpox into Europe 1961–1973. WHO Bulletin, Geneva: WHO/SE/74.62.

Henderson DA. Smallpox – the death of a disease. Amherst: Prometheus Books, 2009.

Henderson DA, Inglesby TV, Bartlett JG et al., for the Working Group on Civil Biodefense. Smallpox as a biological weapon. In Henderson DA, Inglesby TV, O'Toole T (eds). Bioterrorism. Guidelines for medical and public health management. Chicago: AMA Press, 2002.

Henderson GD (ed). Mystics of the North-East. Including: 1. Letters of James Keith MD and others to Lord Deskford. Aberdeen: The Third Spalding Club, 1934.

Herbert EW. Smallpox inoculation in Africa. J African History 1975; XVI:539–59.

Hime TH. Animal vaccination. In Brit Med J, Jenner Centenary Number, 23 May 1896, pp. 1279–89.

Hodge JW. The vaccination superstition: prophylaxis to be realized through the attainment of health, not by the propagation of disease; can vaccination produce syphilis? Niagara Falls, 1901.

Hopkins DR. The greatest killer. Smallpox in history. Chicago: University of Chicago Press, 2002.

Houtton R. Indisputable facts relative to the Suttonian art of inoculation, with observations on its discovery, progress, encouragement, opposition, etc. Dublin: WG Jones, 1768.

Hunter J. Letters from the past from John Hunter to Edward Jenner. Eds EH Cornelius, AJ Harding Rains. London: Royal College of Surgeons of England, 1976.

Ikeda K. The blood in purpuric smallpox. J Amer Med Assoc 1925; 84:1807.

Inhorn MC, Brown PJ. The anthropology of infectious disease. Ann Rev Anthropol 1990; 19:89–117.

Janetta A. The vaccinators: smallpox, medical knowledge and the 'opening' of Japan. Stanford: Stanford University Press, 2007.

Jenner E. Letters to John Baron and others, Royal College of Surgeons of England, London.

Jenner E [Anonymous]. Cursory observations on emetic tartar. Wotton-under-Edge: J Bence, 1780.

Jenner E. Notebook, 1787. MS 372, Royal College of Physicians of London.

Jenner E. Observations on the natural history of the cuckoo. By Mr. Edward Jenner, in a letter to John Hunter, Esq: FRS. Phil Trans Roy Soc 1788; 78:219–37.

Jenner E. A process for preparing pure emetic tartar by re-crystallization. Transactions of a Society for the Improvement of Medical and Chirurgical Knowledge 1793; 1:30–3.
Jenner E. Collection of poems and songs, 1794. MS 3017 (vol. 1) and 3016 (vol. 2), Wellcome Library Archives.
Jenner E. Diary of visits to patients, fees received, expenses, etc. 1794. MS 3018, Wellcome Library Archives.
Jenner E. An inquiry into the natural history of a disease known in Glostershire by the name of the cox-pox. MS in the Royal College of Surgeons of England, London; n.d. (1796).
Jenner E. An Inquiry into the causes and effects of the Variolae Vaccinae: a disease discovered in some of the Western Counties of England, particularly Gloucestershire, and known by the name of the Cow Pox. London: Sampson Low, 1798.
Jenner E. An Inquiry into the causes and effects of the Variolae Vaccinae: a disease discovered in some of the Western Counties of England, particularly Gloucestershire, and known by the name of the Cow Pox, 2nd edition. London: Sampson Low, 1799.
Jenner E. Further observations on the variolae vaccinae or cow-pox. London: Sampson Low, 1799.
Jenner E. A continuation of facts and observations relative to the variolae vaccinae. London: Sampson Low, 1800.
Jenner E. The origin of the vaccine inoculation. London: DN Shury, 1801.
Jenner E. Letter to Dr. John de Carro, 30 March 1803. MS in Royal College of Physicians of London.
Jenner E. Some observations on the migration of birds. By the late Edward Jenner, MD FRS; with an introductory letter to Sir Humphry Davy, Bart. Pres RS. By the Rev. GC Jenner. Phil Trans Roy Soc 1824; 114:11–44.
Johnston RD. The radical middle class. Populist democracy and the question of capitalism in progressive era Portland, Oregon. Princeton: Princeton University Press, 2003.
Jones H. Beauty's triumph: a poem in two cantos. Cited in Deutsch H, Nussbaum F. Defects. Michigan: University of Michigan Press, 2000.
Jurin J. A letter to the learned Dr Caleb Cottesworth, containing a comparison between the danger of the natural smallpox, and of that given by inoculation. Phil Trans Roy Soc 1722; 32:213–17.
Jurin J. Letter to the learned Caleb Cotesworth, containing a comparison between the mortality of the natural small pox, and that given by inoculation. London, 1723.
Kant I. Fragments littéraires. Paris: MV Cousin, 1843.
Kidd BE. Hadwen of Gloucester. Man, medico, martyr. London: John Murray, 1933.
Kirkpatrick J, Barrowby W, Schomberg I. A letter to the real and genuine Pierce Dod, MD, exposing the low absurdity of a late spurious pamphlet falsely ascrib'd to that learned physician: with a full answer to the mistaken case of a natural small-pox, after taking it by inoculation. By Dod Pierce, MS. London: M Cooper, 1746.
Kitson J. Sanitary lessons to working women in Leeds, during the winters of 1872 and 1873. Leeds: Ladies' Council of the Yorkshire Board of Education, 1873.
Kittredge GL. Further notes on Cotton Mather and the Royal Society. Publ Colonial Soc Massachusetts 1913; 14:281–92.
Knaggs HV. Smallpox – a healing crisis and the truth about vaccination. London, 1924.
Koplan JP, Foster SO. Smallpox: clinical types, causes of death and treatment. J Inf Dis 1979; 140:440–1.
Koplow DA. Smallpox. The fight to eradicate a global scourge. Berkeley: University of California Press, 2003.
La Plante E. Salem Witch Judge: the life and repentance of Samuel Sewall. New York: HarperOne, 2007.
Laurance B. Cowpox in man: and its relationship to milkers' nodules. Lancet 1955; 265:764–6.
Le Fanu W. A bibliography of Edward Jenner, 2nd edition. Winchester: St Paul's Bibliographies, 1985.
Lillie W. The history of Middlesbrough. An illustration of the evolution of English industry. Middlesbrough: County Borough of Middlesbrough, 1968.

Little LC. Crimes of the cowpox ring. Some moving pictures thrown on the dead wall of official silence. Minneapolis: The Liberator Publishing Co., 1906.
Li Y, Carroll DS, Gardner SN et al. On the origins of smallpox: correlating variola phylogenetics with historical smallpox records. Proc Natl Acad Sci 2007; 104:15787–92.
Lloyd Davies M, Lloyd Davies TA. The Bible: medicine and myth. Cambridge: Silent Books, 1991.
Loat L. The truth about vaccination and immunization. London, 1951.
Logan JS. A memoir of Dr. Norman Ainley (1924–1962), and a last look at smallpox and vaccination. Ulster Med J 1993; 63:145–52.
Lourie B, Nakano JH, Kemp GE, Setzer HW. Isolation of poxvirus from an African rodent. J Inf Dis 1975; 132:677–81.
Macaulay, Lord Thomas B. The history of England from the Accession of James II, ed. CH Firth. London: Macmillan, 1914.
MacMichael W. The gold-headed cane. London, 1827. Republished in facsimile edition by the Royal College of Physicians of London, 1968.
Mahy BWJ, Esposito JJ, Venter JC. Sequencing the smallpox virus genome: prelude to destruction of a virus species. ASM News 1991; 57:577–80.
Maitland C. Mr. Maitland's account of inoculating the smallpox. London, J Downing, 1722.
Maddox I. A sermon presented before His Grace John Duke of Marlborough, President, the Vice-presidents and Governors of the Hospital for the Small-Pox, and for Inoculation, at the Parish Church of St. Andrew Holborn, on Thursday, March 5th, 1752. London, 1752.
Marsden JP. Variola minor. A personal analysis of 13,686 cases. Bull Hyg 1948; 23:735–46.
Marsden JP. Smallpox. Chapter 5 in Modern practice in infectious fevers. London: Butterworth, 1951.
Marsden JP. On the diagnosis of smallpox. Brit J Clin Pract 1958; 12:1–9.
Marsden JP. Case of malignant smallpox treated with compound 33T57. Brit Med J 1962; 2:524–5.
Massey, E. A sermon against the dangerous and sinful practice of inoculation. Preach'd at St. Andrew's Holborn, on Sunday, July the 8th, 1722.
Mather C. The diary of Cotton Mather. Ed WC Ford, 2 vols. New York: Frederick Ungar (n.d).
Mather C. Remarkable providences relating to witchcrafts and possessions. A faithful account of many wonderful and surprising things. Boston, 1689.
Mather C. The wonders of the invisible world. Observations as well historical as theological, upon the nature, the number, and the operations of the Devils. Boston, 1693.
Mather C. The Negro Christianised. An essay to excite and assist that good work, the instruction of Negro-servants in Christianity. Boston: B Green, 1706.
Mather C. An extract of several letters from Cotton Mather, DD to John Woodward, MD and Richard Waller, Esq; S R Secr. Phil Trans Roy Soc 1714; 29 (no. 339):62–71.
Mather C. The Angel of Bethesda. New London: Timothy Green, 1722.
Mather C. The way of proceeding in the small pox inoculated in New England. Phil Trans Roy Soc 1722; 32 (no. 370):33–5.
Mather C. An account of the method and success of inoculating the small-pox in Boston in New England. London: Jeremiah Dummer, 1724.
Mather C. Christi Magnalia Americana, ed K Murdock, 2 vols. Cambridge MA: Harvard University Press, 1977.
Mather I. Several reasons proving that inoculating or transplanting the small pox is a lawful practice, and that it has been blessed by God for the saving of many a life, ed. GL Kittredge. Cleveland, 1921.
May M. Inoculating the urban poor in the late eighteenth century. Brit J Hist Sci 1997; 30:291–305.
McBean E. The poisoned needle. Suppressed facts about vaccination. Mokelumne Hill, CA: Health Research, 1957. Reprinted by Health Research Books, 1993.
McCormack E. Is vaccination a disastrous illusion? London: The National Anti-Vaccination League, 1909.

McVail JC. Cow-pox and small-pox: Jenner, Woodville and Pearson. Brit Med J 1896; 1:1271–6.
Mead R. A discourse on the small-pox and measles, 1747. In Mead R. The medical works of Richard Mead, MD. Edinburgh, 1775, reprinted New York: AMS Press, 1978.
Meade T. 'Civilising Rio de Janeiro': the public health campaign and the riot of 1904. J Soc Hist 1986; 20:301–22.
Micklos D, Freyer GA, Crotty DA. DNA science: a first course in recombinant DNA technology. Cold Spring Harbor: Cold Spring Harbor Laboratory Press, 2003.
Miller G. Smallpox inoculation in England and America: a reappraisal. The William and Mary Quarterly 1956; series 3, vol. 13:480–1.
Miller G. The adoption of inoculation for smallpox in England and France. Philadelphia: University of Pennsylvania Press, 1957.
Milnes A. Statistics of small-pox and vaccination, with special reference to age-incidence, sex-incidence, and sanitation. J Roy Stat Soc 1897; 60:557.
Montagu, MW [Anonymous]. An account of the inoculating the small pox at Constantinople. The Flying-Post, London, 11–13 September 1722.
Montagu, Lady Mary Wortley. Saturday, The Small Pox [usually known as 'Flavia', after the subject's name]. In Six Town Eclogues. With some other poems. London: M. Cooper, 1747.
Montagu, Lady Mary Wortley. The Lover: a ballad. In R Dodsley (ed) A Collection of Poems by Several Hands. London: R. Dodsley, 1748.
Montagu, Lady Mary Wortley. Letters of the Right Honourable Lady M—y W—y M—e: written during her travels in Europe, Asia and Africa, to persons of distinction, men of letters, & C in different parts of Europe: which contain, among other curious relations, accounts of the policy and manners of the Turks: drawn from sources that have been inaccessible to other travellers. London: T Becket, PS De Hondt, 1763.
Moore J. The history of the smallpox. London: Longman, Hurst, Rees, Orme and Brown, 1815.
Moore W. The Knife Man. London: Bantam Press, 2005.
More LT. Isaac Newton: a biography. New York: Charles Scribner's Sons, 1934.
Morens DM, Littman RJ. 'Thucydides syndrome' reconsidered: new thoughts on the 'Plague of Athens.' Am J Epidemiol 1994; 140:621–8.
Morrison GE. An Australian in China. London, 1895.
Mortimer P. Robert Cory and the vaccine syphilis controversy: a forgotten hero? Lancet 2006; 367:1112–15.
Moseley B. A treatise on Lues Bovilla; or cow pox, 2nd edition. London: J Nichols & Son, 1805.
Nagler FPO, Rake G. The use of the electron microscope in diagnosis of variola, vaccinia and varicella. J Bacteriol 1948; 55:45–51.
Needham J. Science and civilisation in China, vol. 6, part 6. Medicine. Cambridge: Cambridge University Press, 1999.
Nettleton T. An account of the success of inoculating the small-pox. Phil Trans Roy Soc 1722; 32:35–48.
Newman H. The way of proceeding in the small pox inoculated in New England. Communicated by Henry Newman, Esq.: of the Middle Temple. Phil Trans Roy Soc 1722; 32:33–5.
Osler W. The principles and practice of medicine. New York: Appleton, 1892.
Pachiotti X. Sifilide trasmessa per mezzo della vaccinazione in Rivalta presso Acqui. Turin, 1862.
Palmer DD, Palmer BJ. The science of chiropractic: its principles and adjustments. Palmer School of Chiropractic, 1906.
Pankhurst R. The history and traditional treatment of smallpox in Ethiopia. Med Hist 1965; 9:343–55.
Paré A. Deux livres de chirurgie, de la génération de l'homme, & manière d'extraire les enfans hors du ventre de la mère, ensemble ce qu'il faut faire pour la faire mieux, & plus tost accoucher, avec la cure de plusieurs maladies qui luy peuvent survenir. Paris: A Wechel, 1573.

Parker WW. The ancient and modern physician – St. Luke and Jenner. Reprinted from the Transactions of the Medical Society of Virginia, 1891.
Parry CH. An inquiry into the symptoms and causes of the syncope anginosa, commonly called angina pectoris. Bath: Cruttwell, 1799.
Pasteur L. Address at St. James' Hall, London, 8 August 1881. Trans Int Med Congress 1881; i:85.
Patterson KB, Runge T. Smallpox and the American Indian. Am J Med Sci 2002; 323:216–22.
Peachey GC. John Fewster, an unpublished chapter in the history of vaccination. Ann Hist Med (NS) 1929; 1:229–40.
Pead PJ. Benjamin Jesty: new light in the dawn of vaccination. Lancet 2003: 362:2104–9.
Pead PJ. The first vaccinator's 'lost' portrait is found. Wellcome History 2006; Issue 30: 7.
Pead PJ. Vaccination rediscovered. New light in the dawn of man's quest for immunity. Chichester: Timefile Books, 2006.
Pearson GC. An inquiry concerning the history of the cow pox. Principally with a view to supersede and extinguish the small pox. London: J Johnson, 1798.
Pearson GC. On the progress of the variolae vaccinae. Med Phys J 1799; III:97–8, 399.
Pearson G. An examination of the report of the Committee of the House of Commons on the claims of remuneration for the vaccine pock inoculation. London: Johnson, 1802.
Pearson G, Nihell L, Nelson T et al. Report of the Original Vaccine Pock Institution. Edin Med Surg J 1805; 1:513–14.
Pickering J. Anti-vaccination: the statistics of the medical officers of the Leeds Small-pox Hospital exposed and refuted, in a letter to the Leeds Board of Guardians. London: F Pitman, 1876.
Plett PC. Peter Plett und die übrigen Entdecker der Kuhpockenimpfung vor Edward Jenner. Sudhoffs Archiv, Zeitschrift für Wissenschaftgeschichte 2006; 90 (2); S219–32.
Porter R. The greatest benefit to mankind. A medical history of humanity from antiquity to the present. London: Fontana Press, 1997.
Preston R. The demon in the freezer. The New Yorker, 12 July 1999; pp. 44–61.
Prince T. Gentleman's Magazine, London 1753; XXIII.
Probst CO. Smallpox in Ohio. J Amer Med Ass 1899; 33:1589–90.
Pylarinus J. Variolas excitandi per transplantationem methodus. Venice, 1715.
Pylarini J. Nova et tuta variolas excitandi per transplantationem methodus, nuper inventa & in usum tracta. Phil Trans Roy Soc 1716; 24 (347):393–9.
Radford E, Radford MA. Encyclopedia of superstition. Kessinger Publishing, 1949.
Ramsay AM, Emond RTD. Smallpox. Chapter 14 in Infectious Diseases, 2nd edition. London: William Heinemann, 1978.
Reich F. Peter Plett, a forerunner of Jenner. Contribution to the history of vaccination. Medizinische 1953; 21 (15):520–1.
Regulations and Transactions of the Glostershire Medical Society, initiated May 1788. MS 736, Royal College of Physicians of London.
Ricketts TF. The diagnosis of smallpox. London: Cassell, 1908.
Ricketts TH, Byles JB. The red light treatment of smallpox. Lancet 1904; 2:287–90.
Ring J. An answer to Mr. Goldson, proving that vaccination is a permanent security against the small-pox. London: J Murray, 1804.
Roberts J. The madness of Queen Maria. The remarkable life of Maria I of Portugal. Langley Burrell: Templeton Press, 2009.
Robertson RG. Rotting face: smallpox and the American Indian. Caldwell, Idaho: Caxton, 2001.
Roddis LH. Edward Jenner and the discovery of smallpox vaccination. Menasha, WI: George Banta Publishing Company, 1930.
Roland CG. Early history of smallpox in North America. Ann Roy Coll Phys Surg Canada 1992; 25:121–3.
Rowley W. Cox-pox inoculation no security against small-pox infection. London: J Harris, 1805.
Royal College of Physicians. Report of the Royal College of Physicians of London on Vaccination. Med Phys J 1807; 18:97–102.

Royal Vaccination Commission. Final report of the Royal Commission appointed to inquire into the subject of vaccination. London: Her Majesty's Stationery Office, 1896.
Rubin BA. A note on the development of the bifurcated needle for smallpox vaccination. WHO Chron 1980; 34: 180–1.
Rudolf R, Musher DM. Inoculation in the Boston smallpox epidemic of 1721. Arch Int Med 1905; 115:692–6.
Russell A. An account of inoculation in Arabia, in a letter from Dr. Patrick Russell, Physician at Aleppo, to Alexander Russell, MD FRS. Phil Trans Roy Soc 1768; 58:140–50.
Sandwith FM. Smallpox and its early history. Clin J 1910; 36:29–302.
Saunders P. Edward Jenner. The Cheltenham years, 1795–1823. Hanover, NH: University Press of New England, 1982.
Schmidt JS. Wo sind die ersten Kuhblatten inoculiert worden? Schleswig-Holsteinischen Provinzialberichte 1815; 1:77–8.
Shooter RA. Report of the investigation into the cause of the 1978 Birmingham smallpox occurrence. London: Her Majesty's Stationery Office, 1980.
Silverman K. The life and times of Cotton Mather. New York: Harper & Row, 1984.
Simpson HM. The impact of disease on American history. New Engl J Med 1954; 250:679–87.
Sloane H, Birch T. An account of inoculation by Sir Hans Sloane, Bart. Given to Mr. Ranby, to be published. Anno 1736. Phil Trans Roy Soc 1755–56; 49:516–20.
Smith GL. Poxviruses. Chapter 7.10.4 in Warrell DA, Cox TM, Firth JD (eds), Oxford Textbook of Medicine, 4th edition. Oxford: Oxford University Press, 2003, pp. 345–9.
Smith MM. The 'Real Expedición Marítima de la Vacuna' in New Spain and Guatemala. Trans Amer Phil Soc 1974; N.S. 64, part 1:1–74.
Southey R, Southey CC. The life of the Rev. Andrew Bell. 3 vols. London: J Murray, 1844.
Squirrel R. Observations addressed to the public in general on the cow-pox, showing it to originate in the scrophula. London: W Smith & Co., 1805.
Stark RB. Immunization saves Washington's Army. Surg Gynecol Obstetr 1977; 144:425–31.
Stearn EW, Stearn AE. The effect of smallpox on the destiny of the Amerindian. Boston: Humphries, 1945.
Stearn RP. Remarks upon the introduction of inoculation for smallpox in England. Bull Med Hist 1950; 24:103–22.
Steer FW. A relic of an 18th-century isolation hospital. Lancet 1956; 270:200–1.
Stern BJ. Society and medical progress. Princeton: Princeton University Press, 1941. Reprinted McGrath Publishing, 1970.
Stewart I. Reminiscences of a smallpox epidemic. Nursing Record, 12 April 1888.
'The story of the smallpox epidemic in Middlesbrough.' Portraits of the workers. Views of the wards. Supplement to the Northern Weekly Gazette. Middlesbrough, 1898.
Stuart L. Biographical anecdotes of Lady Mary Wortley Montagu. In R Halsband, I Grundy (eds) Lady Mary Wortley Montagu. Essays and poems. Oxford: Clarendon Press, 1977.
Sutton D. The inoculator or Suttonian system of inoculation, fully set forth in a plain and familiar manner. London, 1796.
Sutton R. A new method of inoculating for the small pox, Ipswich Journal, 1 May 1762.
Taylor PA. Speech to the House of Commons, Hansard 11 June 1880.
Taylor PA. The vaccination question. Speech of Mr. P.A. Taylor on Dr. Cameron's resolution respecting animal vaccine. Delivered in the House of Commons, 11 June 1880. From Hansard's Parliamentary Debates, Vol. CCLII. London: Waterlow.
Taylor PA, Hopwood CH. Speeches of Mr. P.A. Taylor and Mr. C.H. Hopwood on vaccination in the House of Commons, 19 June 1883. London: EW Allen, 1883.
Tebb W. Compulsory vaccination in England: with incidental references to foreign states. London: EW Allen, 1884.
Temple R. The genius of China: 3,000 years of science, discovery and invention. New York: Simon and Schuster, 1986.
Thacher J. American medical biography; or, memoirs of eminent physicians who have flourished in America. 2 vols, 1828. Republished, New York: Milford House, 1967.

Thornton J. Second letter to Mr. Tilloch on the cow-pock. Philosoph Magazine 1805; 20:145–6.
Timonius E. An account, or history, of procuring the smallpox by incision, or inoculation, as it has for some time been practised at Constantinople. Phil Trans Roy Soc 1714; 29 (339):72–92.
Tovey D. The Bradford smallpox outbreak in 1962: a personal account. J Roy Soc Med 2004; 97:244–7.
Trevelyan M. Folklore and folk stories of Wales, 1st edition. 1909.
Tucker JB. Scourge: the once and future threat of smallpox. New York: Atlantic Monthly Press, 2001.
Tucker JB, Zilinskas RA. The 1971 smallpox epidemic in Aralsk, Kazakhstan, and the Soviet biological warfare program. Occasional paper no. 9. Monterey, CA: Monterey Institute and Center for Nonproliferation Studies, 2002.
Van de Wettering M. A reconsideration of the inoculation controversy. New Engl Quarterly 1985; 58:46–67.
Van Zwanenberg D. The Suttons and the business of inoculation. Med Hist 1978; 22:71–82.
Vollgnad H. Globus vitellinus, misc. curiosa sive ephem. nat. cur. Jena, 1671.
Wagstaffe W. A letter to Dr. Freind; showing the danger and uncertainty of inoculating the small pox. London, 1722.
Wagstaffe W. Lettre au Docteur Freind; montrant le danger et l'incertitude d'insérer la petite vérole. London, 1722. Also in Journal des Savants, February 1723, pp. 133–41.
Wallace AR. 45 years of registration statistics proving vaccination to be both useless and dangerous. London: EW Allen, 1885.
Wallace AR. Vaccination a delusion. Its penal enforcement a crime. Chapter 2, Much of the evidence adduced for vaccination is worthless. London: Swan Sonnenschein, 1898.
Wallace EM. The first vaccinator – Benjamin Jesty of Yetminster and Worth Matravers and his family. Wareham: Anglebury-Bartlett, 1981.
Waterhouse B. A prospect of exterminating the small pox; being the history of the variolae vaccinae, or kine pox, commonly called the cow pox, as it has appeared in England, with an account of a series of inoculations performed for the kine pox in Massachusetts. Cambridge: Cambridge Press, W Hillard, 1800 (part 1), 1802 (part 2).
Weinstein I. An outbreak of smallpox in New York City. Am J Publ Hlth 1947; 37:1376–84.
Wesley J. Primitive Physic: or, an easy and natural method of curing most diseases. 1747.
Wheeler A. Vaccination in the light of history. London: EW Allen, 1878.
Wheeler A. Vaccination. Opposed to science and a disgrace to English law. An address delivered at Darlington, 10 November 1879. London: EW Allen, 1879.
Wheeler A. Vaccination. London: EW Allen, 1883.
Wilkinson JJG. Pasteur and Jenner. An example and a warning. London, 1881.
Williams J. Several arguments, proving that inoculating the small pox is not contained in the Laws of Physick, natural or divine, and is therefore unlawful. Boston: J Franklin, 1721.
Williams LR. The smallpox epidemic at Niagara Falls. Am J Publ Hlth 1915; 5:423–37.
Williams P. Part of two letters containing a method of procuring the small pox, used in South Wales. Phil Trans Roy Soc 1723; 32:262–4.
Williams P. Part of a letter from the same learned and ingenious gentleman, upon the same subject, to Dr. Jurin, R.S. Sect. Phil Trans Roy Soc 1723; 32:264–6.
Winslow D. A destroying angel: the conquest of smallpox in Colonial Boston. Boston: Houghton Mifflin, 1974.
Wolfe RM, Sharp LK. Anti-vaccinationists past and present. Brit Med J 2002; 325:430–2.
Woodville W. Medical botany, containing systematic and general descriptions, with plates, of all the medicinal plants, indigenous and exotic, comprehended in the Catalogues of the Materia Medica, as published by the Royal Colleges of Physicians of London and Edinburgh. London: J Phillips, 1790–93.
Woodville W. The history of the inoculation of the small-pox in Great Britain, vol. 1. London: J Phillips, 1796.
Woodville W. Report of a series of inoculations for the variolae vaccinae, or cow-pox. London: James Phillips, 1799.

Woodward SB. The story of smallpox in Massachusetts. New Engl J Med 1932; 206:1181–91.
Wright R. A letter on the same subject from Mr. Richard Wright, Surgeon of Haverford West, to Mr. Sylvanus Bevan, Apothecary in London. Phil Trans Roy Soc 1723; 32:267–9.
Wujastuk, D. 'A pious fraud': the Indian claim for pre-Jennerian smallpox vaccination. Chapter 9 in Meulenberg GJ, Wujastuk D (eds), Studies on Indian medical history. Delhi: Motilal Banrsidass, 2001.
Zielonka JS. Inoculation: pro and con. J Hist Med Allied Sci 1972; 27:447.

Index

Compiled by Sue Carlton